Robot Manipulators Modeling, Performance Analysis and Control

Robot Manipulators Modeling, Performance Analysis and Control

Edited by
Etienne Dombre
Wisama Khalil

iSTE

Part of this book adapted from "Analyse et modèlisation des robots manipulateure" and "Commande des robots manipulateurs" published in France by Hermès Science/Lavoisier
First published with revisions in Great Britain and the United States in 2007 by ISTE Ltd

ISTE Ltd
6 Fitzroy Square
London W1T 5DX
UK
www.iste.co.uk

ISTE USA
4308 Patrice Road
Newport Beach, CA 92663
USA

First South Asian Edition 2007

Library of Congress Cataloging-in-Publication Data

Modeling, performance analysis and control of robot manipulators/edited by Etienne Dombre, Wisama Khalil.
p. cm.
Includes index.
ISBN 13: 978-1-905209-10-1
ISBN 10: 1-905209-10-X
1. Robotics. 2. Manipulators (Mechanism) I. Dombre, E. (Etienne) II. Khalil, W. (Wisama)

TJ211.M626 2006
629.8'933—dc22

2006032328

British Library Cataloguing-in-Publication Data
A CIP record for this book is available from the British Library
ISBN 10: 1-905209-10-X
ISBN 13: 978-1-905209-10-1

Printed in Brijbasi Art Press Ltd., I-72, Sector-9, Noida, U.P. India.

Table of Contents

Chapter 1. Modeling and Identification of Serial Robots 1

Wisama KHALIL and Etienne DOMBRE

1.1. Introduction 1
1.2. Geometric modeling 2
1.2.1. Geometric description 2
1.2.2. Direct geometric model 6
1.2.3. Inverse geometric model 7
1.2.3.1. Stating the problem 8
1.2.3.2. Principle of Paul's method 10
1.3. Kinematic modeling 14
1.3.1. Direct kinematic model 14
1.3.1.1 Calculation of the Jacobian matrix by derivation of the DGM 15
1.3.1.2. Kinematic Jacobian matrix 17
1.3.1.3. Decomposition of the kinematic Jacobian matrix into three matrices 19
1.3.1.4. Dimension of the operational space of a robot 20
1.3.2. Inverse kinematic model 21
1.3.2.1. General form of the kinematic model 21
1.3.2.2. Inverse kinematic model for the regular case 22
1.3.2.3. Solution at the proximity of singular positions 23
1.3.2.4. Inverse kinematic model of redundant robots 24

1.4. Calibration of geometric parameters 26
1.4.1. Introduction 26
1.4.2. Geometric parameters 26
1.4.2.1. Geometric parameters of the robot. 26
1.4.2.2. Parameters of the robot's location 27
1.4.2.3. Geometric parameters of the end-effector 28
1.4.3. Generalized differential model of a robot. 29
1.4.4. Principle of geometric calibration 30
1.4.4.1. General form of the calibration model 30
1.4.4.2. Identifying the geometric parameters 31
1.4.4.3. Solving the identification equations 34
1.4.5. Calibration methods of geometric parameters 35
1.4.5.1. Calibration model by measuring the end-effector location. . . . 35
1.4.5.2. Autonomous calibration models 36
1.4.6. Correction of geometric parameters 39
1.5. Dynamic modeling 40
1.5.1. Lagrange formalism 42
1.5.1.1. General form of dynamic equations 43
1.5.1.2. Calculation of energy 44
1.5.1.3. Properties of the dynamic model. 46
1.5.1.4. Taking into consideration the friction 47
1.5.1.5. Taking into account the inertia of the actuator's rotor 48
1.5.1.6. Taking into consideration the forces and moments exerted
by the end-effector on its environment 48
1.5.2. Newton-Euler formalism 50
1.5.2.1. Newton-Euler equations linear in the inertial parameters 50
1.5.2.2. Practical form of Newton-Euler equations 52
1.5.3. Determining the base inertial parameters 53
1.6. Identification of dynamic parameters. 59
1.6.1. Introduction 59
1.6.2. Identification principle of dynamic parameters 60
1.6.2.1. Solving method 60
1.6.2.2. Identifiable parameters. 62
1.6.2.3. Choice of identification trajectories 63
1.6.2.4. Evaluation of joint coordinates. 65
1.6.2.5. Evaluation of joint torques 65

1.6.3. Identification model using the dynamic model 66
1.6.4. Sequential formulation of the dynamic model 68
1.6.5. Practical considerations 69
1.7. Conclusion 70
1.8. Bibliography 71

Chapter 2. Modeling of Parallel Robots 81

Jean-Pierre MERLET and François PIERROT

2.1. Introduction. 81
2.1.1. Characteristics of classic robots 81
2.1.2. Other types of robot structure. 82
2.1.3. General advantages and disadvantages 86
2.1.4. Present day uses. 88
2.1.4.1. Simulators and space applications 88
2.1.4.2. Industrial applications 91
2.1.4.3. Medical applications 93
2.1.4.4. Precise positioning 94
2.2. Machine types 95
2.2.1. Introduction 95
2.2.2. Plane robots with three degrees of freedom 100
2.2.3. Robots moving in space 101
2.2.3.1. Manipulators with three degrees of freedom 101
2.2.3.2. Manipulators with four or five degrees of freedom 107
2.2.3.3. Manipulators with six degrees of freedom 109
2.3. Inverse geometric and kinematic models 113
2.3.1. Inverse geometric model. 113
2.3.2. Inverse kinematics 115
2.3.3. Singular configurations 117
2.3.3.1. Singularities and statics 121
2.3.3.2. State of the art. 121
2.3.3.3. The geometric method 122
2.3.3.4. Maneuverability and condition number. 125
2.3.3.5. Singularities in practice 126

2.4. Direct geometric model 126
2.4.1. Iterative method 127
2.4.2. Algebraic method 128
2.4.2.1. Reminder concerning algebraic geometry 128
2.4.2.2. Planar robots 130
2.4.2.3. Manipulators with six degrees of freedom 133
2.5. Bibliography 134

Chapter 3. Performance Analysis of Robots 141

Philippe WENGER

3.1. Introduction 141
3.2. Accessibility 143
3.2.1. Various levels of accessibility 143
3.2.2. Condition of accessibility 144
3.3. Workspace of a robot manipulator 146
3.3.1. General definition 146
3.3.2. Space of accessible positions 148
3.3.3. Primary space and secondary space 149
3.3.4. Defined orientation workspace 151
3.3.5. Free workspace 152
3.3.6. Calculation of the workspace 155
3.4. Concept of aspect 157
3.4.1. Definition 157
3.4.2. Mode of aspects calculation 158
3.4.3. Free aspects 160
3.4.4. Application of the aspects 161
3.5. Concept of connectivity 163
3.5.1. Introduction 163
3.5.2. Characterization of n-connectivity 165
3.5.3. Characterization of t-connectivity 168
3.6. Local performances 174
3.6.1. Definition of dexterity 174
3.6.2. Manipulability 174
3.6.3. Isotropy index 180

3.6.4. Lowest singular value 181
3.6.5. Approach lengths and angles 181
3.7. Conclusion 183
3.8. Bibliography 183

Chapter 4. Trajectory Generation 189

Moussa HADDAD, Taha CHETTIBI, Wisama KHALIL and Halim LEHTIHET

4.1. Introduction. 189
4.2. Point-to-point trajectory in the joint space under kinematic constraints 190
4.2.1. Fifth-order polynomial model 191
4.2.2. Trapezoidal velocity model 193
4.2.3. Smoothed trapezoidal velocity model 198
4.3. Point-to-point trajectory in the task-space under kinematic constraints 201
4.4. Trajectory generation under kinodynamic constraints 204
4.4.1. Problem statement 205
4.4.1.1. Constraints 206
4.4.1.2. Objective function 207
4.4.2. Description of the method 208
4.4.2.1. Outline 208
4.4.2.2. Construction of a random trajectory profile 209
4.4.2.3. Handling kinodynamic constraints. 212
4.4.2.4. Summary 216
4.4.3. Trapezoidal profiles 218
4.5. Examples 221
4.5.1. Case of a two dof robot 221
4.5.1.1. Optimal free motion planning problem 221
4.5.1.2. Optimal motion problem with geometric path constraint. 223
4.5.2. Case of a six dof robot 224
4.5.2.1. Optimal free motion planning problem 225
4.5.2.2. Optimal motion problem with geometric path constraints 226

4.5.2.3. Optimal free motion planning problem with intermediate points 227
4.6. Conclusion 229
4.7. Bibliography 230
Appendix: Stochastic Optimization Techniques 234

Chapter 5. Position and Force Control of a Robot in a Free or Constrained Space 241

Pierre DAUCHEZ and Philippe FRAISSE

5.1. Introduction 241
5.2. Free space control 242
5.2.1. Hypotheses applying to the whole chapter 242
5.2.2. Complete dynamic modeling of a robot manipulator 243
5.2.3. Ideal dynamic control in the joint space 246
5.2.4. Ideal dynamic control in the operational working space 248
5.2.5. Decentralized control 250
5.2.6. Sliding mode control 251
5.2.7. Robust control based on high order sliding mode 254
5.2.8. Adaptive control 255
5.3. Control in a constrained space 257
5.3.1. Interaction of the manipulator with the environment 257
5.3.2. Impedance control 257
5.3.3. Force control of a mass attached to a spring 258
5.3.4. Non-linear decoupling in a constrained space 262
5.3.5. Position/force hybrid control 263
5.3.5.1. Parallel structure 263
5.3.5.2. External structure 269
5.3.6. Specificity of the force/torque control 271
5.4. Conclusion 275
5.5. Bibliography 275

Chapter 6. Visual Servoing 279

François CHAUMETTE

6.1. Introduction. 279
6.2. Modeling visual features 281
6.2.1. The interaction matrix 281
6.2.2. Eye-in-hand configuration. 282
6.2.3. Eye-to-hand configuration. 283
6.2.4. Interaction matrix 284
6.2.4.1. Interaction matrix of a 2-D point. 284
6.2.4.2. Interaction matrix of a 2-D geometric primitive. 287
6.2.4.3. Interaction matrix for complex 2-D shapes. 290
6.2.4.4. Interaction matrix by learning or estimation 293
6.2.5. Interaction matrix related to 3-D visual features. 294
6.2.5.1. Pose estimation 294
6.2.5.2. Interaction matrix related to θu. 297
6.2.5.3. Interaction matrix related to a 3-D point 298
6.2.5.4. Interaction matrix related to a 3-D plane 300
6.3. Task function and control scheme 301
6.3.1. Obtaining the desired value s* 301
6.3.2. Regulating the task function 302
6.3.2.1. Case where the dimension of s is 6 (k = 6) 304
6.3.2.2. Case where the dimension of s is greater than 6 (k > 6) 312
6.3.3. Hybrid tasks 317
6.3.3.1. Virtual links 317
6.3.3.2. Hybrid task function 319
6.3.4. Target tracking 323
6.4. Other exteroceptive sensors 325
6.5. Conclusion 326
6.6. Bibliography 328

Chapter 7. Modeling and Control of Flexible Robots 337

Frédéric BOYER, Wisama KHALIL, Mouhacine BENOSMAN and George LE VEY

7.1. Introduction. 337
7.2. Modeling of flexible robots 337
7.2.1. Introduction 337
7.2.2. Generalized Newton-Euler model for a kinematically free elastic body 339
7.2.2.1. Definition: formalism of a dynamic model. 339
7.2.2.2. Choice of formalism 340
7.2.2.3. Kinematic model of a free elastic body 341
7.2.2.4. Balance principle compatible with the mixed formalism. 343
7.2.2.5. Virtual power of the field of acceleration quantities 344
7.2.2.6. Virtual power of external forces 346
7.2.2.7. Virtual power of elastic cohesion forces 347
7.2.2.8. Balance of virtual powers 348
7.2.2.9. Linear rigid balance in integral form 349
7.2.2.10. Angular rigid balance in integral form. 349
7.2.2.11. Elastic balances in integral form 350
7.2.2.12. Linear rigid balance in parametric form. 351
7.2.2.13. Intrinsic matrix form of the generalized Newton-Euler model. 353
7.2.3. Velocity model of a simple open robotic chain 356
7.2.4. Acceleration model of a simple open robotic chain. 357
7.2.5. Generalized Newton-Euler model for a flexible manipulator 358
7.2.6. Extrinsic Newton-Euler model for numerical calculus 359
7.2.7. Geometric model of an open chain. 362
7.2.8. Recursive calculation of the inverse and direct dynamic models for a flexible robot 363
7.2.8.1. Introduction 363
7.2.8.2. Recursive algorithm of the inverse dynamic model. 364
7.2.8.3. Recursive algorithm of the direct dynamic model. 368
7.2.8.4. Iterative symbolic calculation. 373
7.3. Control of flexible robot manipulators 373
7.3.1. Introduction 373
7.3.2. Reminder of notations 374

7.3.3. Control methods 375
7.3.3.1. Regulation 375
7.3.3.2. Point-to-point movement in fixed time 375
7.3.3.3. Trajectory tracking in the joint space 380
7.3.3.4. Trajectory tracking in the operational space 383
7.4. Conclusion 388
7.5. Bibliography 389

List of Authors 395

Index 397

Chapter 1

Modeling and Identification of Serial Robots

1.1. Introduction

The design and control of robots require certain mathematical models, such as:

– transformation models between the operational space (in which the position of the end-effector is defined) and the joint space (in which the configuration of the robot is defined). The following is distinguished:

- direct and inverse geometric models giving the location of the end-effector (or the tool) in terms of the joint coordinates of the mechanism and vice versa,

- direct and inverse kinematic models giving the velocity of the end-effector in terms of the joint velocities and vice versa,

– dynamic models giving the relations between the torques or forces of the actuators, and the positions, velocities and accelerations of the joints.

This chapter presents some methods to establish these models. It will also deal with identifying the parameters appearing in these models. We will limit the discussion to simple open structures. For complex structure robots, i.e. tree or closed structures, we refer the reader to [KHA 02].

Chapter written by Wisama KHALIL and Etienne DOMBRE.

Mathematical development is based on (4 × 4) homogenous transformation matrices. The homogenous matrix ${}^{i}\mathbf{T}_j$ representing the transformation from frame R_i to frame R_j is defined as:

$$ {}^{i}\mathbf{T}_j = \begin{bmatrix} & {}^{i}\mathbf{R}_j & & {}^{i}\mathbf{P}_j \\ 0 & 0 & 0 & 1 \end{bmatrix} = \begin{bmatrix} {}^{i}\mathbf{s}_j & {}^{i}\mathbf{n}_j & {}^{i}\mathbf{a}_j & {}^{i}\mathbf{P}_j \\ 0 & 0 & 0 & 1 \end{bmatrix} \qquad [1.1] $$

where ${}^{i}\mathbf{s}_j$, ${}^{i}\mathbf{n}_j$ and ${}^{i}\mathbf{a}_j$ of the orientation matrix ${}^{i}\mathbf{R}_j$ indicate the unit vectors along the axes $\mathbf{x}_j$, $\mathbf{y}_j$ and $\mathbf{z}_j$ of the frame R_j expressed in the frame R_i; and where ${}^{i}\mathbf{P}_j$ is the vector expressing the origin of the frame R_j in the frame R_i.

1.2. Geometric modeling

1.2.1. *Geometric description*

A systematic and automatic modeling of robots requires an appropriate method for the description of their morphology. Several methods and notations have been proposed [DEN 55], [SHE 71], [REN 75], [KHA 76], [BOR 79], [CRA 86]. The most widely used one is that of Denavit-Hartenberg [DEN 55]. However, this method, developed for simple open structures, presents ambiguities when it is applied to closed or tree-structured robots. Hence, we recommend the notation of Khalil and Kleinfinger which enables the unified description of complex and serial structures of articulated mechanical systems [KHA 86].

A simple open structure consists of n+1 links noted C_0, …, C_n and of n joints. Link C_0 indicates the robot base and link C_n, the link carrying the end-effector. Joint j connects link C_j to link C_{j-1} (Figure 1.1). The method of description is based on the following rules and conventions:

– the links are assumed to be perfectly rigid. They are connected by revolute or prismatic joints considered as being ideal (no mechanical clearance, no elasticity);

– the frame R_j is fixed to link C_j;

– axis $\mathbf{z}_j$ is along the axis of joint j;

– axis $\mathbf{x}_j$ is along the common perpendicular with axes $\mathbf{z}_j$ and $\mathbf{z}_{j+1}$. If axes $\mathbf{z}_j$ and $\mathbf{z}_{j+1}$ are parallel or collinear, the choice of $\mathbf{x}_j$ is not unique: considerations of symmetry or simplicity lead to a reasonable choice.

The transformation matrix from the frame R_{j-1} to the frame R_j is expressed in terms of the following four geometric parameters:

– α_j: angle between axes $\mathbf{z}_{j-1}$ and $\mathbf{z}_j$ corresponding to a rotation about $\mathbf{x}_{j-1}$;

– d_j: distance between $\mathbf{z}_{j-1}$ and $\mathbf{z}_j$ along $\mathbf{x}_{j-1}$;

– θ_j: angle between axes $\mathbf{x}_{j-1}$ and $\mathbf{x}_j$ corresponding to a rotation about $\mathbf{z}_j$;

– r_j: distance between $\mathbf{x}_{j-1}$ and $\mathbf{x}_j$ along $\mathbf{z}_j$.

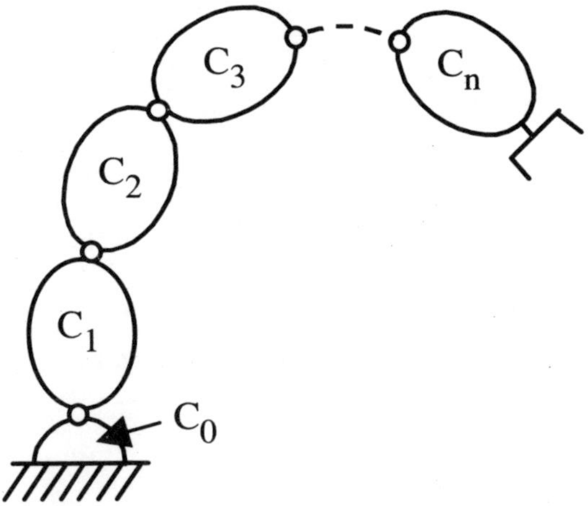

Figure 1.1. *A simple open structure robot*

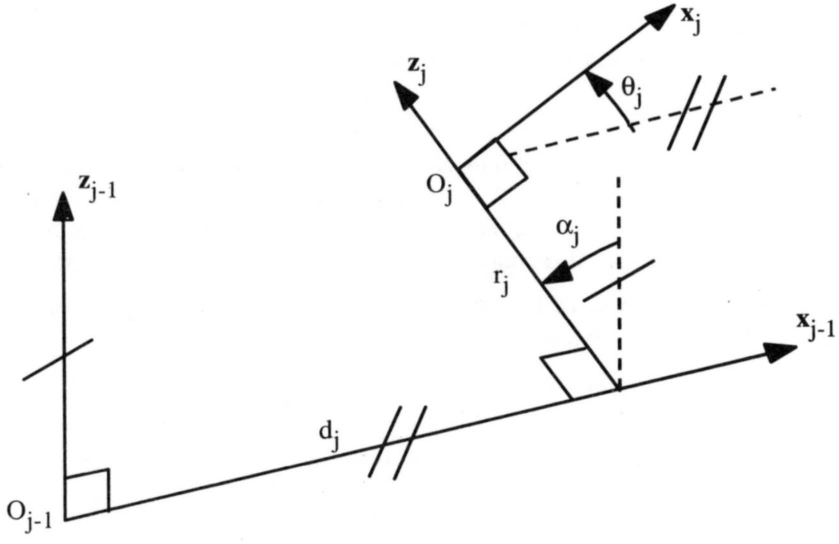

Figure 1.2. *Geometric parameters in the case of a simple open structure*

The joint coordinate q_j associated to the jth joint is either θ_j or r_j, depending on whether this joint is revolute or prismatic. It can be expressed by the relation:

$$q_j = \bar{\sigma}_j \theta_j + \sigma_j r_j \quad [1.2]$$

with:

– $\sigma_j = 0$ if the joint is revolute;

– $\sigma_j = 1$ if the joint is prismatic;

– $\bar{\sigma}_j = 1 - \sigma_j$.

The transformation matrix defining the frame R_j in the frame R_{j-1} is obtained from Figure 1.2 by:

$$^{j-1}\mathbf{T}_j = \mathbf{Rot}(\mathbf{x}, \alpha_j)\ \mathbf{Trans}(\mathbf{x}, d_j)\ \mathbf{Rot}(\mathbf{z}, \theta_j)\ \mathbf{Trans}(\mathbf{z}, r_j)$$

$$= \begin{bmatrix} C\theta_j & -S\theta_j & 0 & d_j \\ C\alpha_j S\theta_j & C\alpha_j C\theta_j & -S\alpha_j & -r_j S\alpha_j \\ S\alpha_j S\theta_j & S\alpha_j C\theta_j & C\alpha_j & r_j C\alpha_j \\ 0 & 0 & 0 & 1 \end{bmatrix} \quad [1.3]$$

where $\mathbf{Rot}(\mathbf{u}, \alpha)$ and $\mathbf{Trans}(\mathbf{u}, d)$ are (4×4) homogenous matrices representing, respectively, a rotation α about the axis **u** and a translation d along **u.**

NOTES.

– for the definition of the reference frame R_0, the simplest choice consists of taking R_0 aligned with the frame R_1 when $q_1 = 0$, which indicates that $\mathbf{z}_0$ is along $\mathbf{z}_1$ and $O_0 \equiv O_1$ when joint 1 is revolute, and $\mathbf{z}_0$ is along $\mathbf{z}_1$ and $\mathbf{x}_0$ is parallel to $\mathbf{x}_1$ when joint 1 is prismatic. This choice renders the parameters α_1 and d_1 zero;

– likewise, the axis $\mathbf{x}_n$ of the frame R_n is taken collinear to $\mathbf{x}_{n-1}$ when $q_n = 0$. This choice makes r_n (or θ_n) zero when $\sigma_n = 1$ (or = 0 respectively);

– for a prismatic joint, the axis $\mathbf{z}_j$ is parallel to the axis of the joint; it can be placed in such a way that d_j or d_{j+1} is zero;

– when $\mathbf{z}_j$ is parallel to $\mathbf{z}_{j+1}$, the axis $\mathbf{x}_j$ is placed in such a way that r_j or r_{j+1} is zero;

– in practice, the vector of joint variables $\mathbf{q}$ is given by:

$$\mathbf{q} = \mathbf{K_c}\, \mathbf{q_c} + \mathbf{q_0}$$

where $\mathbf{q_0}$ represents an offset, $\mathbf{q_c}$ are encoder variables and $\mathbf{K_c}$ is a constant matrix.

EXAMPLE 1.1.– description of the structure of the Stäubli RX-90 robot (Figure 1.3). The robot shoulder is of an RRR anthropomorphic type and the wrist consists of three intersecting revolute axes, equivalent to a spherical joint. From a methodological point of view, firstly the axes $\mathbf{z}_j$ are placed on the joint axes and the axes $\mathbf{x}_j$ are placed according to the rules previously set. Next, the geometric parameters of the robot are determined. The link frames are shown in Figure 1.3 and the geometric parameters are given in Table 1.1.

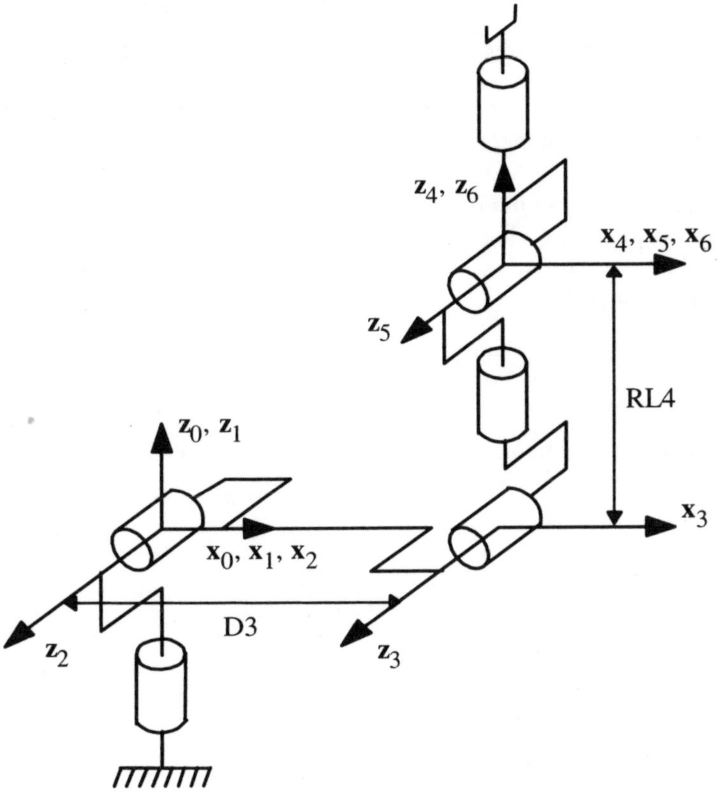

Figure 1.3. *Link frames for the Stäubli RX-90 robot*

j	σ_j	α_j	d_j	θ_j	r_j
1	0	0	0	θ_1	0
2	0	$\pi/2$	0	θ_2	0
3	0	0	D3	θ_3	0
4	0	$-\pi/2$	0	θ_4	RL4
5	0	$\pi/2$	0	θ_5	0
6	0	$-\pi/2$	0	θ_6	0

Table 1.1. *Geometric parameters for the Stäubli RX-90 robot*

1.2.2. *Direct geometric model*

The direct geometric model (DGM) represents the relations calculating the operational coordinates, giving the location of the end-effector, in terms of the joint coordinates. In the case of a simple open chain, it can be represented by the transformation matrix ${}^0\mathbf{T}_n$:

$$ {}^0\mathbf{T}_n = {}^0\mathbf{T}_1(q_1)\ {}^1\mathbf{T}_2(q_2)\ \ldots\ {}^{n-1}\mathbf{T}_n(q_n) \qquad [1.4] $$

The direct geometric model of the robot may also be represented by the relation:

$$ \mathbf{X} = \mathbf{f}(\mathbf{q}) \qquad [1.5] $$

q being the vector of joint coordinates such that:

$$ \mathbf{q} = [q_1\ \ q_2\ \ldots\ q_n]^T \qquad [1.6] $$

The operational coordinates are defined by:

$$ \mathbf{X} = [x_1\ \ x_2\ \ldots\ x_m]^T \qquad [1.7] $$

There are several possibilities to define the vector **X**. For example, with the help of the elements of matrix ${}^{0}\mathbf{T}_{n}$:

$$\mathbf{X} = [P_x \;\; P_y \;\; P_z \;\; s_x \;\; s_y \;\; s_z \;\; n_x \;\; n_y \;\; n_z \;\; a_x \;\; a_y \;\; a_z]^T \qquad [1.8]$$

or otherwise, knowing that **s = nxa**

$$\mathbf{X} = [P_x \;\; P_y \;\; P_z \;\; n_x \;\; n_y \;\; n_z \;\; a_x \;\; a_y \;\; a_z]^T \qquad [1.9]$$

For the orientation, other representations are currently used such as Euler angles, Roll-Pitch-Yaw angles or Quaternions. We can easily derive direction cosines **s**, **n**, **a** from any one of these representations and vice versa [KHA 02].

EXAMPLE 1.2. – direct geometric model for the Stäubli RX-90 robot (Figure 1.3). According to Table 1.1, the relation [1.3] can be used to write the basic transformation matrices ${}^{j-1}\mathbf{T}_{j}$. The product of these matrices gives ${}^{0}\mathbf{T}_{6}$ that has as components:

$$s_x = C1(C23(C4C5C6 - S4S6) - S23S5C6) - S1(S4C5C6 + C4S6)$$
$$s_y = S1(C23(C4C5C6 - S4S6) - S23S5C6) + C1(S4C5C6 + C4S6)$$
$$s_z = S23(C4C5C6 - S4S6) + C23S5C6$$
$$n_x = C1(- C23\,(C4C5S6 + S4C6) + S23S5S6) + S1(S4C5S6 - C4C6)$$
$$n_y = S1(- C23\,(C4C5S6 + S4C6) + S23S5S6) - C1(S4C5S6 - C4C6)$$
$$n_z = - S23(C4C5S6 + S4C6) - C23S5S6$$
$$a_x = - C1(C23C4S5 + S23C5) + S1S4S5$$
$$a_y = - S1(C23C4S5 + S23C5) - C1S4S5$$
$$a_z = - S23C4S5 + C23C5$$
$$P_x = - C1(S23\,RL4 - C2D3)$$
$$P_y = - S1(S23\,RL4 - C2D3)$$
$$P_z = C23\,RL4 + S2D3$$

with C23=cos $(\theta_2 + \theta_3)$ and S23 = sin $(\theta_2 + \theta_3)$.

1.2.3. *Inverse geometric model*

We saw that the direct geometric model of a robot calculates the operational coordinates giving the location of the end-effector in terms of joint coordinates. The inverse problem consists of calculating the joint coordinates corresponding to a given location of the end-effector. When it exists, the explicit form which gives all possible solutions (there is rarely uniqueness of solution) constitutes what we call the inverse geometric model (IGM). We can distinguish three methods for the calculation of IGM:

– Paul's method [PAU 81], which deals with each robot separately and is suitable for most of the industrial robots;

– Pieper's method [PIE 68], which makes it possible to solve the problem for the robots with six degrees of freedom having three revolute joints with intersecting axes or three prismatic joints;

– the general Raghavan and Roth's method [RAG 90] giving the general solution for robots with six joints using at most a 16-degree polynomial.

Whenever calculating an explicit form of the inverse geometric model is not possible, we can calculate a particular solution through numeric procedures [PIE 68], [WHI 69], [FOU 80], [FEA 83], [WOL 84], [GOL 85] [SCI 86].

In this chapter, we present Paul's method; Pieper's method, and Raghavan and Roth's method are detailed in [KHA 02].

1.2.3.1. *Stating the problem*

Let ${}^f\mathbf{T}_E{}^d$ be the homogenous transformation matrix representing the desired location of the end-effector frame R_E with respect to the world frame R_f. In general cases, ${}^f\mathbf{T}_E{}^d$ can be expressed in the following form:

$$ {}^f\mathbf{T}_E{}^d = \mathbf{Z}\, {}^0\mathbf{T}_n(\mathbf{q})\, \mathbf{E} \qquad [1.10] $$

where (see Figure 1.4):

– $\mathbf{Z}$ is the transformation matrix defining the location of the robot frame R_0 in the world reference frame R_f;

– ${}^0\mathbf{T}_n$ is the transformation matrix of the terminal link frame R_n with respect to frame R_0 in terms of the joint coordinates $\mathbf{q}$;

– $\mathbf{E}$ is the transformation matrix defining the end-effector frame R_E in the terminal frame R_n.

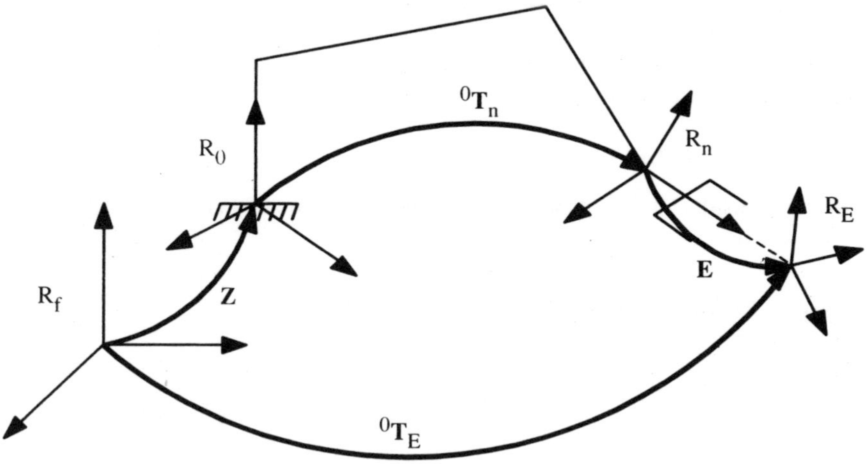

Figure 1.4. *Transformations between the end-effector frame and the world reference frame*

When $n \geq 6$, we can write the following relation by grouping on the right hand side all known terms:

$$^0\mathbf{T}_n(\mathbf{q}) = \mathbf{Z}^{-1}\ {}^f\mathbf{T}_E{}^d\ \mathbf{E}^{-1} \qquad [1.11]$$

When $n < 6$, the robot's operational space is less than six. It is not possible to place the end-effector frame R_E in an arbitrary location $R_E{}^d$ describing the task, except when the frames R_E and $R_E{}^d$ are conditioned in a particular way in order to compensate for the insufficient number of degrees of freedom. Practically, instead of bringing frame R_E onto frame $R_E{}^d$, we will seek to only place some elements of the end-effector (points, straight lines).

In the calculation of IGM, three cases can be distinguished:

a) no solution when the desired location is outside of the accessible zone of the robot. It is limited by the number of degrees of freedom of the robot, the joint limits and the dimension of the links;

b) infinite number of solutions when:

– the robot is redundant with respect to the task,

– the robot is in some singular configuration;

c) a finite number of solutions expressed by a set of vectors $\{\mathbf{q}^1, \ldots, \mathbf{q}^r\}$. A robot is said to be solvable [PIE 68], [ROT 76] when it is possible to calculate all the configurations making it possible to reach a given location. Nowadays, all serial manipulators having up to six degrees of freedom and which are not redundant may be considered as solvable. The number of solutions depends on the structure of the robot and is at most equal to 16.

1.2.3.2. *Principle of Paul's method*

Let us consider a robot whose homogenous transformation matrix has the following form:

$$ {}^{0}\mathbf{T}_{n} = {}^{0}\mathbf{T}_{1}(q_1)\ {}^{1}\mathbf{T}_{2}(q_2) \ldots {}^{n-1}\mathbf{T}_{n}(q_n) \quad [1.12]$$

Let $\mathbf{U}_0$ be the desired location, such that:

$$ U_0 = \begin{bmatrix} s_x & n_x & a_x & P_x \\ s_y & n_y & a_y & P_y \\ s_z & n_z & a_z & P_z \\ 0 & 0 & 0 & 1 \end{bmatrix} \quad [1.13]$$

We seek to solve the following system of equations:

$$ \mathbf{U}_0 = {}^{0}\mathbf{T}_{1}(q_1)\ {}^{1}\mathbf{T}_{2}(q_2) \ldots {}^{n-1}\mathbf{T}_{n}(q_n) \quad [1.14]$$

Paul's method consists of successively pre-multiplying the two sides of equation [1.14] by the matrices ${}^{j}\mathbf{T}_{j-1}$ for j = 1, ..., n–1, operations which make it possible to isolate and identify one after another of the joint coordinates.

For example, in the case of a six degrees of freedom robot, the procedure is as follows:

– left multiplication of both sides of equation [1.14] by ${}^{1}\mathbf{T}_{0}$:

$$ {}^{1}\mathbf{T}_{0}\ \mathbf{U}_0 = {}^{1}\mathbf{T}_{2}\ {}^{2}\mathbf{T}_{3}\ {}^{3}\mathbf{T}_{4}\ {}^{4}\mathbf{T}_{5}\ {}^{5}\mathbf{T}_{6} \quad [1.15]$$

The right hand side is a function of the variables q_2, ..., q_6. The left hand side is only a function of the variable q_1;

– term-to-term identification of the two sides of equation [1.15]. We obtain a system of one or two equations function of q_1 only, whose structure belongs to a particular type amongst a dozen of possible types;

– left multiplication of both sides of equation [1.15] by $^{2}\mathbf{T}_{1}$ and calculation of q_2.

The succession of equations enabling the calculation of all q_j is the following:

$$\begin{aligned}
&\mathbf{U}_0 = {}^{0}\mathbf{T}_1\,{}^{1}\mathbf{T}_2\,{}^{2}\mathbf{T}_3\,{}^{3}\mathbf{T}_4\,{}^{4}\mathbf{T}_5\,{}^{5}\mathbf{T}_6\\
&{}^{1}\mathbf{T}_0\,\mathbf{U}_0 = {}^{1}\mathbf{T}_2\,{}^{2}\mathbf{T}_3\,{}^{3}\mathbf{T}_4\,{}^{4}\mathbf{T}_5\,{}^{5}\mathbf{T}_6\\
&{}^{2}\mathbf{T}_1\,\mathbf{U}_1 = {}^{2}\mathbf{T}_3\,{}^{3}\mathbf{T}_4\,{}^{4}\mathbf{T}_5\,{}^{5}\mathbf{T}_6\\
&{}^{3}\mathbf{T}2\,\mathbf{U}_2 = {}^{3}\mathbf{T}_4\,{}^{4}\mathbf{T}_5\,{}^{5}\mathbf{T}_6\\
&{}^{4}\mathbf{T}_3\,\mathbf{U}_3 = {}^{4}\mathbf{T}_5\,{}^{5}\mathbf{T}_6\\
&{}^{5}\mathbf{T}_4\,\mathbf{U}_4 = {}^{5}\mathbf{T}_6
\end{aligned} \qquad [1.16]$$

with:

$$\mathbf{U}_{j+1} = {}^{j+1}\mathbf{T}_6 = {}^{j+1}\mathbf{T}_j\,\mathbf{U}_j \text{ for } j = 0, \ldots, 4 \qquad [1.17]$$

The use of this method for a large number of industrial robots has shown that only a few types of equations are encountered, and that their solutions are relatively simple.

NOTES.

1) When a robot has more than six degrees of freedom, the system to be solved contains more unknowns than parameters describing the task: it lacks (n–6) relations. Two strategies are possible:

– the first strategy consists of setting arbitrarily (n–6) joint variables. In this case we deal with a problem with six degrees of freedom. The choice of these joints results from the task's specifications and from the structure of the robot;

– the second strategy consists of introducing (n–6) supplementary relations describing the redundancy, like for example in [HOL 84] for robots with seven degrees of freedom.

2) A robot with less than six degrees of freedom cannot place its end-effector at arbitrary positions and orientations. Thus, it is not possible to bring the end-effector frame R_E onto another desired frame $R_E{}^d$ except if certain elements of $^{0}\mathbf{T}_E{}^d$ are imposed in a way that compensates for the insufficient number of degrees of freedom. Otherwise, we have to reduce the number of equations by considering only certain elements (points or axes) of the frames R_E and $R_E{}^d$.

EXAMPLE 1.3.– inverse geometric model of the Stäubli RX-90 robot. After performing all the calculations, we obtain the following solutions:

$$\begin{cases} \theta_1 = \text{atan2}(P_y, P_x) \\ \theta'_1 = \theta_1 + \pi \end{cases}$$

$$\theta_2 = \text{atan2}(S2, C2)$$

with:

$$\begin{cases} C2 = \dfrac{YZ - \varepsilon X\sqrt{X^2 + Y^2 - Z^2}}{X^2 + Y^2} \\ S2 = \dfrac{XZ - \varepsilon Y\sqrt{X^2 + Y^2 - Z^2}}{X^2 + Y^2} \end{cases} \qquad \text{with } \varepsilon = \pm 1$$

$$X = -2P_z D3$$
$$Y = -2\,B1D3$$
$$Z = (RL4)^2 - (D3)^2 - (P_z)^2 - (B1)^2$$
$$B1 = P_x\,C1 + P_y\,S1$$

$$\theta_3 = \text{atan2}\left(\frac{-P_zS2 - B1C2 + D3}{RL4}, \frac{-B1S2 + P_zC2}{RL4}\right)$$

$$\begin{cases} \theta_4 = \text{atan2}[S1\,a_x - C1\,a_y, -C23(C1\,a_x + S1\,a_y) - S23\,a_z] \\ \theta'_4 = \theta_4 + \pi \end{cases}$$

$$\theta_5 = \text{atan2}(S5, C5)$$

with:

$$S5 = -C4\,[C23\,(C1\,a_x + S1\,a_y) + S23a_z] + S4\,(S1\,a_x - C1\,a_y)$$
$$C5 = -S23\,(C1\,a_x + S1\,a_y) + C23\,a_z$$

$$\theta_6 = \text{atan2}(S6, C6)$$

with:

$$S6 = -\,C4\,(S1\,s_x - C1\,s_y) - S4\,[C23\,(C1\,s_x + S1\,s_y) + S23\,s_z]$$
$$C6 = -\,C4\,(S1\,n_x - C1\,n_y) - S4\,[C23\,(C1\,n_x + S1\,n_y) + S23\,n_z]$$

NOTES.

1) Singular positions:

i) when $P_x = P_y = 0$, which corresponds to S23RL4 – C2D3 = 0, the point O_4 is on the axis $\mathbf{z}_0$ (Figure 1.5a). The two arguments used for calculating θ_1 are zero and consequently θ_1 is not determined. We can give any value to θ_1, generally the value of the current position, or, according to optimization criteria, such as maximizing the distance from the mechanical limits of the joints. This means that we can always find a solution, but a small change of the desired position might call for a significant variation of θ_1, which may be impossible to carry out considering the velocity and acceleration limits of the actuators,

ii) when $C23(C1a_x + S1a_y) + S23a_z = 0$ and $S1a_x - C1a_y = 0$, the two arguments of the atan2 function used for the calculation of θ_4 are zero and hence the function is not determined. This configuration happens when axes 4 and 6 are aligned ($C\theta_5 = \pm 1$) and it is the sum ($\theta_4 \pm \theta_6$) which can be obtained (see Figure 1.5b). We can give to θ_4 its current value, then we calculate θ_6 according to this value. We can also calculate the values of θ_4 and θ_6, which move joints 4 and 6 away from their limits,

iii) a third singular position occurring when C3 = 0 will be highlighted along with the kinematic model. This singularity does not pose any problem for the inverse geometric model (see Figure 1.5c).

2) Number of solutions: apart from singularities, the Stäubli RX-90 robot has eight theoretical configurations for the IGM (product of the number of possible solutions on each axis). Some of these configurations may not be accessible due to their joint limits.

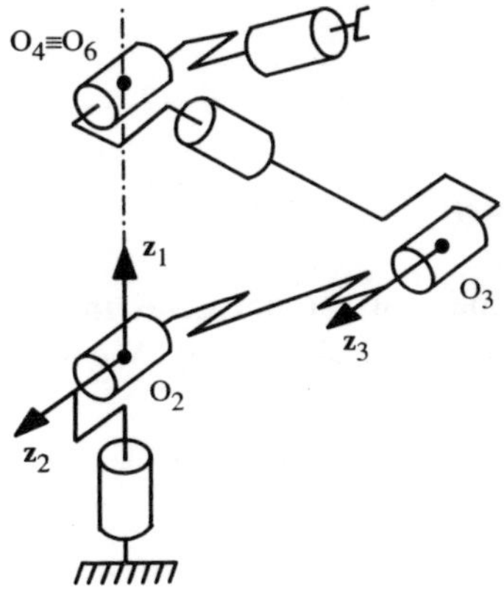

a) Singularity of the shoulder ($P_x = P_y = 0$ and S23RL4 – C2D3 = 0)

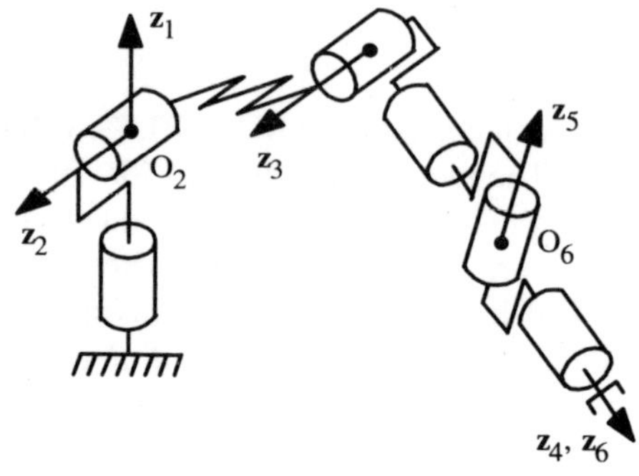

b) Singularity of the wrist (S5 = 0)

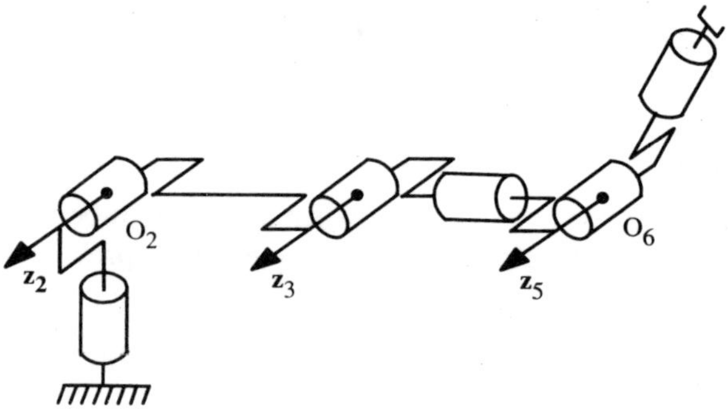

c) Singularity of the elbow (C3 = 0)

Figure 1.5. *Singular positions of the Stäubli RX-90 robot*

1.3. Kinematic modeling

1.3.1. *Direct kinematic model*

The direct kinematic model of a robot gives the velocities of the operational coordinates $\dot{\mathbf{X}}$ in terms of the joint velocities $\dot{\mathbf{q}}$. We write:

$$\dot{\mathbf{X}} = \mathbf{J}(\mathbf{q})\dot{\mathbf{q}} \qquad [1.18]$$

where $\mathbf{J(q)}$ indicates the (m × n) Jacobian matrix of the mechanism, such that:

$$\mathbf{J(q)} = \frac{\partial \mathbf{X}}{\partial \mathbf{q}} \quad [1.19]$$

This Jacobian matrix appears in calculating the direct differential model that gives the differential variations **dX** of the operational coordinates in terms of the differential variations of the joint coordinates **dq**, such as:

$$\mathbf{dX} = \mathbf{J(q)}\ \mathbf{dq} \quad [1.20]$$

The Jacobian matrix has multiple applications in robotics [WHI 69], [PAU 81]:

– it is at the base of the inverse differential model, which can be used to calculate a local solution of joint coordinates **q** corresponding to an operational coordinates **X**;

– in static force model, we use the Jacobian matrix in order to calculate the forces and torques of the actuators in terms of the forces and moments exerted on the environment by the end-effector;

– it facilitates the calculation of singularities and of the dimension of accessible operational space of the robot [BOR 86], [WEN 89].

1.3.1.1. *Calculation of the Jacobian matrix by derivation of the DGM*

The calculation of the Jacobian matrix can be done by differentiating the DGM, $\mathbf{X} = \mathbf{f(q)}$, using the following relation:

$$J_{ij} = \frac{\partial f_i(\mathbf{q})}{\partial q_j} \quad i = 1, \ \ldots, m; j = 1, \ \ldots, n \quad [1.21]$$

where J_{ij} is the (i, j) element of the Jacobian matrix **J**.

This method is easy to apply for robots with two or three degrees of freedom, as shown in the following example. The calculation of the kinematic Jacobian matrix presented in section 1.3.1.2 is more practical for robots with more than three degrees of freedom.

EXAMPLE 1.4.– let us consider the planar robot with three degrees of freedom of parallel revolute axes represented in Figure 1.6. We use L1, L2 and L3 to denote the lengths of the links. We choose as operational coordinates the Cartesian coordinates (P_x, P_y) of point E in the plane ($\mathbf{x}_0$, $\mathbf{y}_0$) and the angle α between $\mathbf{x}_0$ and $\mathbf{x}_3$.

$$P_x = C1\ L1 + C12\ L2 + C123\ L3$$
$$P_y = S1\ L1 + S12\ L2 + S123\ L3$$
$$\alpha = \theta_1 + \theta_2 + \theta_3$$

with $C12 = \cos(\theta_1 + \theta_2)$, $S12 = \sin(\theta_1 + \theta_2)$, $C123 = \cos(\theta_1 + \theta_2 + \theta_3)$ and $S123 = \sin(\theta_1 + \theta_2 + \theta_3)$

The Jacobian matrix is calculated by differentiating these relations with respect to θ1, θ2 and θ3:

$$\mathbf{J} = \begin{bmatrix} -S1L1 - S12L2 - S123L3 & -S12L2 - S123L3 & -S123L3 \\ C1L1 + C12L2 + C123L3 & C12L2 + C123L3 & C123L3 \\ 1 & 1 & 1 \end{bmatrix}$$

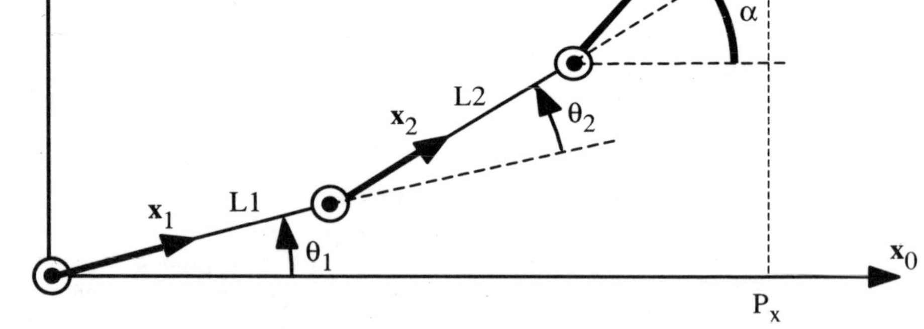

Figure 1.6. *Example of a planar robot with three degrees of freedom*

1.3.1.2. *Kinemat:c Jacobian matrix*

The kinematic Jacobian matrix is obtained through a direct calculation using the relation between the linear and angular velocity vectors $\mathbf{V}_n$ and $\boldsymbol{\omega}_n$ of the frame R_n and the joint velocity $\dot{\mathbf{q}}$:

$$\begin{bmatrix} \mathbf{V}_n \\ \omega_n \end{bmatrix} = \mathbf{J}_n \, \dot{\mathbf{q}} \qquad [1.22]$$

We note that $\mathbf{V}_n$ is the derivative of the position vector $\mathbf{P}_n$ with respect to time. On the other hand, $\boldsymbol{\omega}_n$ is not a derivative of any representation of the orientation.

The expression of the Jacobian matrix is identical if we consider the relation between the differential translational and rotational vectors ($\mathbf{dP}_n$, $\boldsymbol{\delta}_n$) of the frame R_n and the differential increments of the joint coordinates $\mathbf{dq}$:

$$\begin{bmatrix} \mathbf{dP}_n \\ \boldsymbol{\delta}_n \end{bmatrix} = \mathbf{J}_n \; \mathbf{dq} \qquad [1.23]$$

i) Calculation of the kinematic Jacobian matrix

Let us consider the kth joint of an articulated serial chain. The velocity $\dot{q}_k$ produces on the end-effector frame R_n the linear velocity $\mathbf{V}_{k,n}$ and the angular velocity $\omega_{k,n}$. We recall that $\mathbf{a}_k$ is the unit vector along axis $\mathbf{z}_k$ of the joint k and we indicate by $L_{k,n}$ the vector of origin O_k and extremity O_n. By applying the velocity composition law, the linear and angular velocities of the end-effector frame are written as follows:

$$\begin{cases} \mathbf{V}_n = \sum_{k=1}^{n} \mathbf{V}_{k,n} = \sum_{k=1}^{n} [\sigma_k \mathbf{a}_k + \bar{\sigma}_k (\mathbf{a}_k x \mathbf{L}_{k,n}) \, \dot{\mathbf{q}}_k \\ \omega_n = \sum_{k=1}^{n} \omega_{k,n} = \sum_{k=1}^{n} \bar{\sigma}_k \mathbf{a}_k \dot{\mathbf{q}}_k \end{cases} \qquad [1.24]$$

If we write this system in a matrix form and if we identify it with the relation [1.22], we conclude that:

$$\mathbf{J}_n = \begin{bmatrix} \sigma_1 \mathbf{a}_1 + \bar{\sigma}_1 (\mathbf{a}_1 x \mathbf{L}_{1,n}) \cdots \sigma_n \mathbf{a}_n + \bar{\sigma}_n (\mathbf{a}_n x \mathbf{L}_{n,n}) \\ \bar{\sigma}_1 \mathbf{a}_1 \quad \cdots \quad \bar{\sigma}_n \mathbf{a}_n \end{bmatrix} \qquad [1.25]$$

In general, $\mathbf{V}_n$ and $\boldsymbol{\omega}_n$ are expressed either in the frame R_n or in the frame R_0. The corresponding Jacobian matrix is noted either ${}^n\mathbf{J}_n$ or ${}^0\mathbf{J}_n$. These matrices can also be calculated by using a matrix ${}^i\mathbf{J}_n$, i = 0, ..., n, due to the following relation of transformation of the Jacobian matrix between frames:

$$ {}^s\mathbf{J}_n = \begin{bmatrix} {}^s\mathbf{R}_i & \mathbf{0}_3 \\ \mathbf{0}_3 & {}^s\mathbf{R}_i \end{bmatrix} {}^i\mathbf{J}_n \qquad [1.26] $$

where ${}^s\mathbf{R}_i$ is the (3×3) orientation matrix of frame R_i expressed in frame R_s.

The matrix ${}^s\mathbf{J}_n$ can therefore be decomposed into two matrices, the first one always being a full rank matrix.

Since the two matrices ${}^i\mathbf{J}_n$ and ${}^s\mathbf{J}_n$ have the same singular positions, we practically try to use the projection frame R_i which simplifies the elements of the matrix ${}^i\mathbf{J}_n$. In general, we obtain the simplest ${}^i\mathbf{J}_n$ matrix when we consider i = integer(n/2).

ii) Calculation of the ${}^iJ_n$ matrix

We notice that the vector product $\mathbf{a}_k \times \mathbf{L}_{k,n}$ can be transformed[2] into $\hat{\mathbf{a}}_k \mathbf{L}_{k,n}$ and the kth column of ${}^i\mathbf{J}_n$, denoted ${}^i\mathbf{j}_{n;k}$, becomes:

$$ {}^i\mathbf{j}_{n;k} = \begin{bmatrix} \sigma_k \, {}^i\mathbf{a}_k + \bar{\sigma}_k \, {}^i\mathbf{R}_k \, {}^k\hat{\mathbf{a}}_k \, {}^k\mathbf{L}_{k,n} \\ \bar{\sigma}_k \, {}^i\mathbf{a}_k \end{bmatrix} \qquad [1.27] $$

By developing and noting that ${}^k\mathbf{a}_k = [0 \;\; 0 \;\; 1]^T$ and that ${}^k\mathbf{L}_{k,n} = {}^k\mathbf{P}_n$, we obtain:

$$ {}^i\mathbf{j}_{n;k} = \begin{bmatrix} \sigma_k \, {}^i\mathbf{a}_k + \bar{\sigma}_k (-{}^kP_{n_y} \, {}^i\mathbf{s}_k + {}^kP_{n_x} \, {}^i\mathbf{n}_k \\ \bar{\sigma}_k \, {}^i\mathbf{a}_k \end{bmatrix} \qquad [1.28] $$

where ${}^kP_{nx}$ and ${}^kP_{ny}$ are respectively the x and y components of the vector ${}^k\mathbf{P}_n$.

---

2 The skew symmetric matrix $\hat{\mathbf{a}}$ is defined by:

$$ \hat{\mathbf{a}} = \begin{bmatrix} 0 & -a_z & a_y \\ a_z & 0 & -a_x \\ -a_y & a_x & 0 \end{bmatrix} $$

Similarly, the kth column of ${}^{i}\mathbf{J}_n$ can be written as follows:

$$ {}^{i}\mathbf{j}_{n;k} = \begin{bmatrix} \sigma_k\, {}^{i}\mathbf{a}_k + \bar{\sigma}_k\, {}^{i}\hat{\mathbf{a}}_k ({}^{i}\mathbf{P}_n - {}^{i}\mathbf{P}_k) \\ \bar{\sigma}_k\, {}^{i}\mathbf{a}_k \end{bmatrix} \qquad [1.29] $$

When $i = 0$, the elements of column k are obtained from those of the matrix ${}^{0}\mathbf{T}_k$ and the vector ${}^{0}\mathbf{P}_n$. We must then calculate the matrices ${}^{0}\mathbf{T}_k$, for $k = 1, \ldots, n$.

EXAMPLE 1.5.– calculation of the kinematic Jacobian matrix ${}^{3}\mathbf{J}_6$ of the Stäubli RX-90 robot. Column k of the matrix ${}^{3}\mathbf{J}_6$ of a 6R robot with six degrees of freedom can be written as follows:

$$ {}^{3}\mathbf{j}_{6;k} = \begin{bmatrix} -{}^{k}P_{6y}\, {}^{3}\mathbf{s}_k + {}^{k}P_{6x}\, {}^{3}\mathbf{n}_k \\ {}^{3}\mathbf{a}_k \end{bmatrix} $$

Hence:

$$ {}^{3}\mathbf{J}_6 = \begin{bmatrix} 0 & -RL4 + S3D3 & -RL4 & 0 & 0 & 0 \\ 0 & C3D3 & 0 & 0 & 0 & 0 \\ S23\,RL4 - C2D3 & 0 & 0 & 0 & 0 & 0 \\ S23 & 0 & 0 & 0 & S4 & -S5C4 \\ C23 & 0 & 0 & 1 & 0 & C5 \\ 0 & 1 & 1 & 0 & C4 & S5S4 \end{bmatrix} $$

1.3.1.3. *Decomposition of the kinematic Jacobian matrix into three matrices*

With the help of relation [1.26], we have shown that the matrix ${}^{s}\mathbf{J}_n$ could be decomposed into two matrices, the first one always being of full rank and the second one containing simple elements. Renaud [REN 80b] demonstrated that we can also decompose the Jacobian matrix into three matrices: the first two are always of full rank and their inversion is immediate; the third one is of the same rank as ${}^{s}\mathbf{J}_n$ but contains much more simple elements. We obtain [KHA 02]:

$$ {}^{s}\mathbf{J}_n = \begin{bmatrix} {}^{s}\mathbf{R}_i & \mathbf{0}_3 \\ \mathbf{0}_3 & {}^{s}\mathbf{R}_i \end{bmatrix} \begin{bmatrix} \mathbf{I}_3 & -{}^{i}\hat{\mathbf{L}}_{j,n} \\ \mathbf{0}_3 & \mathbf{I}_3 \end{bmatrix} {}^{i}\mathbf{J}_{n,j} \qquad [1.30] $$

the elements of the kth column of ${}^{i}\mathbf{J}_{n,j}$ being expressed in the frame R_i in the following way:

$$ {}^{i}\mathbf{j}_{n,j;k} = \begin{bmatrix} \sigma_k \, {}^{i}\mathbf{a}_k + \bar{\sigma}_k(-{}^{k}P_{jy} \, {}^{i}\mathbf{s}_k + {}^{k}P_{jx} \, {}^{i}\mathbf{n}_k) \\ \bar{\sigma}_k \, {}^{i}\mathbf{a}_k \end{bmatrix} \qquad [1.31] $$

1.3.1.4. *Dimension of the operational space of a robot*

For a given joint configuration $\mathbf{q}$, the rank r of the Jacobian matrix ${}^{i}\mathbf{J}_n$ henceforth denoted $\mathbf{J}$ for simplification of notations, corresponds to the number of degrees of freedom associated with the end-effector. It defines the dimension of the accessible operational space in this configuration. We call number of degrees of freedom M of the operational space of a robot, the maximal rank r_{max} that the Jacobian matrix has in all possible configurations. Two cases are to be examined [GOR 84]:

– if M is equal to the number of degrees of freedom n of the robot, the robot is not redundant: it has just the number of joints enabling it to give M degrees of freedom to its end-effector;

– if $n > M$, the robot is redundant of the order $(n - M)$. It has more joints than are needed to give M degrees of freedom to its end-effector.

Whatever the case, for certain joint configurations, the rank r may be less than M: we say that the robot has a *singularity* of order $(M - r)$. It then loses, locally, the possibility of generating a velocity along or about certain directions. When matrix $\mathbf{J}$ is square, singularities of order one are solution of $\det(\mathbf{J}) = 0$, where $\det(\mathbf{J})$ indicates the determinant of the Jacobian matrix of the robot. They are given by $\det(\mathbf{J}\mathbf{J}^T) = 0$ in the redundant case.

Based on the results obtained in Example 1.5, we can verify that for the Stäubli RX-90 robot, the determinant of ${}^{3}\mathbf{J}_6$ can be written as follows:

$$ \det({}^{3}\mathbf{J}_6) = -\,C3\ D3\ RL4\ S5\ (S23\ RL4 - C2\ D3) $$

The maximal rank is such that $r_{max} = 6$. The robot is non-redundant since it has six degrees of freedom and six joints. However, this rank is equal to five in the following three singular configurations:

$$ \begin{cases} C3 = 0 \\ S23RL4 - C2D3 = 0 \\ S5 = 0 \end{cases} $$

1.3.2. *Inverse kinematic model*

The objective of the inverse kinematic model is to calculate, at a given configuration $\mathbf{q}$, the joint velocity $\dot{\mathbf{q}}$ which provides a desired operational velocity $\dot{\mathbf{X}}$ to the end-effector. This definition is similar to that of the inverse differential model: the latter determines the joint differential variation $\mathbf{dq}$ corresponding to a specified differential variation of the operational coordinates $\mathbf{dX}$. In order to obtain the inverse kinematic model, we inverse the direct kinematic model by solving a system of linear equations. The implementation may be done in an analytical or numerical manner:

– the analytical solution has the advantage of considerably reducing the number of operations, but all singular cases must be treated separately [CHE 87];

– the numeric methods are more general, the most widely used one being based on the pseudo-inverse: the algorithms deal in a unified way with the regular, singular, and redundant cases. They require a relatively significant calculation time.

In this section we will present the techniques to be implemented in order to establish the inverse kinematic model for regular, singular and redundant cases.

1.3.2.1. *General form of the kinematic model*

Let $\mathbf{X} = [\mathbf{X}_p^T \ \mathbf{X}_r^T]^T$ be any representation with respect to R_0 of the location of frame R_n, the elements $\mathbf{X}_p$ and $\mathbf{X}_r$ designating, respectively, the operational position and orientation vectors. The relations between the velocities $\dot{\mathbf{X}}_p$ and $\dot{\mathbf{X}}_r$ and the velocity vectors ${}^0\mathbf{V}_n$ and ${}^0\boldsymbol{\omega}_n$ of the end-effector frame R_n are the following:

$$\begin{bmatrix} \dot{\mathbf{X}}_p \\ \dot{\mathbf{X}} \end{bmatrix} = \begin{bmatrix} \boldsymbol{\Omega}_p & \mathbf{0}_3 \\ \mathbf{0}_3 & \boldsymbol{\Omega}_r \end{bmatrix} \begin{bmatrix} {}^0\mathbf{V}_n \\ {}^0\boldsymbol{\omega}_n \end{bmatrix} = \boldsymbol{\Omega} \begin{bmatrix} {}^0\mathbf{V}_n \\ {}^0\boldsymbol{\omega}_n \end{bmatrix} \quad [1.32]$$

the matrices $\boldsymbol{\Omega}_p$ and $\boldsymbol{\Omega}_r$ depending on the representation having been chosen for position and for orientation respectively [KHA 02].

On the basis of equations [1.22], [1.30] and [1.32], the direct kinematic model has as general form:

$$\dot{\mathbf{X}} = \begin{bmatrix} \boldsymbol{\Omega}_p & \mathbf{0}_3 \\ \mathbf{0}_3 & \boldsymbol{\Omega}_r \end{bmatrix} \begin{bmatrix} {}^0\mathbf{R}_i & \mathbf{0}_3 \\ \mathbf{0}_3 & {}^0\mathbf{R}_i \end{bmatrix} \begin{bmatrix} \mathbf{I}_3 & -{}^i\hat{\mathbf{L}}_{j,n} \\ \mathbf{0}_3 & \mathbf{I}_3 \end{bmatrix} {}^i\mathbf{J}_{n,j}\, \dot{\mathbf{q}} \quad [1.33]$$

or, as a short form:

$$\dot{\mathbf{X}} = \mathbf{J}\,\dot{\mathbf{q}} \qquad [1.34]$$

1.3.2.2. *Inverse kinematic model for the regular case*

In this case, the Jacobian matrix **J** is square and of full rank, hence we can calculate $\mathbf{J}^{-1}$, the inverse matrix of **J**, which makes it possible to determine the joint velocities $\dot{\mathbf{q}}$ with the relation:

$$\dot{\mathbf{X}} = \mathbf{J}^{-1}\dot{\mathbf{q}} \qquad [1.35]$$

When the matrix **J** has the following form:

$$\mathbf{J} = \begin{bmatrix} \mathbf{A} & \mathbf{0} \\ \mathbf{B} & \mathbf{C} \end{bmatrix} \qquad [1.36]$$

the matrices **A** and **C** being inversible and square, it is easy to show that the inverse of this matrix can be written:

$$\mathbf{J} = \begin{bmatrix} \mathbf{A}^{-1} & \mathbf{0} \\ -\mathbf{C}^{-1}\mathbf{B}\mathbf{A}^{-1} & \mathbf{C}^{-1} \end{bmatrix} \qquad [1.37]$$

Thus, the solution of this problem lies on the inversion of two matrices of smaller dimension. When the robot has six degrees of freedom and a spherical wrist, the general form of **J** is that of the relation [1.36], **A** and **C** having the dimension (3×3) [GOR 84].

EXAMPLE 1.6.– calculation of the inverse Jacobian matrix of the Stäubli RX-90 robot. The Jacobian matrix ${}^{3}\mathbf{J}_{6}$ was calculated in Example 1.5. The calculation of the inverse matrices of **A** and **C** gives:

$$\mathbf{A}^{-1} = \begin{bmatrix} 0 & 0 & V1 \\ 0 & V3 & 0 \\ -1/RL4 & V2V3/RL4 & 0 \end{bmatrix}, \quad \mathbf{C}^{-1} = \begin{bmatrix} V4 & 1 & -V5 \\ S4 & 0 & C4 \\ -C4/S5 & 0 & S4/S5 \end{bmatrix}$$

with:

$$V1 = \frac{1}{S23RL4 - C2D3}$$

$$V2 = -RL4 + S3D3$$

$$V3 = \frac{1}{C3D3}$$

$$V4 = C4\ \text{cotg}5$$

$$V5 = S4\ \text{cotg}5$$

By using equation [1.37], we obtain:

$$^{3}\mathbf{J}_{6}^{-1} = \begin{bmatrix} 0 & 0 & V1 & 0 & 0 & 0 \\ 0 & V3 & 0 & 0 & 0 & 0 \\ -1/RL4 & V2V3/RL4 & 0 & 0 & 0 & 0 \\ -S4C5V7 & V5V6 & V8 & V4 & 1 & -V5 \\ C4/RL4 & -C4V6 & -S23S4V1 & S4 & 0 & C4 \\ S4V7 & -S4V6/S5 & S23C4V1/S5 & -C4/S5 & 0 & S4/15 \end{bmatrix}$$

with:

$$V6 = \frac{S3}{C3RL4}$$

$$V7 = \frac{1}{S5RL4}$$

$$V8 = (-S23V4 - C23)V1$$

1.3.2.3. *Solution at the proximity of singular positions*

We have seen that when the robot is not redundant, the singularities of order one are the solution of det(**J**) = 0. In the redundant case, they are given by det($\mathbf{JJ}^{\mathrm{T}}$) = 0. The higher order singularities are determined based on singular configurations of order one. The passage in the proximity of a singular position is however determined in a more precise way by considering the singular values: the decrease of one or several singular values is generally more significant than the decrease of the determinant.

At a singular configuration, the velocity vector $\dot{\mathbf{X}}$ generally consists of a vector of the image space vector $I(\mathbf{J})$ of $\mathbf{J}$, and of an orthogonal vector of degenerated components belonging to $I(\mathbf{J})^{\perp}$; no joint velocity can generate operational velocity following this last direction. In the proximity of singular positions, the use of the classic inverse kinematic model can cause significant joint velocities, incompatible with the characteristics of the actuators.

In order to avoid singularities, one solution consists of increasing the number of degrees of freedom of the mechanism [HOL 84], [LUH 85]. The robot becomes redundant and, with appropriate criteria, it is possible to determine singularity free motion. However, inevitable singularities [BAI 84] exist, which must be taken into account by the designer of the control.

At singular configurations, it is not possible to calculate $\mathbf{J}^{-1}$. It is common to use the pseudo-inverse $\mathbf{J}^{+}$ of the matrix $\mathbf{J}$:

$$\dot{\mathbf{q}} = \mathbf{J}^{+}\,\dot{\mathbf{X}} \qquad [1.38]$$

This solution, proposed by Whitney [WHI 69], [WHI 72], minimizes the Euclidean norm $\|\dot{\mathbf{q}}\|^2$ as well as the error norm $\|\dot{\mathbf{X}} - \mathbf{J}\dot{\mathbf{q}}\|^2$. In a singular configuration, we distinguish the following two particular cases:

– $\dot{\mathbf{X}}$ belongs only to $I(\mathbf{J})$. Thus, solution [1.38] is exact and the error is zero, although the inverse $\mathbf{J}^{-1}$ is not defined;

– $\dot{\mathbf{X}}$ belongs only to $I(\mathbf{J})^{\perp}$. Therefore, solution [1.38] gives $\dot{\mathbf{q}} = \mathbf{0}$.

1.3.2.4. *Inverse kinematic model of redundant robots*

A robot is redundant when the number of joints n is greater than the dimension of the operational space of the end-effector M. Therefore, an infinite number of joint solutions exist in order to carry out a given task. The inverse geometric and kinematic models have in this case an infinite number of solutions; hence the possibility of choosing the solution that meets supplementary optimization criteria, such as:

– avoiding obstacles [MAC 85], [BAI 86];

– avoiding singular configurations [YOS 84];

– avoiding joint limits [FOU 80], [KLE 84];

– minimizing joint torques and forces [BAI 84], [HOL 85].

For such a mechanism, the matrix $\mathbf{J}$ has the dimension $(m \times n)$ where $n > m$, if we suppose that the operational coordinates used are independent $(m = M)$. Several methods of solving the system [1.34] are conceivable. A classic solution consists of using a pseudo-inverse with an optimization term. The general solution of the system of linear equations [1.34] can be written:

$$\dot{\mathbf{q}} = \mathbf{J}^{+}\dot{\mathbf{X}} + \left(\mathbf{I}_n - \mathbf{J}^{+}\mathbf{J}\right)\mathbf{Z} \qquad [1.39]$$

where $\mathbf{J}^{+}$ designates the pseudo-inverse of $\mathbf{J}$ and where $\mathbf{Z}$ represents an arbitrary $(n \times 1)$ vector.

The second term on the right hand side, called the homogenous solution or optimization term, belongs to the zero space of $\mathbf{J}$ and does not affect the value of $\dot{\mathbf{X}}$. It can be used to realize supplementary optimization criteria. Let $\phi(\mathbf{q})$ be a positive definite scalar function of $\mathbf{q}$ and let $\nabla\phi$ be the gradient of this function. We see that taking $\mathbf{Z} = \alpha\nabla\phi$ leads to the decrease of the function $\phi(\mathbf{q})$ when $\alpha < 0$ and leads to the increase of this function when $\alpha > 0$. The solution is thus written:

$$\dot{\mathbf{q}} = \mathbf{J}^{+}\dot{\mathbf{X}} + \alpha\left(\mathbf{I}_n - \mathbf{J}^{+}\mathbf{J}\right)\nabla\phi \qquad [1.40]$$

with:

$$\nabla\phi = \begin{bmatrix} \frac{\partial\phi}{\partial q_1} & \frac{\partial\phi}{\partial q_n} \end{bmatrix}^T \qquad [1.41]$$

The coefficient α makes it possible to find a compromise between the objectives of minimization of $\|\dot{\mathbf{q}}\|^2$ and of optimization of $\phi(\mathbf{q})$. Several choices are possible for the optimization criteria such as the distance of joint limits or the increase of manipulability.

Another approach consists of adding to the vector of operational coordinates $\mathbf{X}$ a vector of (n-m) supplementary linearly independent relations [BAI 85], [CHA 86], [NEN 92]. These relations can represent either physical constraints on the robot, or constraints linked to its environment or simply relations between different joint positions of the robot.

1.4. Calibration of geometric parameters

1.4.1. *Introduction*

The objective of geometric calibration is to identify the exact values of the geometric parameters, which play a role in the calculation of the geometric models of the robot, in order to improve its static accuracy. The nominal values of these parameters are known and generally provided by the manufacturer. It is then a matter of identifying the distance between the nominal values and the real values [HOL 89], [KHA 02].

1.4.2. *Geometric parameters*

The location of the end-effector in the world frame R_f is given by (see section 1.2.3.1):

$$ {}^{f}\mathbf{T}_E = \mathbf{Z}\, {}^{0}\mathbf{T}_n(\mathbf{q})\, \mathbf{E} = \mathbf{Z}\, {}^{0}\mathbf{T}_1\, {}^{1}\mathbf{T}_2 \ldots {}^{n-1}\mathbf{T}_n\, \mathbf{E} \qquad [1.42] $$

with:

– $\mathbf{Z}$, $\mathbf{E}$: transformation matrices defined in section 1.2.3.1;

– ${}^{0}\mathbf{T}_n$: transformation matrix of the robot.

Based on relation [1.42], we distinguish three types of parameters: the geometric parameters of the robot required to calculate the matrix ${}^{0}\mathbf{T}_n$, the parameters defining matrix $\mathbf{Z}$ and those defining matrix $\mathbf{E}$.

1.4.2.1. *Geometric parameters of the robot*

Matrix ${}^{j-1}\mathbf{T}_j$ defined by equation [1.3] is based on (α_j, d_j, θ_j, r_j). In the case where two consecutive axes j-1 and j are parallel, the choice of $\mathbf{x}_{j-1}$ is not unique, which leads to an arbitrary choice of the position of $\mathbf{x}_{j-1}$. When an error takes place such that $\mathbf{z}_{j-1}$ and $\mathbf{z}_j$ are not rigorously parallel, the error Δr_{j-1} for parameter r_{j-1} can become very significant. For this reason, we add a fifth parameter β_j [HAY 83], which is a rotation about $\mathbf{y}_{j-1}$ and which makes it possible to characterize the parallelism error of axes $\mathbf{z}_{j-1}$ and $\mathbf{z}_j$. Relation ${}^{j-1}\mathbf{T}_j$ becomes:

$$ {}^{j-1}\mathbf{T}_j = \mathbf{Rot}(\mathbf{y},\beta_j)\, \mathbf{Rot}(\mathbf{x},\alpha_j)\, \mathbf{Trans}(\mathbf{x},d_j)\, \mathbf{Rot}(\mathbf{z},\theta_j)\, \mathbf{Trans}(\mathbf{z},r_j) \qquad [1.43] $$

The nominal value of β_j is zero. When $\mathbf{z}_{j-1}$ and $\mathbf{z}_j$ are not parallel, β_j is not identifiable. We also note that when $\mathbf{z}_{j-1}$ and $\mathbf{z}_j$ are parallel, we identify either r_{j-1} or r_j. The maximum number of parameters per joint frame is therefore four.

1.4.2.2. *Parameters of the robot's location*

We associate index –1 to the world frame R_f. Since this reference frame can be chosen arbitrarily, six parameters (γ_z, b_z, α_z, d_z, θ_z, r_z) (see Figure 1.7) are needed, in order to define the robot base frame R_0 in R_f. These parameters are defined by [KHA 91]:

$$\mathbf{Z} = {}^{-1}\mathbf{T}_0 = \mathbf{Rot(z,}\gamma_z\mathbf{)}\ \mathbf{Trans(z,b_z)}\ \mathbf{Rot(x,}\alpha_z\mathbf{)}\ \mathbf{Trans(x,d_z)}$$
$$\mathbf{Rot\ (z,}\theta_z\mathbf{)}\ \mathbf{Trans(z,r_z)} \qquad [1.44]$$

and since α_1 and d_1 are zero, we can write that:

$${}^{-1}\mathbf{T}_1 = \mathbf{Rot(x,}\alpha_0\mathbf{)}\ \mathbf{Trans(x,d_0)}\ \mathbf{Rot(z,}\theta_0\mathbf{)}\ \mathbf{Trans(z,r_0)}$$
$$\mathbf{Rot(x,}\alpha'_1\mathbf{)}\ \mathbf{Trans(x,d'_1)}\ \mathbf{Rot(z,}\theta'_1\mathbf{)}\ \mathbf{Trans(z,\ r'_1)} \qquad [1.45]$$

with: $\alpha_0 = 0$, $d_0 = 0$, $\theta_0 = \gamma_z$, $r_0 = b_z$, $\alpha'_1 = \alpha_z$, $d'_1 = d_z$, $\theta'_1 = \theta_1 + \theta_z$, $r'_1 = r_1 + r_z$.

Hence, equation [1.45] represents two transformations of the classic type [1.3].

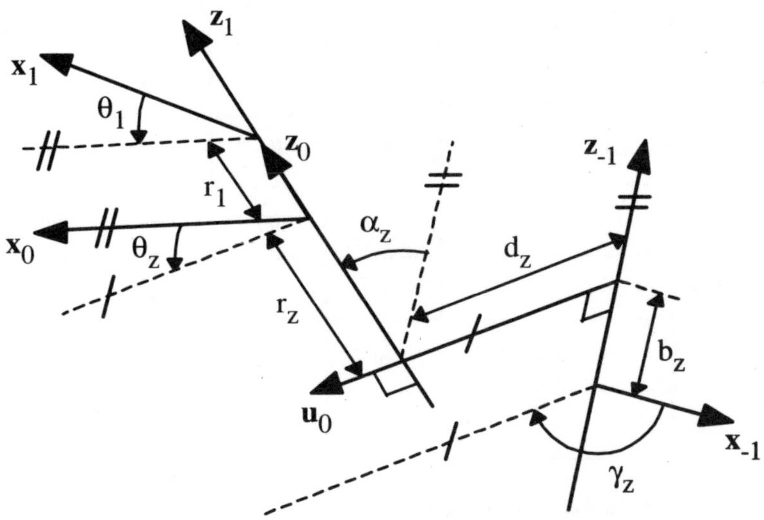

Figure 1.7. *Description of frame R_0 in frame R_{-1}*

1.4.2.3. *Geometric parameters of the end-effector*

We associate index n+1 to the end-effector frame R_E. Six parameters are needed (γ_e, b_e, α_e, d_e, θ_e, r_e) in order to arbitrarily characterize matrix **E** in such a way that:

$$^{n}\mathbf{T}_{n+1} = \mathbf{Rot}(\mathbf{z},\gamma_e)\ \mathbf{Trans}(\mathbf{z},b_e)\ \mathbf{Rot}(\mathbf{x},\alpha_e)\ \mathbf{Trans}(\mathbf{x},d_e)\ \mathbf{Rot}(\mathbf{z},\theta_e)\ \mathbf{Trans}(\mathbf{z},r_e) \quad [1.46]$$

from which:

$$^{n-1}\mathbf{T}_{n+1} = \mathbf{Rot}(\mathbf{x},\alpha_n)\ \mathbf{Trans}(\mathbf{x},d_n)\ \mathbf{Rot}(\mathbf{z},\theta'_n)\ \mathbf{Trans}(\mathbf{z},r'_n)\ \mathbf{Rot}(\mathbf{x},\alpha_{n+1})\ \mathbf{Trans}(\mathbf{x},d_{n+1})\ \mathbf{Rot}(\mathbf{z},\theta_{n+1})\ \mathbf{Trans}(\mathbf{z},r_{n+1}) \quad [1.47]$$

with: $\theta'_n = \theta_n + \gamma_e$, $r'_n = r_n + b_e$, $\alpha_{n+1} = \alpha_e$, $d_{n+1} = d_e$, $\theta_{n+1} = \theta_e$, $r_{n+1} = r_e$.

Finally, in order to describe the location of the end-effector of a robot with n joints in the world frame, a maximum of (4n + 6) independent parameters are required. In fact, since with a prismatic joint only two parameters can be identified, the number of parameters is reduced to a maximum of ($4n_r + 2n_p + 6$), n_r and n_p being the number of revolute and prismatic joints [EVE 89]. Table 1.2 outlines the parameters to be calibrated ($\sigma_j = 2$ indicates a fixed transformation).

j	σ_j	α_j	d_j	θ_j	r_j	β_j
0	2	0	0	γ_z	b_z	0
1	σ_1	α_z	d_z	$\theta_{1'}$	$r_{1'}$	β_1
2	σ_2	α_2	d_2	θ_2	r_2	β_2
...	...	...	...	...	...	...
n	σ_n	α_n	d_n	$\theta_{n'}$	$r_{n'}$	β_n
n+1	2	α_E	d_E	θ_E	r_E	β_{n+1}

Table 1.2. *Definition of complete geometric parameters*

1.4.3. *Generalized differential model of a robot*

The generalized differential model calculates the differential variation of the location of the end-effector in terms of differential variations of geometric parameters. It is expressed by the following relation:

$$\Delta \mathbf{X} = \begin{bmatrix} \mathbf{dP}_{n+1} \\ \boldsymbol{\delta}_{n+1} \end{bmatrix} = \boldsymbol{\Psi}\,\Delta\xi \qquad [1.48]$$

– $\mathbf{dP}_{n+1}$ is the (3×1) vector of the differential variation in position of the origin O_{n+1};

– $\boldsymbol{\delta}_{n+1}$ is the (3×1) vector of the differential variation in rotation of frame R_{n+1};

– $\Delta\xi$ is the $(N_{par} \times 1)$ vector of the differential variation of geometric parameters;

– $\boldsymbol{\Psi}$ is the $(6 \times N_{par})$ generalized Jacobian matrix representing the variation of operational coordinates in terms of variations of the geometric parameters ξ.

The Jacobian matrices of all calibration methods are deduced from the generalized Jacobian matrix $\boldsymbol{\Psi}$. The calculation of the columns of the Jacobian matrix $\boldsymbol{\Psi}$ is similar to the one used for the calculation of the Jacobian matrix of the kinematic model presented in section 1.3.1. According to the type of parameter, we obtain [KHA 89]:

$$\boldsymbol{\Psi}\beta_i = \begin{bmatrix} \mathbf{n}_{i-1} \mathrm{x} \mathbf{L}_{i-1,n+1} \\ \mathbf{n}_{i-1} \end{bmatrix} \qquad [1.49]$$

$$\boldsymbol{\Psi}\alpha_i = \begin{bmatrix} \mathbf{s}_{i-1} \mathrm{x} \mathbf{L}_{i-1,n+1} \\ \mathbf{s}_{i-1} \end{bmatrix} \qquad [1.50]$$

$$\boldsymbol{\Psi}d_i = \begin{bmatrix} \mathbf{s}_{i-1} \\ \mathbf{0}_{3\mathrm{x}1} \end{bmatrix} \qquad [1.51]$$

$$\boldsymbol{\Psi}\theta_i = \begin{bmatrix} \mathbf{a}_i \mathrm{x} \mathbf{L}_{i,n+1} \\ \mathbf{a}_i \end{bmatrix} \qquad [1.52]$$

$$\Psi r_i = \begin{bmatrix} \mathbf{a}_i \\ \mathbf{0}_{3x1} \end{bmatrix} \quad [1.53]$$

The calculation of the vectors playing a role in the calculation of Ψ is done with respect to the reference frame R_{-1}. The vectors ${}^{-1}\mathbf{s}_i$, ${}^{-1}\mathbf{n}_i$ and ${}^{-1}\mathbf{a}_i$ are obtained directly from the matrices ${}^{-1}\mathbf{T}_i$, i = 0, …, n+1. The vector $\mathbf{L}_{i,n+1}$ is obtained by using the same matrices through the relation:

$${}^{-1}\mathbf{L}_{i,n+1} = {}^{-1}\mathbf{P}_{n+1} - {}^{-1}\mathbf{P}_i \quad [1.54]$$

1.4.4. *Principle of geometric calibration*

The numerous calibration methods found in other works all have four steps:

– choice of a calibration model in terms of the geometric parameters of the robot and of the operational coordinates **X**;

– collection of experimental data that guarantee a sufficient number of equations;

– solution of a system of non-linear equations in order to obtain an optimal estimate of the geometric parameters;

– compensation of errors.

1.4.4.1. *General form of the calibration model*

The general form of a geometric calibration model is given by the following non-linear equation:

$$\mathbf{0} = \mathbf{f}(\mathbf{q}, \mathbf{x}, \xi^r) \quad [1.55]$$

where **f** is a vector of p components, **x** represents a vector of external measure such as the Cartesian coordinates of the end-effector, and ξ^r represents the $(N_{par} \times 1)$ vector of real standard geometric parameters of the robot such that:

$$\xi^r = \xi + \Delta\xi \quad [1.56]$$

ξ being the vector of nominal parameters.

When **f** is only a function of **q** and ξ, the method of calibration is called autonomous method since it does not need an external sensor.

By developing equation [1.56] to the first order, we obtain the linear form:

$$\Delta\mathbf{y}(\mathbf{q}, \mathbf{x}, \xi) = \phi(\mathbf{q}, \xi)\, \Delta\xi + \rho \qquad [1.57]$$

By using a sufficiently high number of c configurations, the non-linear system to solve is given by:

$$\mathbf{0} = \mathbf{F}(\mathbf{q}, \mathbf{x}, \xi) + \rho \qquad [1.58]$$

with:

$$\mathbf{F} = \begin{bmatrix} \mathbf{f}(\mathbf{q}(1), \mathbf{x}(1), \xi) \\ \dots \\ \mathbf{f}(\mathbf{q}(e), \mathbf{x}(e), \xi) \end{bmatrix} \qquad [1.59]$$

ρ being the residual vector or the error vector.

The general form of the system of linear equations, for e configurations, is obtained based on equation [1.57] through:

$$\Delta\mathbf{Y} = \mathbf{W}(\mathbf{q}, \xi)\, \Delta\xi + \rho \qquad [1.60]$$

with:

$$\Delta\mathbf{Y} = \begin{bmatrix} \Delta\mathbf{y}(1) \\ \dots \\ \Delta\mathbf{y}(e) \end{bmatrix}, \quad \mathbf{W} = \begin{bmatrix} \phi(1) \\ \dots \\ \phi(e) \end{bmatrix}$$

The matrix **W** is the observation matrix. It is necessary that the r = p.e number of equations is $>>N_{par}$, where p represents the number of equations of the calibration model.

1.4.4.2. *Identifying the geometric parameters*

Before solving the calibration equations, it is necessary to verify that the vector of unknown elements contains only the identifiable parameters in order to obtain an observation matrix **W** of maximum rank. If **W** is not of full rank, there will be an infinite number of solutions. We choose to identify the base parameters, linear combinations of standard parameters $\Delta\xi$, by eliminating the redundant parameters. The loss of rank of the observation matrix can have two origins:

– *structural loss of rank:* it is due to the structure of the robot and to the calibration model used. It results in the existence of linear relations among the columns **W**, irrespective of the robot configurations used. The principle of determining the base parameters is presented in section 1.4.4.2.1;

– *loss of rank due to the choice of configurations* **q**(j)*:* the excitation of the parameters by configurations used is an indispensable condition for the success of any method of identification. This can be measured by the condition number cond(**W**) of the observation matrix of the system [1.60].

1.4.4.2.1. Determining the identifiable parameters

In order to obtain a set of identifiable parameters, we use a general numeric method based on the QR decomposition [GAU 91], [KHA 91]. The research of identifiable parameters comes down to the study of vectorial space generated by the columns of a matrix **W** ($r \times N_{par}$) with $r >> N_{par}$, calculated on the basis of the matrices $\mathbf{\Phi}$ corresponding to a random sequence of configurations **q**(1), …, **q**(e).

We show that:

– the rank b of the matrix **W** gives the number of identifiable parameters;

– the identifiable parameters correspond to b independent columns of **W**, the other columns and their corresponding parameters can be eliminated. The choice as to what parameters to keep is not unique: in practice, we place from left to right the columns of **W** corresponding to the parameters that can be easily updated in analytical direct and inverse geometric models. The identifiable parameters are then determined by implementing a QR decomposition [GAU 91] of the matrix **W**. In fact we show that **W** can have the form [DON 79], [LAW 74], [GOL 96]:

$$\mathbf{W} = \mathbf{Q}\begin{bmatrix} \mathbf{R} \\ \mathbf{0}_{(r-Np)xNp} \end{bmatrix} \qquad [1.61]$$

an expression in which **Q** is an orthogonal ($r \times r$) matrix, **R** is an ($Np \times Np$) upper triangular matrix and $\mathbf{0}_{ixj}$ is the ($I \times j$) matrix of zero elements.

The rank b of matrix **W** corresponds to the number of non-zero elements of the diagonal of matrix **R**. Practically, it is defined using a tolerance τ such that b is equal to the number of elements $|R_{ii}| \geq \tau$ [FOR 77]. The numerical zero may be chosen such that [DON 79]:

$$\tau = r.\varepsilon.\max |R_{ii}| \qquad [1.62]$$

ε being the machine precision and r the number of rows.

The parameter i of vector $\Delta\xi$ is not identifiable if $|R_{ii}| < \tau$. Hence, a new model is built, obtained from [1.60] by keeping only the identifiable parameters, also called the vector of base parameters:

$$\Delta\mathbf{y}(\mathbf{q}, \mathbf{x}, \xi) = \phi_b(\mathbf{q}, \xi)\, \Delta\xi_b + \rho \qquad [1.63]$$

$\mathbf{\Phi}_b$ consisting of b first independent columns of $\mathbf{\Phi}$, and $\Delta\xi_b$ being the vector of identifiable parameters.

1.4.4.2.2. Choice of optimal configurations

This means basically choosing a set of configurations $\mathbf{q}(1), \ldots, \mathbf{q}(e)$, known as optimal configurations, in such a way that the condition number of the associated observation matrix **W** is minimum. We recall that the condition number of norm 2 of a matrix **W** is given by:

$$\mathrm{cond}_2(\mathbf{W}) = \frac{\sum_{max}}{\sum_{min}} \qquad [1.64]$$

where Σ_{max} and Σ_{min} are the maximal and minimal singular values of **W**.

Other observation criteria have been proposed in other works, such as, for example, the smallest singular value [NAH 94] or the product of singular values of **W** [BOR 91], but [HOL 96] showed that condition number was the most efficient criterion.

We note that the majority of calibration methods provide acceptable condition number for arbitrary configurations [KHA 00].

1.4.4.3. *Solving the identification equations*

We define the solution $\hat{\xi}$ of equation [1.58] in the least squares sense by:

$$\hat{\xi} = \min_{\Delta\xi} \|\mathbf{F}\|^2 \qquad [1.65]$$

Solving this non-linear optimization problem can be done either with the help of the Levenberg-Marquard method by using the corresponding function in Matlab or through the iterative solution of the differential system [1.60] and the update of geometric parameters after each iteration. The method using the differential model gives good results for rigid robots.

The least squares solution $\widehat{\Delta\xi}$ of equation [1.60] is given by:

$$\widehat{\Delta\xi} = \min_{\Delta\xi} \|\rho\|^2 = \min_{\Delta\xi} \left\|\Delta\mathbf{Y} - \mathbf{W}\ \widehat{\Delta\xi}\right\|^2 \qquad [1.66]$$

The explicit solution of this problem leads to the relation:

$$\widehat{\Delta\xi} = \mathbf{W}^{+}\ \Delta\mathbf{Y} \qquad [1.67]$$

where $\mathbf{W}^{+}$ indicates the pseudo-inverse matrix of $\mathbf{W}$. If $\mathbf{W}$ is of maximal rank, the explicit form of $\mathbf{W}^{+}$ is given by $(\mathbf{W}^T\ \mathbf{W})^{-1}\ \mathbf{W}^T$.

We calculate an estimate of standard deviation for the values identified by using matrix $\mathbf{W}$ in terms of the estimated parameters. We suppose that $\mathbf{W}$ is deterministic and that ρ is a zero mean additive independent noise vector of standard deviation σ_ρ and of variance-covariance matrix $\mathbf{C}_\rho$ such that:

$$\mathbf{C}_\rho = \mathrm{E}(\rho\ \rho^T) = \sigma_\rho{}^2\ \mathbf{I}_r \qquad [1.68]$$

E indicating the expectation operator and $\mathbf{I}_r$ indicating the $(r \times r)$ unit matrix.

Since the error vector is assumed to be centered, with independent components and equal dispersions, the standard deviation σ_ρ can be calculated with the help of the following unbiased estimator:

$$\sigma_\rho{}^2 = \frac{\left\|\Delta\mathbf{Y} - \mathbf{W}\ \widehat{\Delta\xi}\right\|^2}{(r - c)} \qquad [1.69]$$

Thus, the variance-covariance matrix of estimation error is given by [DEL 77]:

$$\mathbf{C}_{\hat{\xi}} = \mathrm{E}\left[\left(\xi-\hat{\xi}\right)\left(\xi-\hat{\xi}\right)^{\mathrm{T}}\right] = \mathbf{W}^{+}\mathbf{C}_{\rho}\left(\mathbf{W}^{+}\right)^{\mathrm{T}} = \sigma_{\rho}{}^{2}\left(\mathbf{W}^{\mathrm{T}}\mathbf{W}\right)^{-1} \qquad [1.70]$$

The standard deviation for the estimation of the jth parameter is obtained based on the element (j, j) of $\mathbf{C}_{\hat{\xi}}$:

$$\sigma_{\hat{\xi}_j} = \sqrt{\mathbf{C}_{\hat{\xi}}(j,j)} \qquad [1.71]$$

Equations [1.69] and [1.70] are valid when using the Levenberg-Marquard algorithm, but $\sigma_\rho{}^2$ is calculated from the residual of $\|\mathbf{F}\|^2$:

$$\sigma_\rho{}^2 = \frac{\left\|\mathbf{F}(\mathbf{q},\,\mathbf{x},\hat{\xi})\right\|^2}{(r-c)} \qquad [1.72]$$

In order to validate the calibration procedure, we can calculate the residual errors for some configurations that were not used in the identification.

1.4.5. *Calibration methods of geometric parameters*

1.4.5.1. *Calibration model by measuring the end-effector location*

The model is based on the calculation of the distance between the real location of the end-effector ${}^{-1}\mathbf{T}^{r}_{n+1}(\mathbf{x})$, measured by an appropriate external sensor, and the location ${}^{-1}\mathbf{T}^{m}_{n+1}(\mathbf{q},\xi)$ calculated with the help of the direct geometric model:

$$\mathbf{0} = {}^{-1}\mathbf{T}^{r}_{n+1}(\mathbf{x}) - {}^{-1}\mathbf{T}^{m}_{n+1}(\mathbf{q},\xi) + \rho \qquad [1.73]$$

Relation [1.73] contains 12 non-zero elements. In order to reduce the dimension of the problem to six, we use the following relation:

$$\Delta\mathbf{X} = \begin{bmatrix} \Delta\mathbf{X}_p(\mathbf{x},\mathbf{q},\xi) \\ \Delta\mathbf{X}_r(\mathbf{x},\mathbf{q},\xi) \end{bmatrix} = \mathbf{0} \qquad [1.74]$$

where $\Delta\mathbf{X}_p$ is the (3×1) vector representing the error in position:

$$\Delta\mathbf{X}_p = {}^{-1}\mathbf{P}^{r}_{n+1}(\mathbf{x}) - {}^{-1}\mathbf{P}^{m}_{n+1}(\mathbf{q},\xi) \qquad [1.75]$$

and where $\Delta\mathbf{X}_r$ is the (3×1) vector of the error in orientation:

$$\Delta\mathbf{X}_r = \mathbf{u}\,\alpha \qquad [1.76]$$

$\mathbf{u}$ and α being the solution of the following equation:

$${}^{-1}\mathbf{R}^{r}_{n+1}(\mathbf{x}) = \mathbf{rot}(\mathbf{u}, \alpha)\ {}^{-1}\mathbf{R}^{m}_{n+1}(\mathbf{q},\xi) \qquad [1.77]$$

where ${}^{-1}\mathbf{R}^{r}_{n+1}$ represents the measured orientation matrix and ${}^{-1}\mathbf{R}^{m}_{n+1}$ corresponds to the orientation matrix calculated by the direct geometric model.

We can also solve the calibration problem using the linear calibration model given by:

$$\Delta\mathbf{y} = \begin{bmatrix} \Delta\mathbf{X}_p(\mathbf{x},\mathbf{q},\xi) \\ \Delta\mathbf{X}_r(\mathbf{x},\mathbf{q},\xi) \end{bmatrix} = \mathbf{\Psi}(\mathbf{q},\xi)\,\Delta\xi + \rho \qquad [1.78]$$

where $\Delta\mathbf{y}$ is the error in location and orientation of the end-effector.

NOTES.

– since the measures of orientation are more difficult to realize than the measures of position, calibration is very often carried out only by measuring the position of the end-effector terminal point. In this case, three parameters of the last frame will not be identifiable, the matrix $\mathbf{\Psi}$ being reduced to its first three rows;

– the vector to identify contains elements of different units. By using the meter for lengths and the radian for angles, we obtain, however, good results when the robot has standard dimensions (between 1 m and 2 m).

1.4.5.2. *Autonomous calibration models*

The autonomous calibration indicates the procedure for which the only information required to identify the geometric parameters is the vectors of joint coordinates [BEN 91]. We present the equations of two methods, imposing respectively point-point or frame-frame links, as well as point-plane links.

1.4.5.2.1. Calibration using a point-point or a frame-frame link.

We do not need external measures by exploiting the fact that for a given location of the end-effector, a robot can often have several configurations corresponding to multiple solutions of IGM [KHA 95]. By supposing that configurations $\mathbf{q}^a$ and $\mathbf{q}^b$ correspond to the same location of the end-effector, we can write the non-linear calibration model with the help of the following relation:

$$\mathbf{f}(\mathbf{q}^a, \xi) = \mathbf{f}(\mathbf{q}^b, \xi) \qquad [1.79]$$

the function **f** representing the direct geometric model.

System [1.79] is reduced, at the most, to a system of six equations as explained in section 1.4.5.1. The Jacobian matrix of this calibration method is obtained after a first order development:

$$\Delta\mathbf{y} = \left[\boldsymbol{\Psi}(\mathbf{q}^b, \xi) - \boldsymbol{\Psi}(\mathbf{q}^a, \xi)\right] \Delta\xi + \rho \qquad [1.80]$$

In the case of a frame-frame link, $\Delta\mathbf{y}$ is given by $\mathbf{f}(\mathbf{q}^a, \xi) - \mathbf{f}(\mathbf{q}^b, \xi)$:

$$\Delta\mathbf{y} = \begin{bmatrix} {}^{-1}\mathbf{dP}_{n+1} \\ {}^{-1}\boldsymbol{\delta}_{n+1} \end{bmatrix} \qquad [1.81]$$

In the case of a point-point link, $\Delta\mathbf{y}$ is given by:

$$\Delta\mathbf{y} = {}^{-1}\mathbf{dP}_{n+1} = {}^{-1}\mathbf{P}_{n+1}(\mathbf{q}^a) - {}^{-1}\mathbf{P}_{n+1}(\mathbf{q}^b) \qquad [1.82]$$

The identifiable parameters through this method are different from those of classical methods using external sensors. It is shown, for example, that if a column from matrix $\boldsymbol{\Psi}$ is constant for all configurations, the corresponding column in the observation matrix **W** is zero and the corresponding parameter is not identifiable. This is what occurs in the case of parameters $\Delta\gamma_z$, Δb_z, $\Delta\alpha_z$, Δd_z, $\Delta\theta_{1'}$, $\Delta r_{1'}$.

1.4.5.2.2. Calibration by using a set of points in a plane

In this method, we use a set of configurations that force the terminal point of the end-effector to remain in the same plane. The observation matrix is built by writing the equation of this plane [TAN 94], [ZHO 95], [KHA 96], [IKI 97]. By indicating

P_{xt}, P_{yt}, P_{zt} as the Cartesian coordinates of the terminal point, the general equation of the plane that does not pass through the origin of the reference frame –1 is given by:

$$a' P_{xt} + b' P_{yt} + c' P_{zt} + d' = 0 \quad [1.83]$$

where again after normalization:

$$a P_{xt} + b P_{yt} + c P_{zt} + 1 = 0 \quad [1.84]$$

where P_{xt}, P_{yt} and P_{zt} are a function of $\mathbf{q}$ and ξ.

Hence, equation [1.84] represents the non-linear equation of calibration. As far as solving by linearization is concerned, if we suppose that coefficients a, b, and c of the plane are known, we obtain:

$$a [P_{xt}(\mathbf{q}) + \boldsymbol{\Psi}_x(\mathbf{q}) \Delta\xi] + b [P_{yt}(\mathbf{q}) + \boldsymbol{\Psi}_y(\mathbf{q}) \Delta\xi] + c [P_{zt}(\mathbf{q}) + \boldsymbol{\Psi}_z(\mathbf{q}) \Delta\xi] + 1 = 0 \quad [1.85]$$

$\boldsymbol{\Psi}_x$, $\boldsymbol{\Psi}_y$ and $\boldsymbol{\Psi}_z$ being respectively the first, second, and third rows of the generalized Jacobian matrix [1.48]. This relation is also written:

$$[a \boldsymbol{\Psi}_x(\mathbf{q}) + b \boldsymbol{\Psi}_y(\mathbf{q}) + c \boldsymbol{\Psi}_z(\mathbf{q})] \Delta\xi = - a P_{xt}(\mathbf{q}) - b P_{yt}(\mathbf{q}) - c P_{zt}(\mathbf{q}) - 1 \quad [1.86]$$

Equation [1.86] gives a single linear equation in which the vector of unknown parameters is $\Delta\xi$.

NOTES.

– we can build an observation matrix which is not dependent on the coefficient of the plane d' by the subtraction of two equations of type [1.83] [MAU 96], [ZHO 95]. The coefficients a', b' and c' can be easily obtained due to an inclinometer;

– the concatenation of the data of two or three known non-parallel planes increases the number of identifiable parameters [ZHU 99]. The use of three planes gives the same identifiable parameters as the position measuring method if the robot has at least one prismatic joint, whereas four planes are necessary if the robot has only revolute joints [BES 00].

1.4.6. *Correction of geometric parameters*

The last step of calibration consists of introducing the results of the identification into control algorithms. Basically, the model identified is generally much more complex than the implemented analytical model because certain parameters supposed to be nominally zero are not like that on real robot(s).

A solution is then to implement the following principles (see Figure 1.8):

– all parameters identified can be taken into account in a "complete" direct geometric model;

– for the inverse geometric model, we only consider the parameters which appear explicitly in this model;

– the effect of the other parameters are corrected by using the differential model in the following way:

1) calculating the joint coordinates **q** corresponding to the desired position $\mathbf{X}^d$ by using the nominal inverse geometric model (or with partially adjusted parameters),

2) calculating the differential errors $\Delta\mathbf{X}$ between $\mathbf{X}^d$ and $\mathbf{f}(\xi, \mathbf{q})$ by using the "complete" direct geometric model or by using the "complete" differential model taking into account the non-adjustable errors,

3) calculating vector $\Delta\mathbf{q}$ corresponding to vector $\Delta\mathbf{X}$ by using the inverse differential model,

4) the result is the sum of vectors $\mathbf{q} + \Delta\mathbf{q}$.

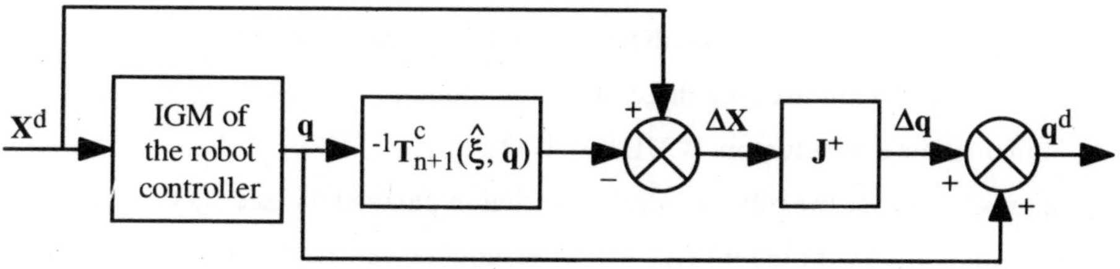

Figure 1.8. *Principle of the correction of errors*

1.5. Dynamic modeling

The dynamic model is the relation between the torques (and/or forces) applied to the actuators and the joint positions, velocities and accelerations. The dynamic model is represented by a relation of the form:

$$\mathbf{\Gamma} = \mathbf{f}(\mathbf{q}, \dot{\mathbf{q}}, \ddot{\mathbf{q}}, \mathbb{f}_{\mathbf{e}}) \quad [1.87]$$

with:

– $\mathbf{\Gamma}$: vector of torques/forces of the actuators, depending on whether the joint is revolute or prismatic. From now on, we will simply write torques;

– $\mathbf{q}$: vector of joint positions;

– $\dot{\mathbf{q}}$: vector of joint velocities;

– $\ddot{\mathbf{q}}$: vector of joint accelerations;

– $\mathbb{f}_{\mathbf{e}}$: vector representing the external forces and moments that the robot exert on the environment.

It is common to call relation [1.87] the inverse dynamic model, or quite simply the dynamic model.

The direct dynamic model expresses the joint accelerations in terms of joint positions, velocities and torques. Hence, it is represented by the relation:

$$\ddot{\mathbf{q}} = \mathbf{g}(\mathbf{q}, \dot{\mathbf{q}}, \mathbf{\Gamma}, \mathbb{f}_{\mathbf{e}}) \quad [1.88]$$

Among the applications of the dynamic model, we can mention:

– simulation, which uses the direct dynamic model;

– dimensioning of actuators [CHE 90], [POT 86];

– identification of the robot inertial and friction parameters (see section 1.5.4);

– control, which uses the inverse dynamic model (see Chapter 5).

Several formalisms were used to obtain the dynamic model of the robots [REN 75], [COI 81], [VUK 82]. The most frequently used formalisms are:

a) Lagrange formalism [UIC 69], [REN 80a], [HOL 80], [PAU 81], [MEG 84], [REN 85];

b) Newton-Euler formalism [HOO 65], [ARM 79], [LUH 80], [ORI 79], [KHA 85], [KHO 86], [KHA 87], [REN 87].

In this section, we present these two formalisms for serial robots (for complex chain robots, see [KHA 02]). We will also deal with the determination of minimum inertial parameters as well as their identification.

The principal notations used are the following:

$\mathbf{a}_j$ unit vector along axis $\mathbf{z}_j$;

$\mathbf{F}_j$ resultant of external forces on link C_j;

$\mathbf{f}_j$ forces applied on link C_j by link C_{j-1};

$\mathbf{f}_{ej}$ forces applied by link C_j on the environment;

F_{sj} Coulomb friction parameter of joint j;

F_{vj} viscous friction parameter of joint j;

$\mathbf{g}$ acceleration of gravity;

G_j center of gravity of link C_j;

$\mathbf{I}_{Gj}$ inertial tensor of link C_j with respect to a frame parallel to R_j and of origin G_j;

Ia_j moment of inertia of the rotor of actuator j and of its transmission system referred to the joint side;

${}^j\mathbf{J}_j$ inertial tensor of link C_j with respect to frame R_j , which is expressed by:

$${}^j\mathbf{J}_j = \begin{bmatrix} \int(y^2+z^2)dm & -\int xydm & -\int xzdm \\ -\int xydm & \int(x^2+z^2)dm & -\int yzdm \\ -\int xydm & -\int yzdm & \int(x^2+y^2)dm \end{bmatrix} = \begin{bmatrix} XX_j & XY_j & XZ_j \\ XY_j & YY_j & YZ_j \\ XZ_j & YZ_j & ZZ_j \end{bmatrix} \quad [1.89]$$

$\mathbf{L}_j$ vector connecting the origin of frame R_{j-1} and the origin of frame R_j. It is equal to $\mathbf{O}_{j-1}\mathbf{O}_j$;

M_j mass of link C_j;

$\mathbf{MS}_j$ first moments of inertia of link C_j about the origin of the frame R_j, equal to $M_j\,\mathbf{S}_j$. Let $[\, MX_j \;\; MY_j \;\; MZ_j \,]^T$ be the components of ${}^j\mathbf{MS}_j$;

$\mathbf{M}_{Gj}$ total moment exerted on link C_j about G_j;

$\mathbf{M}_j$ total moment exerted on link C_j about O_j;

$\mathbf{m}_j$ moment about O_j exerted on link C_j by link C_{j-1};

$\mathbf{m}_{ej}$ moment about O_j exerted by link C_j on the environment;

$\mathbf{S}_j$ vector whose origin is O_j and whose extremity is the mass center of link C_j. It is equal to $\mathbf{O}_j\mathbf{G}_j$;

$\mathbf{V}_j$ velocity of point O_j;

$\dot{\mathbf{V}}_j$ linear acceleration of point O_j;

$\mathbf{V}_{Gj}$ velocity of the center of gravity of link C_j;

$\dot{\mathbf{V}}_{Gj}$ acceleration of the center of gravity of link C_j;

$\boldsymbol{\omega}_j$ angular velocity of link C_j;

$\dot{\boldsymbol{\omega}}_j$ angular acceleration of link C_j.

1.5.1. *Lagrange formalism*

The goal of this section is to study the general form of the dynamic model, and to present its main properties. The method presented does not give the most competitive model from the point of view of the number of operations, but it is the simplest method considering these objectives. We will consider first an ideal robot without friction or elasticity which does not exert any external forces on the environment. These phenomena will be covered in section 1.5.1.4.

Lagrange formalism describes the equations of motion, when the external forces of the end-effector are supposed zero, through the following equation:

$$\Gamma_i = \frac{d}{dt}\frac{\partial L}{\partial \dot{q}_i} - \frac{\partial L}{\partial q_i} \qquad i = 1, \ldots, n \qquad [1.90]$$

with:

– L: Lagrangian of the system, equal to E – U;

– E: total kinetic energy of the system;

– U: total potential energy of the system.

1.5.1.1. *General form of dynamic equations*

The kinetic energy of the system is a quadratic function of joint velocities:

$$E = \frac{1}{2}\dot{\mathbf{q}}^T \mathbf{A}\,\dot{\mathbf{q}} \qquad [1.91]$$

where $\mathbf{A}$ is the $(n \times n)$ matrix of kinetic energy, of generic element A_{ij}, also called inertia matrix of the robot, which is symmetric and positive definite. Its elements are a function of the joint variables $\mathbf{q}$.

Since the potential energy is a function of joint variables $\mathbf{q}$, the joint torque or force components of $\mathbf{\Gamma}$ can be obtained using equations [1.90] and [1.91] in the following form:

$$\mathbf{\Gamma} = \mathbf{A}(\mathbf{q})\,\ddot{\mathbf{q}} + \mathbf{C}(\mathbf{q},\dot{\mathbf{q}})\,\dot{\mathbf{q}} + \mathbf{Q}(\mathbf{q}) \qquad [1.92]$$

with:

– $\mathbf{C}(\mathbf{q},\dot{\mathbf{q}})\,\dot{\mathbf{q}}$: $(n \times 1)$ vector representing Coriolis and centrifugal forces/torques such that:

$$\mathbf{C}\dot{\mathbf{q}} = \dot{\mathbf{A}}\,\dot{\mathbf{q}} - \frac{\partial E}{\partial \mathbf{q}} \qquad [1.93]$$

– $\mathbf{Q} = [Q_1 \; \dots \; Q_n]^T$: vector of torques/forces of gravity.

Several forms are possible for the matrix $\mathbf{C}$. For example, we can calculate its elements from the Christophel symbol $c_{i,jk}$ such that:

$$\begin{cases} C_{ij} = \sum_{k=1}^{n} c_{i,jk}\,\dot{q}_k \\ c_{i,jk} = \frac{1}{2}\left[\frac{\partial A_{ij}}{\partial q_k} + \frac{\partial A_{ik}}{\partial q_j} - \frac{\partial A_{jk}}{\partial q_i}\right] \end{cases} \qquad [1.94]$$

The elements of vector $\mathbf{Q}$ are calculated by:

$$Q_i = \frac{\partial U}{\partial q_i} \qquad [1.95]$$

The elements of **A**, **C** and **Q** are functions of the geometric and inertial parameters of the robot. Hence, the dynamic equations of an articulated mechanism form a system of n coupled and non-linear second order differential equations.

1.5.1.2. *Calculation of energy*

The kinetic energy of the system is given by the relation:

$$E = \sum_{j=1}^{n} E_j \qquad [1.96]$$

where E_j indicates the kinetic energy of link C_j, which is expressed by:

$$E_j = \frac{1}{2}\left(\boldsymbol{\omega}_j^{T}\mathbf{I}_{Gj}\boldsymbol{\omega}_j + M_j\mathbf{V}_{Gj}^{T}\mathbf{V}_{Gj}\right) \qquad [1.97]$$

Given that (Figure 1.9):

$$\mathbf{V}_{Gj} = \mathbf{V}_j + \boldsymbol{\omega}_j \text{ x } \mathbf{S}_j \qquad [1.98]$$

and knowing that:

$$\mathbf{J}_j = \mathbf{I}_{Gj} - M_j\ \hat{\mathbf{S}}_j\hat{\mathbf{S}}_j \qquad [1.99]$$

relation [1.97] becomes:

$$E_j = \frac{1}{2}\ [\boldsymbol{\omega}_j^{T}\ \mathbf{J}_j\ \boldsymbol{\omega}_j + M_j\ \mathbf{V}_j^{T}\ \mathbf{V}_j + 2\ \mathbf{MS}_j^{T}\ (\mathbf{V}_j \text{ x } \boldsymbol{\omega}_j)] \qquad [1.100]$$

Relation [1.97] is non-linear in the parameters of vector $\mathbf{S}_j$, contrary to relation [1.100] which is linear with respect to the elements of M_j, $\mathbf{MS}_j$ and $\mathbf{J}_j$ called standard inertial parameters.

The calculation of $\mathbf{V}_j$ and of $\boldsymbol{\omega}_j$ is carried out by the velocity composition equations (see Figure 1.9):

$$\boldsymbol{\omega}_j = \boldsymbol{\omega}_{j-1} + \bar{\sigma}_j\dot{q}_j\ \mathbf{a}_j \qquad [1.101]$$

$$\mathbf{V}_j = \mathbf{V}_{j-1} + \boldsymbol{\omega}_{j-1} \times \mathbf{L}_j + \sigma_j \dot{q}_j \mathbf{a}_j \qquad [1.102]$$

For a robot with a fixed base, the initial conditions are $\mathbf{V}_0 = \mathbf{0}$ and $\boldsymbol{\omega}_0 = \mathbf{0}$.

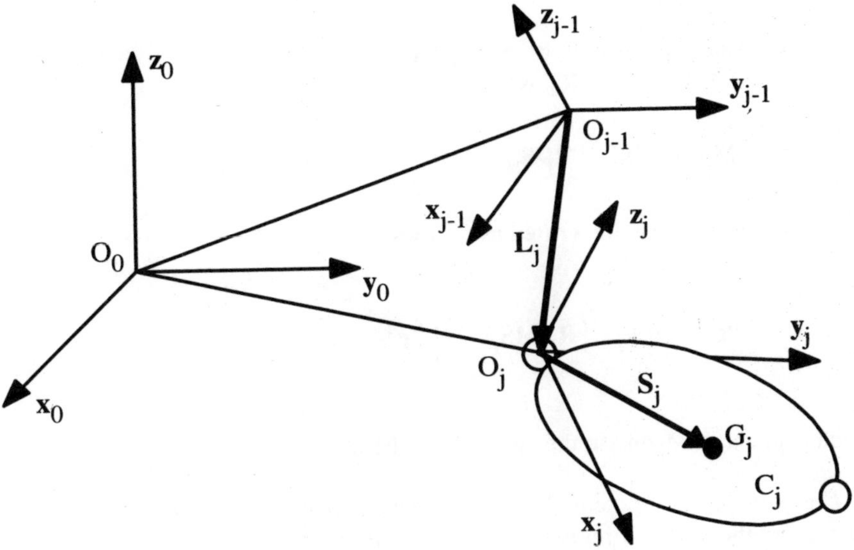

Figure 1.9. *Composition of velocities*

In equation [1.100], all elements must be expressed in the same frame. The simplest way is to express them in frame R_j. Therefore, we rewrite equations [1.100], [1.101] and [1.102] giving E_j, ${}^j\boldsymbol{\omega}_j$ and ${}^j\mathbf{V}_j$ as follows:

$$E_j = \frac{1}{2}\left[{}^j\boldsymbol{\omega}_j^T\, {}^j\mathbf{J}_j\, {}^j\boldsymbol{\omega}_j + M_j\, {}^j\mathbf{V}_j^T\, {}^j\mathbf{V}_j + 2\, {}^j\mathbf{MS}_j^T ({}^j\mathbf{V}_j \times {}^j\boldsymbol{\omega}_j)\right] \qquad [1.103]$$

$${}^j\boldsymbol{\omega}_j = {}^j\mathbf{R}_{j-1}\, {}^{j-1}\boldsymbol{\omega}_{j-1} + \bar{\sigma}_j \dot{q}_j\, {}^j\mathbf{a}_j = {}^j\boldsymbol{\omega}_{j-1} + \bar{\sigma}_j \dot{q}_j\, {}^j\mathbf{a}_j \qquad [1.104]$$

$${}^j\mathbf{V}_j = {}^j\mathbf{R}_{j-1} ({}^{j-1}\mathbf{V}_{j-1} + {}^{j-1}\boldsymbol{\omega}_{j-1} \times {}^{j-1}\mathbf{P}_j) + \sigma_j \dot{q}_j\, {}^j\mathbf{a}_j \qquad [1.105]$$

The elements ${}^j\mathbf{J}_j$ and ${}^j\mathbf{MS}_j$ are constant. In order to simplify matters, they are noted $\mathbf{J}_j$ and $\mathbf{MS}_j$.

Potential energy can be written:

$$U = \sum_{j=1}^{n} U_j = \sum_{j=1}^{n} -M_j \mathbf{g}^T \left(\mathbf{L}_{0,j} + \mathbf{S}_j \right) \qquad [1.106]$$

$\mathbf{L}_{0,j}$ indicates the vector of origin 0_0 and of extremity O_j. By projecting the vectors of this relation into R_0, we obtain:

$$U_j = -M_j \, {}^0\mathbf{g}^T \left({}^0\mathbf{P}_j + {}^0\mathbf{R}_j \, {}^j\mathbf{S}_j \right) \qquad [1.107a]$$

This expression can be written as follows:

$$U_j = -{}^0\mathbf{g}^T \left(M_j \, {}^0\mathbf{P}_j + {}^0\mathbf{R}_j \, {}^j\mathbf{MS}_j \right) = -\begin{bmatrix} {}^0\mathbf{g}^T & 0 \end{bmatrix} {}^0\mathbf{T}_j \begin{bmatrix} {}^j\mathbf{MS}_j \\ M_j \end{bmatrix} \qquad [1.107b]$$

which is linear in M_j and in the elements of ${}^j\mathbf{MS}_j$.

The kinetic and potential energy being linear with respect to the inertial parameters, the dynamic model is linear as well in these parameters.

1.5.1.3. *Properties of the dynamic model*

a) matrix **A** is symmetric and positive definite, thus $A_{ij} = A_{ji}$;

b) the energies of link C_j are a function of $(q_1, \ldots, q_j)$ and $(\dot{q}_1, \ldots, \dot{q}_j)$;

c) based on property b and on relation [1.90], we can prove that Γ_i is a function of inertial parameters of link C_i and links $C_{i+1}, \ldots, C_n$;

d) we show that, **C** being defined according to relation [1.94], the matrix $[\frac{d}{dt} \mathbf{A} - 2\mathbf{C}(\mathbf{q}, \dot{\mathbf{q}})]$ is antisymmetric [KOD 84], [ARI 84], which is an interesting property for the control;

e) the dynamic model is linear with respect to the elements of the inertial parameters M_j, ${}^j\mathbf{MS}_j$ and ${}^j\mathbf{J}_j$, called standard inertial parameters. This property will be used in order to identify the inertial parameters and reduce the number of operations of the dynamic model.

1.5.1.4. *Taking into consideration the friction*

The non-compensated frictions at the joints, reduction gears and transmissions produce static errors, delays and limit cycles [CAN 90]. Various friction models have been proposed in other works, like for example those of [DAH 77], [CAN 89], [ARM 88], [ARM 91], [ARM 94].

In many applications, the friction model relies on a constant term for dry friction (or Coulomb friction) and a term function of velocity for the viscous friction (see Figure 1.10). The expression of the friction torque Γ_{fi} of the joint i is therefore written:

$$\Gamma_{fi} = F_{si}\,\mathrm{sign}(\dot{q}_i) + F_{vi}\,\dot{q}_i \qquad [1.108]$$

F_{si} and F_{vi} indicate the parameters of Coulomb and viscous friction and sign(.) represents the sign function.

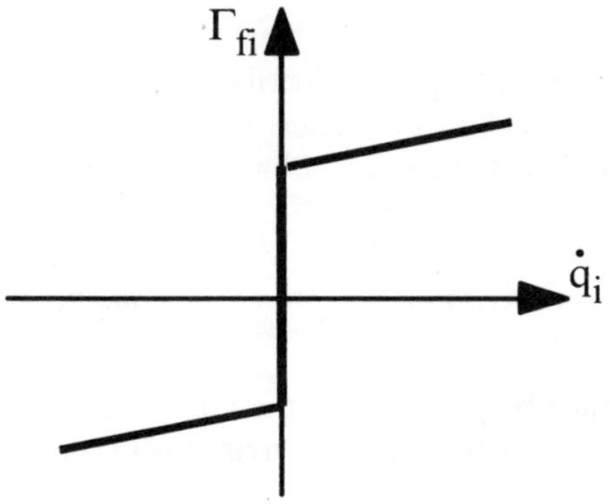

Figure 1.10. *Friction model*

We can therefore take into consideration the friction torques/forces by adding vector $\mathbf{\Gamma}_f$ to the right hand side of the expression [1.92] so that:

$$\mathbf{\Gamma}_f = \mathbf{diag}(\dot{\mathbf{q}})\mathbf{F}_s + \mathbf{diag}[\mathrm{sign}(\dot{\mathbf{q}})]\mathbf{F}_v \qquad [1.109]$$

with

– $\mathbf{F}_s = \begin{bmatrix} F_{s1} & \dots & F_{sn} \end{bmatrix}^T$

– $\mathbf{F}_V = [F_{v1} \;\; \ldots \;\; F_{vn}]^T$

– **diag**(.): $(n \times n)$ diagonal matrix

The non-linearity of this model can also be approximated by a piecewise linear model.

1.5.1.5. *Taking into account the inertia of the actuator's rotor*

We represent the kinetic energy of the rotor of actuator j by a term having the form $\frac{1}{2}\, Ia_j\, \dot{q}_j^{\,2}$. The inertial parameter Ia_j can be written:

$$Ia_j = N_j^2\, J_{mj} \qquad [1.110]$$

where J_{mj} is the moment of inertia of the actuator's rotor j, N_j is the transmission ratio of the axis j equal to $\dot{q}_{mj} / \dot{q}_j$, $\dot{q}_{mj}$ indicating the velocity of the actuator's rotor j.

We conclude that element A_{jj} of the inertia matrix **A** of equation [1.92] must be increased by Ia_j. This modeling of inertias of actuators neglects their gyroscopic effect. We find more complete modeling of actuators and transmissions in [LLI 83], [CHE 86], [SCI 94].

1.5.1.6. *Taking into consideration the forces and moments exerted by the end-effector on its environment*

The torques that must be provided by the joint actuators so that its end-effector can apply a force and a moment $\mathbb{f}_{en}$ on the environment can be written as follows:

$$\mathbf{\Gamma}_e = \mathbf{J}_n^{\,T}\, \mathbb{f}_{en} \qquad [1.111]$$

We then add the term $\mathbf{\Gamma}_e$ to the right hand side of the expression [1.92].

EXAMPLE 1.7.– finding the elements of matrices **A** and **Q** of a robot with three degrees of freedom, having the same structure as the shoulder of the Stäubli RX-90 robot described in Example 1.1. We suppose that:

$${}^{j}\mathbf{MS}_j = [MX_j \;\; MY_j \;\; MZ_j]^T$$

$$ {}^{i}\mathbf{J}_{j}=\begin{bmatrix} XX_j & XY_j & XZ_j \\ XY_j & YY_j & YZ_j \\ XZ_j & YZ_j & ZZ_j \end{bmatrix},\ \mathbf{I}_a=\begin{bmatrix} I_{a1} & 0 & 0 \\ 0 & I_{a2} & 0 \\ 0 & 0 & I_{a3} \end{bmatrix} $$

All calculations having been done, we obtain:

$$ A_{11} = Ia_1 + ZZ_1 + SS2\ XX_2 + 2CS2\ XY_2 + CC2\ YY_2 + SS23\ XX_3 + 2CS23\ XY_3 + CC23\ YY_3 + 2C2\ C23\ D3\ MX_3 - 2C2\ S23\ D3\ MY_3 + CC2\ D3^2\ M_3 $$

$$ A_{12} = S2\ XZ_2 + C2\ YZ_2 + S23\ XZ_3 + C23\ YZ_3 - S2\ D3\ MZ_3 $$

$$ A_{13} = S23\ XZ_3 + C23\ YZ_3 $$

$$ A_{22} = Ia_2 + ZZ_2 + ZZ_3 + 2C3\ D3\ MX_3 - 2S3\ D3\ MY_3 + D3^2\ M_3 $$

$$ A_{23} = ZZ_3 + C3\ D3\ MX_3 - S3\ D3\ MY_3 $$

$$ A_{33} = Ia_3 + ZZ_3 $$

with $SSj = (\sin\theta_j)^2$, $CCj = (\cos\theta_j)^2$ and $CSj = \cos\theta_j \sin\theta_j$. The elements of the Coriolis and centrifugal matrix **C** can be obtained from these expressions due to relation [1.94].

For the calculation of the forces of gravity, we suppose that:

$$ {}^{0}\mathbf{g} = [0 \quad 0 \quad G3]^T $$

The potential energy is obtained by using relation [1.107]:

$$ U = -G3\ (MZ_1 + S2MX_2 + C2MY_2 + S23MX_3 + C23MY_3 + D3S2M_3) $$

Hence:

$$ Q_1 = 0 $$

$$ Q_2 = -G3\ (C2\ MX_2 - S2\ MY_2 + C23\ MX_3 - S23\ MY_3 + D3\ C2\ M_3) $$

$$ Q_3 = -G3\ (C23\ MX_3 - S23\ MY_3) $$

1.5.2. *Newton-Euler formalism*

The Newton-Euler equations express the total external forces and moments acting in the center of mass G_j of link j by equations:

$$\mathbf{F}_j = M_j\ \dot{\mathbf{V}}_{Gj} \qquad [1.112]$$

$$\mathbf{M}_{Gj} = \mathbf{I}_{Gj}\ \dot{\omega}_j + \omega_j \text{ x } (\mathbf{I}_{Gj}\ \omega_j) \qquad [1.113]$$

The Luh, Walker and Paul method [LUH 80], considered as a significant progress towards the possibility of on-line calculation of the dynamic model of robots, uses these equations and is based on a double recurrence. The forward recurrence, from the base of the robot towards the end-effector, calculates the velocities and accelerations of the links, and then the total forces and moments about the links. A backward recurrence, from the end-effector towards the base, enables the calculation of actuator torques by expressing for each link the resultant of external forces and moments.

This method makes it possible to directly obtain the inverse dynamic model [1.90] without having to explicitly calculate the matrices **A**, **C** and **Q**. The inertial parameters used are M_j, $\mathbf{S}_j$ and $\mathbf{I}_{Gj}$. The model obtained in this way is not linear in the inertial parameters.

1.5.2.1. *Newton-Euler equations linear in the inertial parameters*

In this section, we will present a Newton-Euler algorithm based on the double recurrence of the Luh *et al.* [LUH 80] method, but expressing the external forces and moments about O_j rather than about G_j, by using the inertial parameters M_j, $\mathbf{MS}_j$ and $\mathbf{J}_j$ [KHA 87], [KHO 86]. In this way, the generated model is linear in the inertial parameters. It can be calculated by using the base inertial parameters according to the property of linearity.

The modified Newton-Euler equations giving the external forces and moments about O_j are written:

$$\mathbf{F}_j = M_j\ \dot{\mathbf{V}}_j + \dot{\omega}_j \text{ x } \mathbf{MS}_j + \omega_j \text{ x } (\omega_j \text{ x } \mathbf{MS}_j) \qquad [1.114]$$

$$\mathbf{M}_j = \mathbf{J}_j\,\dot{\omega}_j + \omega_j \text{ x } (\mathbf{J}_j\ \omega_j) + \mathbf{MS}_j \text{ x } \dot{\mathbf{V}}_j \qquad [1.115]$$

where $\dot{\boldsymbol{\omega}}_j$ and $\dot{\mathbf{V}}_j$ are obtained by the differentiation with respect to time of equations [1.101] and [1.102], leading to:

$$\dot{\boldsymbol{\omega}}_j = \dot{\boldsymbol{\omega}}_{j-1} + \overline{\sigma}_j \left(\ddot{q}_j \mathbf{a}_j + \boldsymbol{\omega}_{j-1} \times \dot{q}_j \mathbf{a}_j \right) \qquad [1.116]$$

$$\dot{\mathbf{V}}_j = \dot{\mathbf{V}}_{j-1} + \dot{\boldsymbol{\omega}}_{j-1} \times \mathbf{L}_j + \boldsymbol{\omega}_{j-1} \times (\boldsymbol{\omega}_{j-1} \times \mathbf{L}_j) + \sigma_j \left(\ddot{q}_j \mathbf{a}_j + 2\, \boldsymbol{\omega}_{j-1} \times \dot{q}_j \mathbf{a}_j \right) \qquad [1.117]$$

– *forward recurrence:* this calculates $\mathbf{F}_j$ and $\mathbf{M}_j$ due to relations [1.114] and [1.115]. We initialize this recurrence by $\boldsymbol{\omega}_0 = \mathbf{0}$, $\dot{\boldsymbol{\omega}}_0 = \mathbf{0}$ and $\dot{\mathbf{V}}_0 = \mathbf{0}$.

– *backward recurrence:* the equations are obtained by calculating the total forces and the total moments at O_j. We obtain (Figure 1.11):

$$\mathbf{F}_j = \mathbf{f}_j - \mathbf{f}_{j+1} + M_j\, \mathbf{g} - \mathbf{f}_{ej} \qquad [1.118]$$

$$\mathbf{M}_j = \mathbf{m}_j - \mathbf{m}_{j+1} - \mathbf{L}_{j+1} \times \mathbf{f}_{j+1} + \mathbf{S}_j \times M_j\, \mathbf{g} - \mathbf{m}_{ej} \qquad [1.119]$$

We can consider the effect of gravity without having to take it into account explicitly as seen in equation [1.118]. For that, we set:

$$\dot{\mathbf{V}}_0 = -\,\mathbf{g} \qquad [1.120]$$

hence the following equilibrium equations are derived from [1.118] and [1.119]:

$$\mathbf{f}_j = \mathbf{F}_j + \mathbf{f}_{j+1} + \mathbf{f}_{ej} \qquad [1.121]$$

$$\mathbf{m}_j = \mathbf{M}_j + \mathbf{m}_{j+1} + \mathbf{L}_{j+1} \times \mathbf{f}_{j+1} + \mathbf{m}_{ej} \qquad [1.122]$$

recurrence initialized by $\mathbf{f}_{n+1} = \mathbf{0}$ and $\mathbf{m}_{n+1} = \mathbf{0}$.

We thus obtain the torques for the actuators Γ_j by projecting, according to the type of joint j, the vectors $\mathbf{f}_j$ or $\mathbf{m}_j$ on the axis $\mathbf{z}_j$ of joint j. We add the corrective terms representing the effect of friction and inertias of the actuators' rotors, which leads to:

$$\Gamma_j = (\sigma_j\, \mathbf{f}_j + \overline{\sigma}_j\, \mathbf{m}_j)^T\, \mathbf{a}_j + F_{sj}\, \mathrm{sign}\,(\dot{q}_j) + F_{vj}\, \dot{q}_j + Ia_j\, \ddot{q}_j \qquad [1.123]$$

Based on equations [1.121] and [1.122], we conclude that the terms $\mathbf{f}_j$ and $\mathbf{m}_j$ depend only on the inertial parameters of link j, j+1, …, n. Therefore, we have here property (c) expressed in section 1.5.1.3.

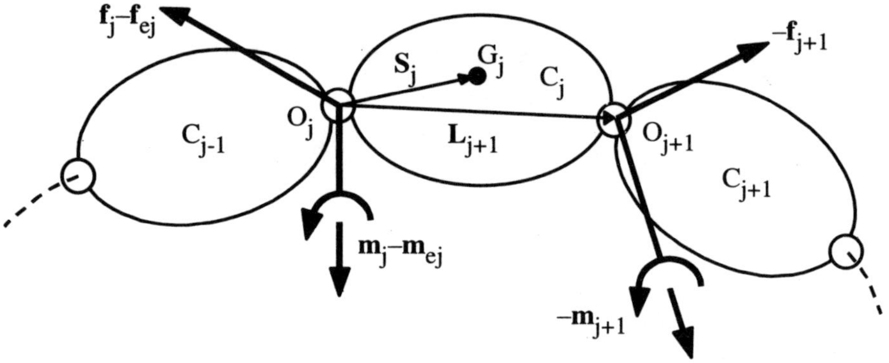

Figure 1.11. *Forces and moments on link j*

1.5.2.2. *Practical form of Newton-Euler equations*

In order to make practical use of the Newton-Euler algorithm shown above, we must project the vectors and tensors which appear in the same equation with respect to a convenient frame. We will use here the choice of Luh *et al.* [LUH 80], which consists of projecting the elements of a link in its frame. The equations of the forward recurrence become, for j = 1, ..., n:

$$ {}^{j}\omega_{j-1} = {}^{j}\mathbf{R}_{j-1}\, {}^{j-1}\omega_{j-1} \qquad [1.124]$$

$$ {}^{j}\omega_{j} = {}^{j}\omega_{j-1} + \bar{\sigma}_j\, \dot{q}_j\, {}^{j}\mathbf{a}_j \qquad [1.125]$$

$$ {}^{j}\dot{\omega}_{j} = {}^{j}\mathbf{R}_{j-1}\, {}^{j-1}\dot{\omega}_{j-1} + \bar{\sigma}_j\, (\ddot{q}_j\, {}^{j}\mathbf{a}_j + {}^{j}\omega_{j-1} \times \dot{q}_j\, {}^{j}\mathbf{a}_j) \qquad [1.126]$$

$$ {}^{j}\mathbf{U}_{j} = {}^{j}\hat{\dot{\omega}}_j + {}^{j}\hat{\omega}_j\, {}^{j}\hat{\omega}_j \qquad [1.127]$$

$$ {}^{j}\dot{\mathbf{V}}_{j} = {}^{j}\mathbf{R}_{j-1}\, ({}^{j-1}\dot{\mathbf{V}}_{j-1} + {}^{j-1}\mathbf{U}_{j-1}\, {}^{j-1}\mathbf{P}_j) + \sigma_j\, (\ddot{q}_j\, {}^{j}\mathbf{a}_j + 2\, {}^{j}\omega_{j-1} \times \dot{q}_j\, {}^{j}\mathbf{a}_j) \qquad [1.128]$$

$$ {}^{j}\mathbf{F}_j = M_j\,{}^{j}\dot{\mathbf{V}}_j + {}^{j}\mathbf{U}_j\,{}^{j}\mathbf{MS}_j \quad [1.129] $$

$$ {}^{j}\mathbf{M}_j = {}^{j}\mathbf{J}_j\,{}^{j}\dot{\boldsymbol{\omega}}_j + {}^{j}\boldsymbol{\omega}_j \times ({}^{j}\mathbf{J}_j\,{}^{j}\boldsymbol{\omega}_j) + {}^{j}\mathbf{MS}_j \times {}^{j}\dot{\mathbf{V}}_j \quad [1.130] $$

with $\boldsymbol{\omega}_0 = \mathbf{0}$, $\dot{\boldsymbol{\omega}}_0 = \mathbf{0}$, $\dot{\mathbf{V}}_0 = -\mathbf{g}$.

For the backward recurrence, for j = n, ..., 1:

$$ {}^{j}\mathbf{f}_j = {}^{j}\mathbf{F}_j + {}^{j}\mathbf{f}_{j+1} + {}^{j}\mathbf{f}_{ej} \quad [1.131] $$

$$ {}^{j-1}\mathbf{f}_j = {}^{j-1}\mathbf{R}_j\,{}^{j}\mathbf{f}_j \quad [1.132] $$

$$ {}^{j}\mathbf{m}_j = {}^{j}\mathbf{M}_j + {}^{j}\mathbf{R}_{j+1}\,{}^{j+1}\mathbf{m}_{j+1} + {}^{j}\mathbf{P}_{j+1} \times {}^{j}\mathbf{f}_{j+1} + {}^{j}\mathbf{m}_{ej} \quad [1.133] $$

$$ \Gamma_j = (\sigma_j\,{}^{j}\mathbf{f}_j + \bar{\sigma}_j\,{}^{j}\mathbf{m}_j)^T\,{}^{j}\mathbf{a}_j + F_{sj}\,\mathrm{sign}(\dot{q}_j) + F_{vj}\,\dot{q}_j + Ia_j\,\ddot{q}_j \quad [1.134] $$

The previous algorithm can be calculated numerically. However, in order to reduce the number of operations, it is better to implement a technique of iterative symbolic calculation and to use the base inertial parameters.

1.5.3. *Determining the base inertial parameters*

In this section we present a symbolic method of calculating a minimal set of inertial parameters, also called base inertial parameters [MAY 90], which fully characterize the dynamic model. The use of these parameters in the calculation of the dynamic model reduces its complexity. Apart from that, this stage is an indispensable step for the identification of inertial parameters, the base parameters being the only identifiable parameters. They are obtained starting from standard parameters by eliminating those which do not have any effect on the dynamic model and grouping some others together.

The calculation of base parameters using the dynamic model proves to be long and fastidious. We present here a symbolic method leading to general rules without having to calculate either the dynamic model or the energy [GAU 90b], [KHA 94]. The demonstration of this method is based on the recursive calculation of the energy.

The total energy of link j, which is linear in the inertial parameters, is written as follows:

$$H_j = E_j + U_j = \mathbf{h}_j \, \mathbf{K}_j = (\mathbf{e}_j + \mathbf{u}_j) \, \mathbf{K}_j \qquad [1.135]$$

$$\mathbf{K}_j = [XX_j \;\; XY_j \;\; XZ_j \;\; YY_j \;\; ZZ_j \;\; MX_j \;\; MY_j \;\; MZ_j \;\; M_j]^T \qquad [1.136]$$

$$\mathbf{h}j = [h_{XXj} \;\; h_{XYj} \;\; h_{XZj} \;\; h_{YYj} \;\; h_{ZZj} \;\; h_{MXj} \;\; h_{MYj} \;\; h_{MZj} \;\; h_{Mj}] \qquad [1.137]$$

with:

– $\mathbf{K}_j$: vector of standard inertial parameters of link j (we do not consider parameter I_{aj});

– $\mathbf{h}_j$: (1 × 10) row matrix of the energy functions corresponding to the inertial parameters of link j;

– $\mathbf{e}_j$: (1 × 10) row matrix of the kinetic energy functions corresponding to the inertial parameters of link j;

– $\mathbf{u}_j$: (1 × 10) row matrix of the potential energy functions corresponding to the inertial parameters of link j.

The elements of $\mathbf{h}_j$ are obtained from relations [1.100] and [1.107] and are written as follows:

$$\begin{cases} h_{XXj} = \frac{1}{2}\omega_{1,j}\omega_{1,j} \\ h_{Xyj} = \omega_{1,j}\omega_{2,j} \\ h_{XZj} = \omega_{1,j}\omega_{3,j} \\ h_{YYj} = \frac{1}{2}\omega_{2,j}\omega_{2,j} \\ h_{YZj} = \omega_{2,j}\omega_{3,j} \\ h_{ZZj} = \frac{1}{2}\omega_{3,j}\omega_{3,j} \\ h_{MXj} = \omega_{3,j}V_{2,j} - \omega_{2,j}V_{3,j} - {}^0\mathbf{g}^T\,{}^0\mathbf{s}_j \\ h_{MYj} = \omega_{1,j}V_{3,j} - \omega_{3,j}V_{1,j} - {}^0\mathbf{g}^T\,{}^0\mathbf{n}_j \\ h_{MZj} = \omega_{2,j}V_{1,j} - \omega_{1,j}V_{2,j} - {}^0\mathbf{g}^T\,{}^0\mathbf{a}_j \\ h_{Mj} = \frac{1}{2}\,{}^j\mathbf{V}_j^T\,{}^j\mathbf{V}_j - {}^0\mathbf{g}^T\,{}^0\mathbf{P}_j \end{cases} \qquad [1.138]$$

with ${}^{j}\omega_j = [\omega_{1,j}\ \ \omega_{2,j}\ \ \omega_{3,j}]^T$ and ${}^{j}\mathbf{V}_j = [V_{1,j}\ \ V_{2,j}\ \ V_{3,j}]^T$.

Based on relations [1.100] and [1.107], we can write the general relation of recurrence between the energy functions of links j and j–1:

$$\mathbf{h}_j = \mathbf{h}_{j-1}\, {}^{j-1}\boldsymbol{\lambda}_j + \dot{q}_j\, \boldsymbol{\eta}_j \qquad [1.139]$$

${}^{j-1}\boldsymbol{\lambda}_j$ being a (10×10) matrix that is a function of the geometric parameters of frame j [GAU 90a]. It makes it possible to express the inertial parameters of link j with respect to frame j–1 according to the relation:

$$ {}^{j-1}\mathbf{K}_j = {}^{j-1}\boldsymbol{\lambda}_j\, {}^{j}\mathbf{K}_j \qquad [1.140]$$

The term $\boldsymbol{\eta}_j$ is written:

$$\begin{aligned}\boldsymbol{\eta}_j = {} & \bar{\sigma}_j\, [0\ \ 0\ \ \omega_{1,j}\ \ 0\ \ \omega_{2,j}\ \ (\omega_{3,j} - \frac{1}{2}\dot{q}_j)\ \ V_{2,j}\ \ -V_{1,j}\ \ 0\ \ 0] \\ & + \sigma_j\, [0\ \ 0\ \ 0\ \ 0\ \ 0\ \ 0\ \ -\omega_{2,j}\ \ \omega_{1,j}\ \ 0\ \ (V_{3,j} - \frac{1}{2}\dot{q}_j)]\end{aligned} \qquad [1.141]$$

The search for parameters that do not affect the dynamic model is based on the following rule:

h_j = constant $\Leftrightarrow$ K_j does not affect the model, we then note that $K_j \equiv 0$ [1.142]

The use of the recurrence relations of velocities [1.104] and [1.105] and of expressions [1.138] of $\mathbf{h}_j$, makes it possible to establish general rules in order to determine the parameters which do not effect the dynamic model without having to calculate the energy [KHA 94].

The grouping of a parameter K_j with other parameters consists of searching for linear combinations between the energy functions of this parameter h_j and those of the other inertial parameters such that:

$$h_j = \sum_{p=1}^{r} t_{jp} h_{jp} + \text{constant} \qquad [1.143]$$

Starting from the recurrence relation [1.139], between the energy functions $\mathbf{h}_j$ and $\mathbf{h}_{j-1}$, we deduce according to the joint type:

– *revolute joint j:* we obtain three linear combinations between the energy functions $\mathbf{h}_j$ and $\mathbf{h}_{j-1}$:

$$h_{XXj} + h_{YYj} = \mathbf{h}_{j-1} \left({}^{j-1}\boldsymbol{\lambda}_j^1 + {}^{j-1}\boldsymbol{\lambda}_j^4 \right) \quad [1.144]$$

$$h_{MZj} = \mathbf{h}_{j-1} \, {}^{j-1}\boldsymbol{\lambda}_j^9 \quad [1.145]$$

$$h_{Mj} = \mathbf{h}_{j-1} \, {}^{j-1}\boldsymbol{\lambda}_j^{10} \quad [1.146]$$

where ${}^{j-1}\boldsymbol{\lambda}_j^k$ is the kth column of the matrix ${}^{j-1}\boldsymbol{\lambda}_j$.

Consequently, we can group three inertial parameters. By choosing to group parameters YY_j, MZ_j and M_j with the other parameters, we obtain:

$$XXR_j = XX_j - YY_j \quad [1.147]$$

$$\mathbf{KR}_{j-1} = \mathbf{K}_{j-1} + YY_j \left({}^{j-1}\boldsymbol{\lambda}_j^1 + {}^{j-1}\boldsymbol{\lambda}_j^4 \right) + MZ_j \, {}^{j-1}\boldsymbol{\lambda}_j^9 + M_j \, {}^{j-1}\boldsymbol{\lambda}_j^{10} \quad [1.148]$$

where the letter "R" indicates the inertial parameters with which other parameters are grouped.

The development of these relations gives the following theorem:

THEOREM 1.1.– when joint j is revolute, parameters YY_j, MZ_j and M_j can be grouped with the parameters of the links C_j and C_{j-1}. The corresponding relations are as follows:

$$\begin{aligned}
XXR_j &= XX_j - YY_j \\
XXR_{j-1} &= XX_{j-1} + YY_j + 2\, r_j\, MZ_j + r_j^2\, M_j \\
XYR_{j-1} &= XY_{j-1} + d_j\, S\alpha_j\, MZ_j + d_j\, r_j\, S\alpha_j\, M_j \\
XZR_{j-1} &= XZ_{j-1} - d_j\, C\alpha_j\, MZ_j - d_j\, r_j\, C\alpha_j\, M_j \\
YYR_{j-1} &= YY_{j-1} + CC\alpha_j\, YY_j + 2\, r_j\, CC\alpha_j\, MZ_j + (d_j^2 + r_j^2\, CC\alpha_j)\, M_j \\
YZR_{j-1} &= YZ_{j-1} + CS\alpha_j\, YY_j + 2\, r_j\, CS\alpha_j\, MZ_j + r_j^2\, CS\alpha_j\, M_j \\
ZZR_{j-1} &= ZZ_{j-1} + SS\alpha_j\, YY_j + 2\, r_j SS\alpha_j\, MZ_j + (d_j^2 + r_j^2\, SS\alpha_j)\, M_j \\
MXR_{j-1} &= MX_{j-1} + d_j\, M_j \\
MYR_{j-1} &= MY_{j-1} - S\alpha_j\, MZ_j - r_j\, S\alpha_j\, M_j \\
MZR_{j-1} &= MZ_{j-1} + C\alpha_j\, MZ_j + r_j\, C\alpha_j\, M_j \\
MR_{j-1} &= M_{j-1} + M_j
\end{aligned} \quad [1.149]$$

with SS(.) = S(.) S(.), CC(.) = C(.) C(.) and CS(.) = C(.) S(.).

– *prismatic joint j:* in this case, $\eta_{1,j} = ... = \eta_{6,j} = 0$ and columns 1 to 6 of ${}^{j-1}\boldsymbol{\lambda}_j$ are constant. Hence, we obtain six relations between the energy functions:

$$h_{XXj} = \mathbf{h}_{j-1}\,{}^{j-1}\boldsymbol{\lambda}_j^1, \ldots, h_{ZZj} = \mathbf{h}_{j-1}\,{}^{j-1}\boldsymbol{\lambda}_j^6 \qquad [1.150]$$

These relations make it possible to group the parameters of the inertia matrix of link j with those of link j–1 through the following general relation:

$$\mathbf{KR}_{j-1} = \mathbf{K}_{j-1} + {}^{j-1}\boldsymbol{\lambda}_j^1\, XX_j + {}^{j-1}\boldsymbol{\lambda}_j^2\, XY_j + \ldots + {}^{j-1}\boldsymbol{\lambda}_j^6\, ZZ_j \qquad [1.151]$$

The development of relation [1.151] leads to the following theorem:

THEOREM 1.2.– when joint j is prismatic, the parameters of the inertia matrix of link j can be grouped with those of link j–1 according to the following relations:

$$\begin{aligned}
XXR_{j-1} &= XX_{j-1} + CC\theta_j\, XX_j - 2\, CS\theta_j\, XY_j + SS\theta_j\, YY_j \\
XYR_{j-1} &= XY_{j-1} + CS\theta_j\, C\alpha_j\, XX_j + (CC\theta_j - SS\theta_j)\, C\alpha_j\, XY_j - C\theta_j\, S\alpha_j\, XZ_j \\
&\quad - CS\theta_j\, C\alpha_j\, YY_j + S\theta_j\, S\alpha_j\, YZ_j \\
XZR_{j-1} &= XZ_{j-1} + CS\theta_j\, S\alpha_j\, XX_j + (CC\theta_j - SS\theta_j)\, S\alpha_j\, XY_j + C\theta_j\, C\alpha_j\, XZ_j \\
&\quad CS\theta_j\, S\alpha_j\, YY_j - S\theta_j\, C\alpha_j\, YZ_j \\
YYR_{j-1} &= YY_{j-1} + SS\theta_j\, CC\alpha_j\, XX_j + 2CS\theta_j\, CC\alpha_j\, XY_j - 2S\theta_j\, CS\alpha_j\, XZ_j \\
&\quad + CC\theta_j\, CC\alpha_j\, YY_j - 2C\theta_j\, CS\alpha_j\, YZ_j + SS\alpha_j\, ZZ_j \qquad [1.152] \\
YZR_{j-1} &= YZ_{j-1} + SS\theta_j\, CS\alpha_j\, XX_j + 2CS\theta_j\, CS\alpha_j\, XY_j + S\theta_j\, (CC\alpha_j - SS\alpha_j)\, XZ_j \\
&\quad + CC\theta_j\, CS\alpha_j\, YY_j + C\theta_j\, (CC\alpha_j - SS\alpha_j)\, YZ_j - CS\alpha_j\, ZZ_j \\
ZZR_{j-1} &= ZZ_{j-1} + SS\theta_j\, SS\alpha_j\, XX_j + 2CS\theta_j\, SS\alpha_j\, XY_j + 2S\theta_j\, CS\alpha_j\, XZ_j \\
&\quad + CC\theta_j\, SS\alpha_j\, YY_j + 2C\theta_j\, CS\alpha_j\, YZ_j + CC\alpha_j\, ZZ_j
\end{aligned}$$

The relations [1.152] are equivalent to the following relation:

$${}^{j-1}\mathbf{JR}_{j-1} = {}^{j-1}\mathbf{J}_{j-1} + {}^{j-1}\mathbf{R}_j\, {}^{j}\mathbf{J}_j\, {}^{j}\mathbf{R}_{j-1} \qquad [1.153]$$

The demonstration of this relation is immediate in such a way that when joint j is prismatic, the angular velocity of link C_j is the same as that of link C_{j-1}.

NOTE.– let r1 be the first revolute joint from the base and r2 be the first revolute joint after r1 and whose axis is not parallel to that of r1. Relations [1.42] and [1.152] lead to obtaining all the grouped parameters of all links of the robot, except those of the links whose joint is prismatic and placed between r1 and r2. Readers who are interested in this aspect can find more information on the particular grouping relations in [KHA 02]. The determination of the parameters that have no effect on the dynamic model of links $C_1, \ldots, C_{r2-1}$ complete the algorithm of determination of the base parameters.

EXAMPLE 1.8.– by using the relations [1.149] for j = n, …, 1, the set of base parameters of the Stäubli RX-90 robot is as follows:

– the parameters that have no effect in the model are XX_1, XY_1, XZ_1, YY_1, YZ_1, MX_1, MY_1, MZ_1, M_1, MZ_2 and M_2; axis 1 being vertical, all parameters of link 1 except ZZ1 are eliminated;

– the parameters eliminated by grouping are Ia_1, YY_2, Ia_2, YY_3, MZ_3, M_3, YY_4, MZ_4, M_4,YY_5, MZ_5, M_5, YY_6, MZ_6 and M_6;

– the grouping relations are:

$$ZZR_1 = ZZ_1 + Ia_1 + YY_2 + D3^2\,(M_3 + M_4 + M_5 + M_6) + YY_3$$
$$XXR_2 = XX_2 - YY_2 - D3^2\,(M_3 + M_4 + M_5 + M_6)$$
$$XZR_2 = XZ_2 - D3\,MZ_3$$
$$ZZR_2 = ZZ_2 + Ia_2 + D3^2\,(M_3 + M_4 + M_5 + M_6)$$
$$MXR_2 = MX_2 + D3\,(M_3 + M_4 + M_5 + M_6)$$
$$XXR_3 = XX_3 - YY_3 + YY_4 + 2\,RL4\,MZ_4 + RL4^2\,(M_4 + M_5 + M_6)$$
$$ZZR_3 = ZZ_3 + YY_4 + 2RL4\,MZ_4 + RL4^2\,(M_4 + M_5 + M_6)$$
$$MYR_3 = MY_3 + MZ_4 + RL4\,(M_4 + M_5 + M_6)$$
$$XXR_4 = XX_4 + YY_5 - YY_4$$
$$ZZR_4 = ZZ_4 + YY_5$$
$$MYR_4 = MY_4 - MZ_5$$
$$XXR_5 = XX_5 + YY_6 - YY_5$$
$$ZZR_5 = ZZ_5 + YY_6$$
$$MYR_5 = MY_5 + MZ_6$$
$$XXR_6 = XX_6 - YY_6$$

Table 1.3 provides the base inertial parameters of the Stäubli RX-90 robot. It has 40 parameters. By using this set of parameters, the cost in calculating the dynamic model of the Stäubli RX-90 robot through an iterative symbolic procedure is of 253 multiplications and of 238 additions, which no longer leads to any real-time calculation problem. If we suppose, moreover, that the links are symmetric, only 160 multiplications and 113 additions are needed.

j	XX_j	XY_j	XZ_j	YY_j	YZ_j	ZZ_j	MX_j	MY_j	MZ_j	M_j	Ia_j
1	0	0	0	0	0	ZZR_1	0	0	0	0	0
2	XXR_2	XY_2	XZR_2	0	YZ_2	ZZR_2	MXR_2	MY_2	0	0	0
3	XXR_3	XY_3	XZ_3	0	YZ_3	ZZR_3	MX_3	MYR_3	0	0	Ia_3
4	XXR_4	XY_4	XZ_4	0	YZ_4	ZZR_4	MX_4	MYR_4	0	0	Ia_4
5	XXR_5	XY_5	XZ_5	0	YZ_5	ZZR_5	MX_5	MYR_5	0	0	Ia_5
6	XXR_6	XY_6	XZ_6	0	YZ_6	ZZ_6	MX_6	MY_6	0	0	Ia_6

Table 1.3. *Base inertial parameters of the Stäubli RX-90 robot*

1.6. Identification of dynamic parameters

1.6.1. Introduction

In the last few years, several control laws have been proposed based on the dynamic model, as we will see in Chapter 5. However, whether we deal with the implementation of these laws in a robot controller or with the simulation of dynamic equations, it is necessary to have a good knowledge of the numeric values of the dynamic parameters of the robot. This section will show how to exploit that the dynamic model is linear in these parameters in order to identify them. Hence, the principle returns to the solution of an over-determined linear system using least squares techniques.

We suppose that the geometric parameter values are known (see section 1.4). The dynamic parameters of link j correspond to the inertial parameters of the link and of the actuator's rotor j as well as to the friction parameters. Thus, we write them as a vector χ^j:

$$\chi^j = [XXj\ \ XYj\ \ XZj\ \ YYj\ \ YZj\ \ ZZj\ \ MXj\ \ MYj\ \ MZj\ \ M_j\ \ Iaj\ \ F_{sj}\ \ F_{vj}]^T \qquad [1.154]$$

The dynamic parameters of the robot are represented by vector $\boldsymbol{\chi}$ as follows:

$$\boldsymbol{\chi} = [\boldsymbol{\chi}^{1T} \ \boldsymbol{\chi}^{2T} \cdots \boldsymbol{\chi}^{nT}]^T \qquad [1.155]$$

n being the number of joints of the robot.

1.6.2. *Identification principle of dynamic parameters*

The identification of the dynamic parameters of robots has been the subject of plenty of research: [FER 84], [MAY 84], [AN 85], [ATK 86], [KAW 88], [KHO 85], [GAU 86], [OLS 86], [BOU 89], [RAU 90], [AUB 91], [PRU 94], [RES 96]. The proposed methods have numerous common points, such as:

– use of a model linear in the unknown dynamic parameters (dynamic model, energetic model, power model or model of the forces and movements exerted by the robot on its base);

– construction of an over-determined linear system of equations by calculating the model at a sufficient number of points along some trajectories;

– estimation of parameters through linear regression techniques (ordinary least squares or other variants).

1.6.2.1. *Solving method*

The system to be solved is given by the following general relation:

$$\mathbf{Y}(\boldsymbol{\Gamma}, \dot{\mathbf{q}}) = \mathbf{W}(\mathbf{q}, \dot{\mathbf{q}}, \ddot{\mathbf{q}})\, \boldsymbol{\chi} + \boldsymbol{\rho} \qquad [1.156]$$

where $\mathbf{W}$ is the (r × c) observation matrix with r >> c, c being the number of parameters and $\boldsymbol{\rho}$ being the residual vector or the error vector.

The theory of the estimation offers a sufficiently wide range of methods, for both the deterministic case and the stochastic case. The use of classic methods for solving over-determined linear systems, and notably those using the SVD or QR decomposition, gives good results provided the elements of $\mathbf{W}$ are properly treated, as we will see in this section. We will notice that the implementation of these methods by scientific software such as Matlab or Mathematica is immediate.

We define the solution $\hat{\chi}$ of equation [1.156] in the least squares sense as follows:

$$\hat{\chi} = \min_{\chi} \|\rho\|_2 \qquad [1.157]$$

If **W** is of full rank, the explicit solution of this problem leads to the relation:

$$\hat{\chi} = (\mathbf{W}^T \mathbf{W})^{-1} \mathbf{W}^T \mathbf{Y} = \mathbf{W}^+ \mathbf{Y} \qquad [1.158]$$

where $\mathbf{W}^+$ indicates the pseudo-inverse matrix of **W**.

One difficulty in identifying the dynamic parameters comes from the fact that the observation matrix **W** is not deterministic but random. In addition, **W** and ρ are the result of correlated random variables [GAU 86], which can bias the least squares estimator [DEL 77], [EYK 74], [MEN 73]. An additional difficulty arises from the non-linearity of **W** in terms of $\mathbf{q}$ and $\dot{\mathbf{q}}$, which makes the calculation of the bias and the estimation error variance difficult, unless we admit certain independence hypotheses for the noises [ARM 89], [RAU 90]. That is why it is important to validate the results obtained through appropriate experimentation.

We basically calculate an estimation of the standard deviation for the identified values in a similar way to the one described in section 1.4.4.3, by considering that **W** is deterministic and that ρ is a zero mean additive independent noise with a standard deviation σ_ρ and with a variance-covariance matrix $\mathbf{C}_\rho$ such that:

$$\mathbf{C}_\rho = E(\rho\, \rho^T) = \sigma_\rho^2\, \mathbf{I}_r \qquad [1.159]$$

E indicating the expectation operator and $\mathbf{I}_r$ the $(r \times r)$ unit matrix.

The standard deviation σ_ρ can be calculated by the following unbiased estimator:

$$\sigma_\rho^2 = \frac{\|\mathbf{Y} - \mathbf{W}\hat{\chi}\|^2}{(r - c)} \qquad [1.160]$$

Hence, the matrix of variance-covariance of the estimation error is [DEL 77]:

$$\mathbf{C}_{\hat{\chi}} = E[(\chi - \hat{\chi})(\chi - \hat{\chi})^T] = \mathbf{W}^+ \mathbf{C}_\rho (\mathbf{W}^+)^T = \sigma_\rho^2 (\mathbf{W}^T \mathbf{W})^{-1} \qquad [1.161]$$

Hence, the standard deviation on the jth parameter:

$$\sigma_{\hat{\chi}_j} = \sqrt{\mathbf{C}_{\hat{\chi}}(j,j)} \qquad [1.162]$$

This interpretation was used in [RAU 90] but must be considered with caution because, in our case, the hypotheses are not satisfied, since **W** is not deterministic.

The relative standard deviation of the parameter χ_j is estimated by:

$$\sigma_{\hat{\chi}jr}\% = 100\frac{\sigma_{\hat{\chi}j}}{|\hat{\chi}_j|} \qquad [1.163]$$

The relative standard deviation can be used as a criterion in order to determine the quality of the estimation for each identified parameter.

1.6.2.2. *Identifiable parameters*

The identification of standard inertial parameters according to [1.156] cannot provide a unique solution when the observation matrix **W** is not of full rank. Hence, we have to determine a set of identifiable parameters called base parameters or minimal parameters.

This problem can be solved for the inertial parameters, either through the symbolic method (see section 1.5.3), or through a numerical method based on the QR decomposition similar to the one used in section 1.4.4.2.1 for geometric calibration. As far as the columns associated with friction parameters are concerned, we easily show that they are independent.

Determining identifiable parameters is a necessary preliminary step in identifying the dynamic parameters. In this case, the grouping relations are not necessary, but only the knowledge of the parameters to eliminate or to group is indispensable. The following algorithm can be used to determine the identifiable parameters by eliminating the parameters that have no effect on the dynamic model or because they can be grouped. We make use of r1 and r2 as defined in section 1.5.3.

For $j = n, ..., 1$:

1) the following parameters are eliminated due to general grouping relations:

- YY_j, MZ_j and M_j if joint j is revolute ($\sigma_j = 0$),
- XX_j, XY_j, XZ_j, YY_j, YZ_j and ZZ_j if joint j is prismatic ($\sigma_j = 1$);

2) if joint j is prismatic and $\mathbf{a}_j$ is parallel to $\mathbf{a}_{r1}$, for $r1 < j < r2$, then eliminate MZ_j, MX_j and MY_j;

3) if joint j is prismatic and $\mathbf{a}_j$ is not parallel to $\mathbf{a}_{r1}$, for $r1 < j < r2$, then:

- if $\mathbf{a}_j$ is not perpendicular to $\mathbf{a}_{r1}$, i.e., the z component of ${}^j\mathbf{a}_{r1}$ is not equal to zero, then eliminate MZ_j,

- if $\mathbf{a}_j$ is perpendicular to $\mathbf{a}_{r1}$, i.e., the z component of ${}^j\mathbf{a}_{r1}$ is equal to zero, then eliminate MX_j if the y component of ${}^j\mathbf{a}_{r1}$ is equal to zero, otherwise eliminate MY_j;

4) if joint j is revolute, for $r1 \leq j < r2$, then eliminate XX_j, XY_j, XZ_j and YZ_j. Note that the axis of this joint is parallel to the axis of joint r1, and that the parameter YY_j has been eliminated by rule 1;

5) if joint j is revolute, for $r1 \leq j < r2$, and the $\mathbf{a}_j$ axis is along $\mathbf{a}_{r1}$, and if $\mathbf{a}_{r1}$ is parallel to $\mathbf{a}_i$ and to gravity $\mathbf{g}$, for all $i < j$, then eliminate the parameters MX_j, MY_j. Note that MZ_j and M_j have been eliminated by rule 1;

6) if joint j is prismatic and $j < r1$, then eliminate the parameters MX_j, MY_j, MZ_j;

7) concerning the rotor inertia parameter Ia_j it can be eliminated by grouping in the following cases (where $j \leq r2$):

- if $j = r1$,

- if $j = r2$ and its axis is orthogonal to that of r1,

- if the axis of joint j is orthogonal to gravity and $j = p1$, with p1 the axis of the first prismatic joint, and if $p1 = 1$ or its axis is aligned with the revolute axes preceding it.

To simplify matters, we will suppose that χ contains only the base inertial parameter (or identifiable inertial parameters) as well as the friction parameters. $\mathbf{W}$ consists of the columns associated with these parameters.

1.6.2.3. *Choice of identification trajectories*

The choice of appropriate trajectories is very important in identifying the dynamic parameters. This problem is directly linked to the condition number of the linear system [1.156]. Thus, it is necessary to find an exciting trajectory with which all parameters will be identified.

We note that the elements of the observation matrix are affected by the modeling errors, the errors in the geometric parameters, and the measurement noise. We show that a robust solution with respect to these errors is obtained when the condition number of matrix $\mathbf{W}$ is minimum and when its last singular value is not too small [PRE 93].

To obtain an exciting trajectory, two strategies can be implemented:

– calculation of a trajectory leading to a condition number close to unity;

– identification of parameters by small groups with simple motion that affect only certain joints, the others being blocked (sequential trajectories).

The exciting strategy consists of calculating a trajectory that minimizes a criterion by non-linear optimization methods whose degrees of freedom are the initial point, the end point, the intermediate points, the maximum joint velocities and accelerations, etc., and whose criterion is a function of the condition number [ARM 89], [LU 93], [GAU 92], [BEN 93]. Various criteria have been proposed, such as:

– condition number of matrix **W** defined by:

$$\text{cond}(\mathbf{W}) = \frac{\sum_{\max}}{\sum_{\min}} \quad [1.164]$$

where Σ_{max} and Σ_{min} indicate the maximum and minimum singular values of **W**;

– if the order of magnitude of the parameters to be identified is known *a priori*, the following criterion is meant to balance the contribution of each parameter to the identification model, which results in equilibrating the relative standard deviation of the different parameters [PRE 93]:

$$C = \text{cond}(\mathbf{W}.\mathbf{diag}(\mathbf{Z})) \quad [1.165]$$

where **diag(Z)** is the diagonal matrix consisting of the elements of **Z** representing the vector of *a priori* values of the dynamic parameters (orders of magnitude).

The synthesis of an exciting trajectory by an optimization program presents the following difficulties:

– there is no analytical expression of the criteria;

– the algorithm must take into consideration the joint limits, maximum velocities and accelerations. The velocities must be high enough for the friction model to be valid;

– the degrees of freedom are very numerous.

The most current approaches suggest using several trajectories: each trajectory affects certain joints, the others being blocked; each test enables the identification of a small number of parameters [MAY 84], [OLS 86], [ATK 86], [HA 89], [AUB 91], [GAU 92]. This strategy makes it possible to simplify the identification. However, it may generate an accumulation of errors when the parameters already identified are used again in order to identify others. To remedy this, a weighted least-squares technique can be used [GAU 01].

1.6.2.4. *Evaluation of joint coordinates*

If the dynamic model is used as the identification model, the observation matrix $\mathbf{W}$ depends on $\mathbf{q}$ and $\dot{\mathbf{q}}$, but also on $\ddot{\mathbf{q}}$. In general, industrial robots are equipped with position sensors whose data are accurate and acceptable. On the other hand, the measures of $\dot{\mathbf{q}}$ and of $\ddot{\mathbf{q}}$ are problematic. The same goes if we want to determine these variables starting from position data. In fact, the measure of position with the help of encoders introduces a quantification noise. The velocity estimated by numeric derivation of $\mathbf{q}$ is distorted. In the same way, the double derivation for calculating $\ddot{\mathbf{q}}$ leads to unusable data because the differentiation increases the high frequency noises.

One solution is to filter the position signal with a low-pass filter before derivation [KHO 86], [GAU 95], [GAU 01]. Filtering can be done with a 4th order Butterworth two-way filter, by using Matlab's "butter" function, while the numeric deviations are made by a central difference to avoid phase difference according to the relation:

$$\dot{\mathbf{q}}(k) = [\mathbf{q}(k+1) - \mathbf{q}(k-1)]/2T \qquad [1.166]$$

where T is the sampling time.

1.6.2.5. *Evaluation of joint torques*

The use of torque sensors or force sensors is not widespread, at least in the industrial field, and often only the reference current at the input of power amplifiers are available. Thus, the relation between these values and the torques must be estimated. In general, owing to the high bandwidth of the current loop, this relation can be taken as a simple constant gain. For joint j, this relation is written (see Figure 1.12):

$$\Gamma_{mj} = G_{Tj}\, u_j \qquad [1.167]$$

with:

$$G_{Tj} = N_j\, K_{aj}\, K_{Tj} \qquad [1.168]$$

where G_{Tj} is the torque gain of the drive chain j, u_j is the reference current, N_j is the gear transmission ratio, K_{aj} is the amplifier gain and K_{Tj} is the actuator torque constant.

In the case of a numerical closed loop control, the reference current u_j is equal to the control calculated by the computer. Various electrical engineering techniques make it possible to identify each one of the terms necessary for the calculation of the overall gain G_T [RES 95]. However, they can be obtained from the manufacturer's data sheets.

Other identification methods of dynamic parameters, which do not explicitly require the evaluation of joint torques, have been proposed [KHA 93], [GAU 94]. They consist of including in the identification model the gains of the drive chains.

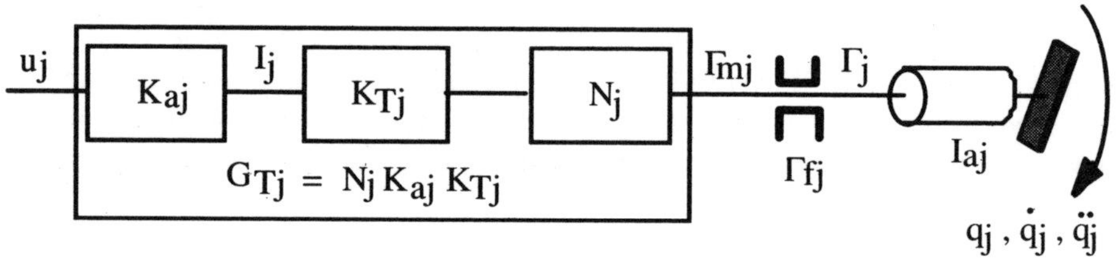

Figure 1.12. *Representation of the drive chain*

1.6.3. *Identification model using the dynamic model*

The dynamic model being linear in the dynamic parameters, the identification model is given by the following relation:

$$\boldsymbol{\Gamma} = \boldsymbol{\Phi}(\mathbf{q}, \dot{\mathbf{q}}, \ddot{\mathbf{q}})\, \boldsymbol{\chi} \qquad [1.169]$$

$\boldsymbol{\Phi}$ being a $(n \times b)$ matrix and b being the number of base dynamic parameters.

For the calculation of the ith column of $\boldsymbol{\Phi}$, written $\boldsymbol{\Phi}^i$, we can show that:

$$\boldsymbol{\Phi}^i = \boldsymbol{\Gamma}(\mathbf{q}, \dot{\mathbf{q}}, \ddot{\mathbf{q}} \text{ with } \chi_i = 1, \chi_j = 0 \text{ for } j \neq i) \qquad [1.170]$$

This relation indicates that column $\boldsymbol{\Phi}^i$ can be calculated with the help of the algorithm of Newton-Euler inverse dynamic model by using as dynamic parameters the values $\chi_i = 1$, $\chi_j = 0$ for $j \neq i$. To improve the performance of such a procedure, we can use the iterative symbolic technique by taking into consideration the fact that calculating the velocities and accelerations is common for all columns. Moreover, this technique is suitable to an efficient vector calculation of the observation matrix.

When the measures $(\mathbf{q}, \dot{\mathbf{q}}, \ddot{\mathbf{q}})_{(i)}$ and $\Gamma(i)$ (i = 1, ..., e, e being the number of samples) are sufficiently numerous for a given trajectory, we create an over-determined linear system of r equations that has b unknown elements:

$$\mathbf{Y}(\Gamma) = \mathbf{W}(\mathbf{q}, \dot{\mathbf{q}}, \ddot{\mathbf{q}})\, \chi + \rho \qquad [1.171]$$

with:

$$\mathbf{Y} = \begin{bmatrix} \Gamma(1) \\ \cdots \\ \Gamma(e) \end{bmatrix}, \quad \mathbf{W} = \begin{bmatrix} \mathbf{\Phi}(1) \\ \cdots \\ \mathbf{\Phi}(e) \end{bmatrix} \qquad [1.172]$$

and such that:

– $r = nxe >> b$

– $\mathbf{\Phi}(i) = \mathbf{\Phi}[(\mathbf{q}, \dot{\mathbf{q}}, \ddot{\mathbf{q}})_{(i)}$

– $(\mathbf{q}, \dot{\mathbf{q}}, \ddot{\mathbf{q}})_{(i)} = \mathbf{q}(t_i), \dot{\mathbf{q}}(t_i), \ddot{\mathbf{q}}(t_i)$

– $\Gamma(i) = \Gamma(t_i)$

It is in general easier to arrange this data by grouping together the equations of each joint. Vector **Y** and the matrix **W** are thus written:

$$\mathbf{Y} = \begin{bmatrix} \mathbf{Y}_1 \\ \cdots \\ \mathbf{Y}_n \end{bmatrix}, \quad \mathbf{W} = \begin{bmatrix} \mathbf{W}1 \\ \cdots \\ \mathbf{W}_n \end{bmatrix} \qquad [1.173]$$

where $\mathbf{Y}_i$ and $\mathbf{W}_i$ represent the equations of joint i for all samples, in such a way that:

$$\mathbf{Y}_i = \begin{bmatrix} \mathbf{\Gamma}_i(1) \\ \cdots \\ \mathbf{\Gamma}_i(e) \end{bmatrix}, \quad \mathbf{W}_i = \begin{bmatrix} \mathbf{\Phi}(i,:)(1) \\ \cdots \\ \mathbf{\Phi}(i,:)(e) \end{bmatrix} \qquad [1.174]$$

where $\mathbf{\Phi}$(i,:)(j) indicates the row corresponding to joint i in the identification model for the jth sample.

In order to eliminate the high frequency ripples that are not modeled in the torque vector $\mathbf{\Gamma}$, we filter vector $\mathbf{Y}$ as well as the columns of the observation matrix $\mathbf{W}$. This process is called parallel filtering [RIC 98]. It can be carried out by Matlab's "decimate" function (Signal Processing ToolBox).

1.6.4. *Sequential formulation of the dynamic model*

Since the dynamics of the joint j depends only on the dynamic parameters of link j and links j+1, ..., n (property (c), see section 1.5.1.3), we can write the dynamic model [1.169] in such a way that matrix $\mathbf{\Phi}$ has a upper triangular form:

$$\begin{bmatrix} \Gamma_1 \\ \Gamma_2 \\ \vdots \\ \Gamma_n \end{bmatrix} = \begin{bmatrix} \mathbf{\Phi}_{11} & \mathbf{\Phi}_{12} \cdots \mathbf{\Phi}_{1n-1} & \mathbf{\Phi}_{1n} \\ \mathbf{0} & \mathbf{\Phi}_{22} \cdots \mathbf{\Phi}_{2n-1} & \mathbf{\Phi}_{2n} \\ \vdots & \mathbf{0} \quad \cdots \quad \vdots & \vdots \\ \mathbf{0} & \mathbf{0} \quad \cdots \quad \mathbf{0} & \mathbf{\Phi}_{nn} \end{bmatrix} \begin{bmatrix} \chi^1 \\ \chi^2 \\ \vdots \\ \chi^n \end{bmatrix} \qquad [1.175]$$

where χ^j contains the base dynamic parameters of the link and actuator j, and where $\mathbf{\Phi}_{ij}$ is the row vector corresponding to χ^j in the Γ_i equation. The system of equation [1.171] is thus written for e samples:

$$\begin{bmatrix} \mathbf{Y}_1 \\ \cdots \\ \mathbf{Y}_n \end{bmatrix} = \begin{bmatrix} \mathbf{W}_{11} & \mathbf{W}_{12} \cdots \mathbf{W}_{1n-1} & \mathbf{W}_{1n} \\ \mathbf{0} & \mathbf{W}_{22} \cdots \mathbf{W}_{2n-1} & \mathbf{W}_{2n} \\ \vdots & \mathbf{0} \quad \cdots \quad \vdots & \vdots \\ \mathbf{0} & \mathbf{0} \quad \cdots \quad \mathbf{0} & \mathbf{W}_{nn} \end{bmatrix} \begin{bmatrix} \chi^1 \\ \chi^2 \\ \vdots \\ \chi^n \end{bmatrix} \qquad [1.176]$$

This structure suggests the idea of identifying sequentially the parameters of the robot, link by link, beginning with link n and ending with link 1. We proceed in the following manner:

– we first identify the parameters of joint n, χ^n, by using the dynamic equation of axis n:

$$\mathbf{Y}_n = \mathbf{W}_{nn}(\mathbf{q}, \dot{\mathbf{q}}, \ddot{\mathbf{q}})\, \chi^n \qquad [1.177]$$

– then, we identify the parameters of χ^{n-1} considering that those of χ^n are known, which gives:

$$\mathbf{Y}_{n-1} - \mathbf{W}_{n-1,n}(\mathbf{q}, \dot{\mathbf{q}}, \ddot{\mathbf{q}})\hat{\chi}^{n} = \mathbf{W}_{n-1,n-1}(\mathbf{q}, \dot{\mathbf{q}}, \ddot{\mathbf{q}})\chi^{n-1} \quad [1.178]$$

– and so on, up to link 1.

With the help of the sequential method, the search for exciting trajectories enabling a good condition number of the observation matrix $\mathbf{W}_{jj}$ is much simpler than using a general method and this is due to two reasons:

– the dimension of the vector to be identified in each step is small, which reduces the dimension of the observation matrix $\mathbf{W}_{jj}$;

– matrix $\mathbf{W}_{jj}$ depends only on positions, velocities and accelerations of the joints 1, 2, …, j. It is therefore possible to estimate the parameters of axis j by blocking the movement of the axes j+1, …, n.

The major disadvantage of the sequential identification is the accumulation of errors during each step, but this problem can be solved by using a weighting least squares method [GAU 01].

1.6.5. *Practical considerations*

Before ending this section, we have to highlight the following points:

– the identification model can be created by using the energetic model, which is based only on the joint positions and velocities. We have presented only the method based on the inverse dynamic model which is structurally more exciting than the energetic model. This is due to the fact that the dynamic model projects its information on the components of torque $\mathbf{\Gamma}$, whereas the energetic model groups all information in a scalar form;

– after filtering the position, the joint velocities are determined by central difference of the positions and the joint accelerations by central difference of velocities;

– when we use the dynamic model, it is necessary to filter the columns of the observation matrix and the torque measure vector in order to eliminate the high frequency ripples. This operation is called parallel filtering and can be carried out by Matlab's "decimate" function;

– if the robot includes a significant number of parameters, it is advisable to carry out the identification in two steps: firstly, the parameters of the wrist links are identified by using only the equations of the wrist joints; and secondly, we identify the parameters of the shoulder by blocking the axes of the links already identified. This procedure is all the more relevant because the order of magnitude of the parameter values of the wrist links is not the same as the ones of the shoulder;

– the number of equations must be higher than the number of inertial parameters by at least 500 times;

– the relative standard deviation estimated by equation [1.163] is used as an identification quality criterion. A parameter is considered as well identified if its relative standard deviation is less than 10. In an opposite case, this parameter may be considered as poorly identified. This means that it is not excited enough by the trajectory used or that it does not have a significant effect on the model. If, with other trajectories, the standard deviation does not decrease below the threshold, we can assume that this parameter has no effect on this model. Eliminating such parameters, we then define a set of essential and better identifiable parameters [PHA 91];

– after identifying the inertial parameters, the values obtained must be validated. Several procedures can be carried out:

- direct validation by calculating the error of prediction for the trajectories used in the identification,

- crossed validation for a different trajectory,

- identification of the robot parameters without load, then re-doing the identification with a load for which the inertial parameters are known, and comparing between the values identified and the values *a priori*,

- validation by elaborating a dynamic control and realization of a simulator.

1.7. Conclusion

We have presented in this chapter the geometric, kinematic and dynamic modeling of serial robots. The identification of geometric and dynamic parameters of these models was also covered, with the most efficient methods from an experimental point of view. The geometric description of the structure is based on a method which makes it possible to generalize the various models for tree-structured robots and closed loop robots [KHA 02]. The symbolic calculation of these models on the computer was dealt with in numerous works [DIL 73], [KHA 76], [ZAB 78], [KRE 79], [ALD 82], [CES 84], [MEG 84], [MUR 84], [KIR 85], [BUR 86], [IZA 86], [KHA 89]. The SYMORO+ [KHA 97] software, which contains all the algorithms of this chapter, is certainly the best performing and the only one able to deal with all of the models mentioned.

1.8. Bibliography

[ALD 82] ALDON M.J., "Elaboration automatique de modèles dynamiques de robots en vue de leur conception et de leur commande", PhD Thesis, USTL, Montpellier, October 1982.

[AN 85] AN C.H., ATKESON C.G., HOLLERBACH J.M., "Estimation of inertial parameters of rigid body links of manipulators", *Proc. 24th IEEE Conf. on Decision and Control*, Fort Lauderdale, 1985, p. 990-995.

[ARI 84] ARIMOTO S., MIYAZAKI F., "Stability and robustness of PID feedback control for robots manipulators of sensory capability", *1st Int. Symp. of Robotics Research*, MIT Press, Cambridge, 1984.

[ARM 79] ARMSTRONG W.W., "Recursive solution to the equation of motion of an N-links manipulator", *Proc. 5th World Congress on Theory of Machines and Mechanisms*, Montreal, 1979, p. 1343-1346.

[ARM 88] ARMSTRONG B., "Dynamics for robot control: friction modeling and ensuring excitation during parameter identification", PhD Thesis, Dept. of Electrical Engineering, Stanford University, May 1988.

[ARM 89] ARMSTRONG B., "On finding exciting trajectories for identification experiments involving systems with non linear dynamics", *The Int. J. of Robotics Research*, Vol. 8(6), 1989, p. 28-48.

[ARM 91] ARMSTRONG B., *Control of Machines with Frictions*, Kluwer Academic Publishing, 1991.

[ARM 94] ARMSTRONG-HÉLOUVRY B., DUPONT P., CANUDAS DE WIT C., "A survey of analysis tools and compensation methods for the control of machines with friction", *Automatica*, Vol. 30(10), 1994, p. 1083-1138.

[ATK 86] ATKESON C.G., AN C.H., HOLLERBACH J.M., "Estimation of inertial parameters of manipulator loads and links", *Int. J. of Robotics Research*, Vol. 5(3), 1986, p. 101-119.

[AUB 91] AUBIN A., "Modélisation, identification et commande du bras manipulateur TAM", PhD Thesis, INPG, Grenoble, 1991.

[BAI 84] BAILLIEUL J., HOLLERBACH J.M., BROCKETT R., "Programming and control of kinematically redundant manipulators", *Proc. 23rd IEEE Conf. on Decision and Control*, Las Vegas, December 1984, p. 768-774.

[BAI 85] BAILLIEUL J., "Kinematic programming alternatives for redundant manipulators", *Proc. IEEE Int. Conf. on Robotics and Automation*, St. Louis, March 1985, p. 722-728.

[BAI 86] BAILLIEUL J., "Avoiding obstacles and resolving kinematic redundancy", *Proc. IEEE Int. Conf. on Robotics and Automation*, San Francisco, April 1986, p. 1698-1704.

[BEN 93] BENHLIMA S., "Identification des paramètres dynamiques des systèmes mécaniques articulés complexes", PhD Thesis, ENSAM, Paris, France, 1993.

[BEN 91] BENNETT D.J., HOLLERBACH J.M., "Autonomous calibration of single-loop closed kinematic chains formed by manipulators with passive endpoint constraints", *IEEE Trans. on Robotics and Automation*, Vol. RA-7(5), 1991, p. 597-606.

[BES 00] BESNARD S., KHALIL W., GARCIA G., "Robot Calibration Using Multiple Plane Constraints", *Advances in Robotic kinematics*, Kluwer Academic Publishing, 2000, p. 61-70.

[BOR 79] BORREL P., "Modèle de comportement de manipulateurs; application à l'analyse de leurs performances et à leur commande automatique", PhD Thesis, USTL, Montpellier, December 1979.

[BOR 86] BORREL P., "Contribution à la modélisation géométrique des robots-manipulateurs; application à la conception assistée par ordinateur", PhD Thesis, USTL, Montpellier, July 1986.

[BOR 91] BORM J.H., MENQ C.H., "Determination of optimal measurement configurations for robot calibration based on observability measure", *The Int. J. of Robotics Research*, Vol. 10(1), p. 51-63, 1991.

[BOU 89] BOUZOUIA B., "Commande des robots manipulateurs: identification des paramètres et étude des stratégies adaptatives", PhD Thesis, UPS, Toulouse, May 1989.

[BUR 86] BURDICK J.W., "An algorithm for generation of efficient manipulator dynamic equations", *Proc. IEEE Int. Conf. on Robotics and Automation*, San Francisco, April 1986, p. 212-218.

[CAN 89] CANUDAS DE WIT C., NOËL P., AUBIN A., BROGLIATO B., DREVET P., "Adaptive Friction compensation in robot manipulators: low-velocities", *Proc. Int. Conf. on Robotics and Automation*, Scottsdale, May 1989, p. 1352-1357.

[CAN 90] CANUDAS DE WIT C., SERONT V., "Robust adaptive friction compensation", *Proc. IEEE Int. Conf. on Robotics and Automation*, Cincinnati, May 1990, p. 1383-1389.

[CES 84] CESAREO G., NICOLO F., NICOSIA S., "DYMIR: a code for generating dynamic models of robots", *Proc. IEEE Int. Conf. on Robotics*, Atlanta, March 1984, p. 115-120.

[CHA 86] CHANG P.H., "A closed form solution for the control of manipulators with kinematic redundancy", *Proc. IEEE Int. Conf. on Robotics and Automation*, San Francisco, April 1986, p. 9-14.

[CHE 86] CHEDMAIL P., GAUTIER M., KHALIL W., "Automatic modelling of robots including parameters of links and actuators", *Proc. IFAC Symp. on Theory of Robots*, Vienna, Austria, December 1986, p. 295-299.

[CHE 88b] CHEVALLEREAU C., "Contribution à la commande des robots-manipulateurs dans l'espace opérationnel", PhD Thesis, ENSM, Nantes, May 1988.

[CHE 90] CHEDMAIL P., GAUTIER M., "Optimum choice of robot actuators", *Trans. of ASME, J. of Engineering for Industry*, Vol. 112(4), 1990, p. 361-367.

[COI 81] COIFFET P., *Les Robots*; *Tome 1: Modélisation et commande*, Hermès, Paris, 1981.

[CRA 86] CRAIG J.J., *Introduction to Robotics: Mechanics and Control*, Addison Wesley Publishing Company, Reading, 1986.

[DAH 77] DAHL P.R., "Measurements of solid friction parameters of ball bearings", *Proc. of the 6th Annual Symp. on Incremental Motion Control Systems and Devices*, University of Illinois, 1977.

[DEL 77] DE LARMINAT P., THOMAS Y., *Automatique des systèmes linéaires; Tome 2: Identification*, Editions Flammarion, 1977.

[DEN 55] DENAVIT J., HARTENBERG R.S., "A kinematic notation for lower pair mechanism based on matrices", *Trans. of ASME, J. of Applied Mechanics*, Vol. 22, June 1955, p. 215-221.

[DIL 73] DILLON S.R., "Computer assisted equation generation in linkage dynamics", PhD Thesis, Ohio State University, August 1973.

[DON 79] DONGARRA J.J., MOLER C.B., BUNCH J.R., STEWART G.W., *LINPACK User's Guide*, Philadelphia, SIAM, 1979.

[EVE 89] EVERETT L.J., "Forward calibration of closed loop jointed manipulators", *Int. J. of Robotics Research*, Vol. 8(4), August 1989, p. 85-91.

[EYK 74] EYKKOF P., *System identification: parameter and state estimation*, John Wiley & Sons, London, 1974.

[FEA 83] FEATHERSTONE R., "Position and velocity transformations between robot end-effector coordinates and joint angles", *The Int. J. of Robotics Research*, Vol. 2(2), 1983, p. 35-45.

[FER 84] FERREIRA E.P., "Contribution à l'identification de paramètres et à la commande des robots manipulateurs", PhD Thesis, UPS, Toulouse, July 1984.

[FOR 77] FORSYTHE G.E., MALCOM M.A., MOLER C.B., *Computer methods for mathematical computations*, Prentice-Hall, Englewood Cliffs, 1977.

[FOU 80] FOURNIER A., "Génération de mouvements en robotique; application des inverses généralisées et des pseudo-inverses", PhD Thesis, USTL, Montpellier, April 1980.

[GAU 86] GAUTIER M., "Identification of robots dynamics", *Proc. IFAC Symp. on Theory of Robots*, Vienne, Austria, December 1986, p. 351-356.

[GAU 90a] GAUTIER M., "Contribution à la modélisation et à l'identification des robots", PhD Thesis, ENSM, Nantes, May 1990.

[GAU 90b] GAUTIER M., KHALIL W., "Direct calculation of minimum set of inertial parameters of serial robots", *IEEE Trans. on Robotics and Automation*, Vol. RA-6(3), 1990, p. 368-373.

[GAU 91] GAUTIER M., "Numerical calculation of the base inertial parameters", *J. of Robotic Systems*, Vol. 8(4), August 1991, p. 485-506.

[GAU 92] GAUTIER M., KHALIL W., "Exciting trajectories for inertial parameters identification", *The Int. J. of Robotics Research*, Vol. 11(4), 1992, p. 362-375.

[GAU 94] GAUTIER M., VANDANJON P.-O., PRESSE C., "Identification of inertial and drive gain parameters of robots", *Proc. IEEE 33rd Conf. on Decision and Control*, Orlando, USA, December 1994, p. 3764-3769.

[GAU 95] GAUTIER M., KHALIL W., RESTREPO P.P., "Identification of the dynamic parameters of a closed loop robot", *Proc. IEEE Int. Conf. on Robotics and Automation*, Nagoya, May 1995, p. 3045-3050.

[GAU 01] GAUTIER M., KHALIL W., "Identification des paramètres des modèles", in *Traité IC2 : Analyse et modélisation des robots manipulteurs*, E. Dombre (ed.), Hermès, 2001, p. 145-197.

[GOL 85] GOLDENBERG A.A., BENHABIB B., FENTON R.G., "A complete generalized solution to inverse kinematics of robots", *IEEE J. of Robotics and Automation*, Vol. RA-1(1), 1985, p. 14-20.

[GOL 96] GOLUB G.H., VAN LOAN C.F., *Matrix computations*, 3rd edition, John Hopkins University Press, Baltimore, 1996.

[GOR 84] GORLA B., RENAUD M., *Modèles des robots-manipulateurs; application à leur commande*, Cepadues Editions, Toulouse, 1984.

[HA 89] HA I.J., KO M.S., KWON S.K., "An efficient estimation algorithm for the model parameters of robotic manipulators", *IEEE Trans. on Robotics and Automation*, Vol. RA-5(6), 1989, p. 386-394.

[HAY 83] HAYATI S.A., "Robot arm geometric link parameter estimation", *Proc. 22nd IEEE Conf. Decision and Control*, San Antonio, USA, December 1983, p. 1477-1483.

[HOL 80] HOLLERBACH J.M., "An iterative lagrangian formulation of manipulators dynamics and a comparative study of dynamics formulation complexity", *IEEE Trans. on Systems, Man, and Cybernetics*, Vol. SMC-10(11), 1980, p. 730-736.

[HOL 84] HOLLERBACH J.M., "Optimum kinematic design for a seven degree of freedom manipulator", *Proc. 2nd Int. Symp. of Robotics Research*, Kyoto, August 1984, p. 349-356.

[HOL 85] HOLLERBACH J.M., SUH K.C., "Redundancy resolution of manipulators through torque optimization", *Proc. IEEE Int. Conf. on Robotics and Automation*, St. Louis, March 1985, p. 1016-1021.

[HOL 89] HOLLERBACH J.M., "A survey of kinematic calibration", in *The Robotics Review n°1*, MIT Press, Cambridge, USA, 1989, p. 207-242.

[HOL 96] HOLLERBACH J.M., WAMPLER C.W., "The calibration index and taxonomy of kinematic calibration methods", *The Int. J. of Robotics Research*, Vol. 14, 1996, p. 573-591.

[HOO 65] HOOKER W.W., MARGULIES G., "The dynamical attitude equations for a n-body satellite", *The Journal of the Astronautical Sciences*, Vol. 12(4), 1965, p. 123-128.

[IKI 97] IKITS M., HOLLERBACH J.M., "Kinematic calibration using a plane constraint", *Proc. IEEE Int. Conf. on Robotics and Automation*, Albuquerque, April 1997, p. 3191-3196.

[IZA 86] IZAGUIRRE A., PAUL R.C.P., "Automatic generation of the dynamic equations of the robot manipulators using a LISP program", *Proc. IEEE Int. Conf. on Robotics and Automation*, San Francisco, April 1986, p. 220-226.

[KAW 88] KAWASAKI H., NISHIMURA K., "Terminal-link parameter estimation and trajectory control of robotic manipulators", *IEEE J. of Robotics and Automation*, Vol. RA-4(5), p. 485-490, 1988.

[KHA 76] KHALIL W., "Modélisation et commande par calculateur du manipulateur MA-23; extension à la conception par ordinateur des manipulateurs", PhD Thesis, USTL, Montpellier, September 1976.

[KHA 85] KHALIL W., KLEINFINGER J.-F., "Une modélisation performante pour la commande dynamique de robots", *Revue RAIRO, APII*, Vol. 6, 1985, p. 561-574.

[KHA 86] KHALIL W., KLEINFINGER J.-F., "A new geometric notation for open and closed-loop robots", *Proc. IEEE Int. Conf. on Robotics and Automation*, San Francisco, April 1986, p. 1174-1180.

[KHA 87] KHALIL W., KLEINFINGER J.-F., "Minimum operations and minimum parameters of the dynamic model of tree structured robots", *IEEE J. of Robotics and Automation*, Vol. RA-3(6), December 1987, p. 517-526.

[KHA 89] KHALIL W., BENNIS F., CHEVALLEREAU C., KLEINFINGER J.-F., "SYMORO: a software package for the symbolic modelling of robots", *Proc. 20th Int. Symp. on Industrial Robots*, Tokyo, October 1989, p. 1023-1030.

[KHA 91] KHALIL W., GAUTIER M., "Calculation of the identifiable parameters for robot calibration", *Proc. IFAC Symp. on Identification and System Parameter Estimation*, Budapest, 1991, p. 888-892.

[KHA 93] KHALIL W., GAUTIER M., "Computed current control of robots", *Proc. IFAC 12th World Congress*, Sydney, Australia, July 1993, Vol. IV, p. 129-132.

[KHA 94] KHALIL W., BENNIS F., "Comments on direct calculation of minimum set of inertial parameters of serial robots", *IEEE Trans. on Robotics and Automation*, Vol. RA-10(1), 1994, p. 78-79.

[KHA 95] KHALIL W., GARCIA G., DELAGARDE J.-F., "Calibration of the geometric parameters of robots without external sensors", *Proc. IEEE Int. Conf. on Robotics and Automation*, Nagoya, May 1995, p. 3039-3044.

[KHA 96] KHALIL W., LEMOINE P., GAUTIER M., "Autonomous calibration of robots using planar points", *Proc. 6th Int. Symp. on Robotics and Manufacturing, WAC'96*, Vol. 3, Montpellier, May 1996, p. 383-388.

[KHA 97] KHALIL W., CREUSOT D., "SYMORO+: a system for the symbolic modelling of robots", *Robotica*, Vol. 15, 1997, p. 153-161.

[KHA 00] KHALIL W., BESNARD S., LEMOINE P., "Comparison study of the geometric parameters calibration methods", *Int. Journal of robotics and Automation*, Vol. 15(2), 2000, p. 56-67.

[KHA 02] KHALIL W., DOMBRE E., *Modelling, Identification and Control of Robots*, Hermès Penton, London, 2002.

[KHO 85] KHOSLA P.K., KANADE T., "Parameter identification of robot dynamics", *Proc. 24th IEEE Conf. on Decision and Control*, Fort Lauderdale, December 1985, p. 1754-1760.

[KHO 86] KHOSLA P.K., "Real-time control and identification of direct drive manipulators", PhD Thesis, Carnegie Mellon University, Pittsburgh, 1986.

[KIR 85] KIRCÁNSKI M., VUKOBRATOVIC M., "Computer-aided generation of manipulator kinematic models in symbolic form", *Proc. 15th Int. Symp. on Industrial Robots*, Tokyo, September 1985, p. 1043-1049.

[KLE 84] KLEIN C.A., "Use of redundancy in the design of robotic systems", *Proc. 2nd Int. Symp. of Robotic Research*, Kyoto, August 1984, p. 58-65.

[KOD 84] KODITSCHEK D.E., "Natural motion for robot arms", *Proc. 23rd IEEE Conf. on Decision and Control*, Las Vegas, December 1984, p. 737-735.

[KRE 79] KREUZER E.J., "Dynamical analysis of mechanisms using symbolical equation manipulation", *Proc. 5th World Congress on Theory of Machines and Mechanisms*, Montreal, 1979, p. 599-602.

[LAW 74] LAWSON C.L., HANSON R.J., *Solving least squares problems*, Prentice-Hall, Englewood Cliffs, 1974.

[LLI 83] LLIBRE M., MAMPEY R., CHRETIEN J.P., "Simulation de la dynamique des robots manipulateurs motorisés", *Congrès AFCET: Productique et robotique intelligente*, Besançon, November 1983, p. 197-207.

[LU 93] LU Z., SHIMOGA K.B., GOLDBERG A., "Experimental determination of dynamic parameters of robotic arms", *J. of Robotic Systems*, Vol. 10(8), 1993, p. 1009-1029.

[LUH 80] LUH J.Y.S., WALKER M.W., PAUL R.C.P., "On-line computational scheme for mechanical manipulators", *Trans. of ASME, J. of Dynamic Systems, Measurement, and Control*, Vol. 102(2), 1980, p. 69-76.

[LUH 85] LUH J.Y.S., GU Y.L., "Industrial robots with seven joints", *Proc. IEEE Int. Conf. on Robotics and Automation*, St. Louis, March 1985, p. 1010-1015.

[MAC 85] MACIEJEWSKI A.A., KLEIN C.A., "Obstacle avoidance for kinematically redundant manipulators in dynamically varying environments", *The Int. J. of Robotics Research*, Vol. 4(3), Fall 1985, p. 109-117.

[MAU 96] MAURINE P., "Développement et mise en œuvre de méthodologies d'étalonnage de robots manipulateurs industriels", PhD Thesis, Montpellier University II, December 1996.

[MAY 84] MAYEDA H., OSUKA K., KANGAWA A., "A new identification method for serial manipulator arms", *Proc. IFAC 9th World Congress*, Budapest, July 1984, p. 74-79.

[MAY 90] MAYEDA H., YOSHIDA K., OSUKA K., "Base parameters of manipulator dynamic models", *IEEE Trans. on Robotics and Automation*, Vol. RA-6(3), 1990, p. 312-321.

[MEG 84] MEGAHED S., "Contribution à la modélisation géométrique et dynamique des robots manipulateurs ayant une structure de chaîne cinématique simple ou complexe; application à leur commande", PhD Thesis, UPS, Toulouse, July 1984.

[MEN 73] MENDEL J.M., *Discrete techniques of parameter estimation: the equation error formulation*, Marcel Dekker Inc. New York, 1973.

[MUR 84] MURRAY J.J., NEWMAN C.P., "ARM: an algebraic robot dynamic modeling program", *Proc. IEEE Int. Conf. on Robotics and Automation*, Atlanta, March 1984, p. 103-104.

[NAH 94] NAHVI A., HOLLERBACH J.M., HAYWARD V., "Calibration of parallel robot using multiple kinematic closed loops", *Proc. IEEE Int. Conf. on Robotics and Automation*, San Diego, May 1994, p. 407-412.

[NEN 92] NENCHEV D.N., "Restricted Jacobian matrices of redundant manipulators in constrained motion tasks", *The Int. J. of Robotics Research*, Vol. 11(6), 1992, p. 584-597.

[OLS 86] OLSEN H.B., BEKEY G.A., "Identification of robot dynamics", *Proc. IEEE Int. Conf. on Robotics and Automation*, San Francisco, 1986, p. 1004-1010.

[ORI 79] ORIN D.E., MCGHEE R.B., VUKOBRATOVIC M., HARTOCH G., "Kinematic and kinetic analysis of open-chain linkages utilizing Newton-Euler methods", *Mathematical Biosciences*, Vol. 43, 1979, p. 107-130.

[PAU 81] PAUL R.C.P., *Robot Manipulators: Mathematics, Programming and Control*, MIT Press, Cambridge, 1981.

[PIE 68] PIEPER D.L., "The kinematics of manipulators under computer control", PhD Thesis, Stanford University, 1968.

[POT 86] POTKONJAK V., "Thermal criterion for the selection of DC drives for industrial robots", *Proc. 16th Int. Symp. on Industrial Robots*, Brussels, September-October 1986, p. 129-140.

[PRE 93] PRESSÉ C., GAUTIER M., "New criteria of exciting trajectories for robot identification", *Proc. IEEE Int. Conf. on Robotics and Automation*, Atlanta, May 1993, p. 907-912.

[PRU 94] PRÜFER M., SCHMIDT C., WAHL F., "Identification of robots dynamics with differential and integral models: a comparison", *Proc. IEEE Int. Conf. on Robotics and Automation*, San Diego, May 1994, p. 340-345.

[RAG 90] RAGHAVAN M., ROTH B., "Inverse kinematics of the general 6R manipulator and related linkages", *Trans. of the ASME, J. of Mechanical Design*, Vol. 115, 1990, p. 502-508.

[RAU 90] RAUCENT B., "Identification des paramètres dynamiques des robots manipulateurs", PhD Thesis, University of Louvain, Belgium, 1990.

[REN 75] RENAUD M., "Contribution à l'étude de la modélisation et de la commande des systèmes mécaniques articulés", PhD Thesis, UPS, Toulouse, December 1975.

[REN 80a] RENAUD M., "Contribution à la modélisation et à la commande dynamique des robots manipulateurs", PhD Thesis, UPS, Toulouse, September 1980.

[REN 80b] RENAUD M., "Calcul de la matrice jacobienne nécessaire à la commande coordonnée d'un manipulateur", *J. of Mechanism and Machine Theory*, Vol. 15(1), 1980, p. 81-91.

[REN 85] RENAUD M., "A near minimum iterative analytical procedure for obtaining a robot-manipulator dynamic model", *IUTAM/IFToMM Symp. on Dynamics of Multi-body Systems*, Udine, 1985.

[REN 87] RENAUD M., "Quasi-minimal computation of the dynamic model of a robot manipulator utilizing the Newton-Euler formalism and the notion of augmented body", *Proc. IEEE Int. Conf. on Robotics and Automation*, Raleigh, March-April 1987, p. 1677-1682.

[RES 95] RESTREPO P.P., GAUTIER M., "Calibration of drive chain of robot joints", *Proc. 4th IEEE Conf. on Control Applications, CCA'95*, Albany, 1995, p. 526-531.

[RES 96] RESTREPO P.P., "Contribution à la modélisation, identification et commande des robots à structures fermées: application au robot Acma SR400", PhD Thesis, University of Nantes, October 1996.

[RIC 98] RICHALET J., *Pratique de l'identification*, 2nd edition, Hermès, Paris, 1998.

[ROT 76] ROTH B., "Performance evaluation of manipulators from a kinematic viewpoint", Lecture on Robotics at IRIA, Toulouse, 1976, p. 233-263.

[SCI 86] SCIAVICCO L., SICILIANO B., "Coordinate transformation; a solution algorithm for one class of robots", *IEEE Trans. on Systems, Man, and Cybernetics*, Vol. SMC-16(4), 1986, p. 550-559.

[SCI 94] SCIAVICCO L., SICILIANO B., VILLANI L., "On dynamic modelling of gear-driven rigid robot manipulators", *Proc. 4th IFAC Symp. on Robot Control, SYROCO'94*, Capri, September 1994, p. 543-549.

[SHE 71] SHETH P.N., UICKER J.J., "A generalized symbolic notation for mechanism", *Trans. of ASME, J. of Engineering for Industry*, Vol. 93, 1971, p. 102-112.

[TAN 94] TANG G.R., LIEU L.S., "A study of three robot calibration methods based on flat surfaces", *J. of Mechanism and Machine Theory*, Vol. 29(2), 1994, p. 195-206.

[UIC 69] UICKER J.J., "Dynamic behavior of spatial linkages", *Trans. of ASME, J. of Engineering for Industry*, Vol. 91, 1969, p. 251-258.

[VUK 82] VUKOBRATOVIC M., POTKONJAK V., *Dynamics of Manipulation Robots;* Vol. 1: *Theory and Applications*, Springer-Verlag, New York, 1982.

[WEN 89] WENGER P., "Aptitude d'un robot manipulateur à parcourir son espace de travail en présence d'obstacles", PhD Thesis, ENSM, Nantes, September 1989.

[WHI 69] WHITNEY D.E., "Resolved motion rate control of manipulators and human prostheses", *IEEE Trans. on Man Machine Systems*, Vol. MMS-10(2), June 1969, p. 47-53.

[WHI 72] WHITNEY D.E., "The mathematics of coordinated control of prosthetic arms and manipulators", *Trans. of ASME, J. of Dynamic Systems, Measurement, and Control*, Vol. 94, December 1972, p. 303-309.

[WOL 84] WOLOVICH W.A., ELLIOTT H., "A computational technique for inverse kinematics", *Proc. 23rd IEEE Conf. on Decision and Control*, Las Vegas, December 1984, p. 1359-1363.

[YOS 84] YOSHIKAWA T., "Analysis and control of robot manipulators with redundancy", *1st Int. Symp. of Robotics Research*, MIT Press, Cambridge, 1984, p. 735-748.

[ZAB 78] ZABALA ITURRALDE J., "Commande des robots-manipulateurs à partir de la modélisation de leur dynamique", PhD Thesis, UPS, Toulouse, July 1978.

[ZHO 95] ZHONG X.L., LEWIS J.M., "A new method for autonomous robot calibration", *Proc. IEEE Int. Conf. on Robotics and Automation,* Nagoya, May 1995, p. 1790-1795.

[ZHU 99] ZHUANG H., MOTAGHEDI S.H., ROTH Z.S., "Robot calibration with planar constraints", *Proc. IEEE Int. Conf. on Robotics and Automation*, Detroit, USA, May 1999, p. 805-810.

Chapter 2

Modeling of Parallel Robots

2.1. Introduction

Over the course of the past few years we have witnessed a large rise in the use of robots in the industrial world, mainly due to their flexibility. However, the mechanical structure of the most commonly used robots is inappropriate for certain tasks. Hence, other types of structures have been explored and have begun to find their place in the world of industrial robotics, and most recently in the field of machine tools. This particularly holds true for parallel manipulators, which we shall define in this chapter.

2.1.1. *Characteristics of classic robots*

To date, the majority of manipulators present an evident anthropomorphic character with a strong resemblance to the human arm. In fact, they consist of a series of segments, each connected to its predecessor and to its successor using a one degree of freedom joint (a revolute joint or a prismatic joint), a structure that we term of the serial robot by analogy with electrical systems. The driving elements, that is to say the actuators, make it possible to change the motion of these joints.

Chapter written by Jean-Pierre MERLET and François PIERROT.

For serial robots, the payload to robot weight ratio is never more than 0.3, which means that when a manipulator is required to carry heavy loads, it will itself become very heavy (in this respect, one of the best existing manipulators to date is probably the Mitsubishi Pa 10 robot, which offers a payload of 10 kg for its own weight in the amount of 30 kg, even if this means sacrificing some rigidity). Another interesting factor concerns robot accuracy. In this domain there are two types of values: absolute accuracy, which is the difference between the set point and the true location of the end-effector, and repeatability, which is the distance measured between the successive locations of the end-effector, when the same set point has been required for different starting locations. Repeatability is in general the measure of accuracy that manufacturers supply and it is far better than absolute accuracy, even though this measure is of far greater interest to users. However, for most industrial robots, even the measure of repeatability is not sufficient. As for absolute accuracy, this measure is sometimes completely wrong.

Low payload and mediocre accuracy of serial robots are intrinsic to the mechanical structure of these manipulators: the segments are submitted to high forces and bending moments requiring them to be very rigid, and thus very heavy (which is detrimental to fast motion), and errors of the internal sensors of the robot travel in an amplified manner to the end-effector.

Hence, we see that a serial robot is inappropriate for tasks requiring either the handling of heavy loads, an adequate level of positioning accuracy, or the ability to move fast.

2.1.2. *Other types of robot structure*

The anthropomorphic aspect of serial robots is unquestionably the driving force that propelled early engineers in their development of manipulation systems (in the sixties) with such a structure. However, there are other possible types of mechanical structures which are used to make robots and some of them are quite surprising.

In 1942, Pollard [POL 42] patented the mechanical structure represented in Figure 2.1, intended for painting automobiles. In this manipulator three revolute actuators move three arms whose extremity is connected to the nacelle by three joint segments. It should be noted at this point that this is no longer a serial structure since the end-effector is connected to the base of the robot through three separate and independent kinematic chains.

In 1947, an engineer by the name of Gough [GOU 57] established the basic principles for a mechanism (see Figure 2.2) making it possible to position and to steer a mobile platform with the aim of testing tire wear, a machine whose prototype he built in 1955 [GOU 55]. In this structure the mobile element is a hexagonal plate, whose top parts are each connected by spherical joints to a segment. The other end of the segment is connected to the base using a universal joint. A linear actuator makes it possible to change the total length of the segment. When the length of the segments varies, the position and orientation of the mobile platform change, and this is possible, as we shall see, for all six degrees of freedom of the platform.

In 1965, Stewart [STE 65] proposed using the mechanism presented in Figure 2.3 as a manipulator for simulators. In this structure, the mobile element is a triangular plate, whose top parts are each connected through a spherical joint to a sub-manipulator consisting of two linear actuators (1,2), also mounted in a triangular manner. One end of each of the jacks is connected via a revolute joint to the segment of a vertical axis, which is able to rotate around its own axis. The other end of one of the jacks is connected to the spherical joint of the mobile plate, while the end of the other jack is connected via a revolute joint to the opposite side. It turns out that one of the reviewers of the Stewart paper was Gough, who reported the existence of his own structure. This structure, however, was also cited by other reviewers of Stewart's publication, who suggested using it as a platform for drilling and milling (which happened to be an excellent prediction of the future, as we shall see). As far as we know Stewart's manipulator never found an application, but the use of Gough's is, on the contrary, very frequent. Ironically, however, Gough's manipulator, which far precedes that of Stewart, is often recognized under the name of Stewart's platform.

The common feature among the manipulators of Pollard, Gough and Stewart is the fact that the end-effector is connected to the base of the manipulator using separate and independent kinematic chains. This is termed closed-chain manipulator, or *parallel* robot, which is how we will refer to it in this chapter, whereas in certain communities Gough's platform is referred to as a *hexapod*. Certain theoretical issues linked to this type of structure have been highlighted previously, way before the existence of the first robot (e.g. as early as 1645 by Christopher Wren, followed by Cauchy (1813) Bricard (1897) and Borel (1908) to name just a few). This interest, rooted in the past, and renewed in the present, arises in connection with the natural advantages of this type of manipulator, which we shall examine in greater detail in the following section.

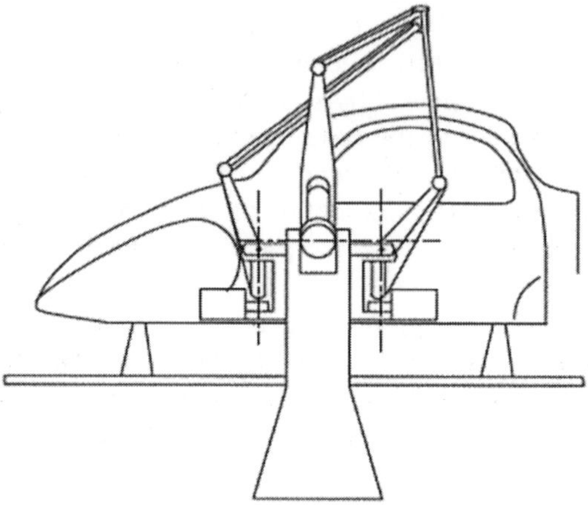

Figure 2.1. *Pollard's manipulator (according to Pollard) [POL 42]*

Figure 2.2. *Gough's machine (1957): the mobile plate (1947) linked to a tire is connected to the ground by six segments of variable length. At one end of each segment there is a universal joint and at the other end a spherical joint. The wheel is driven by a conveyor belt, and hence tire wear may be measured under different conditions [GOU 57]*

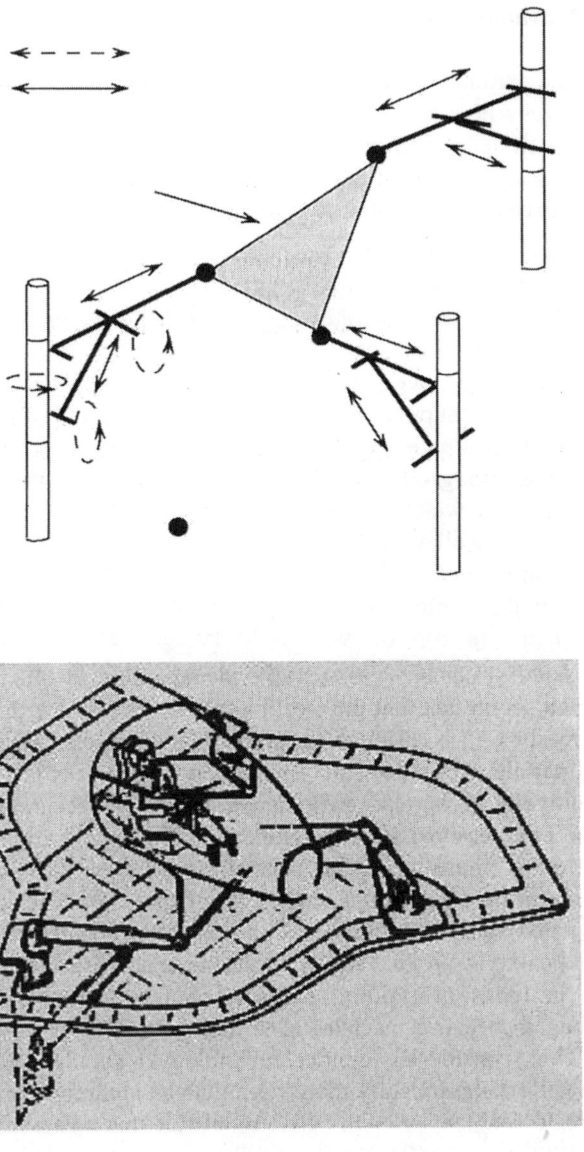

Figure 2.3. *Stewart's manipulator (1965). The motions of the mobile plate are obtained by changing the length of the six joint segments*

2.1.3. *General advantages and disadvantages*

In view of the extreme diversity of existing manipulators, considering both their structures and sizes (there are micro parallel robots, and there are parallel robots the size of a crane), it is of course very difficult to outline a panorama of all the advantages and disadvantages of parallel and serial robots, in general. However, it is possible to put forward a few of the main trends at various levels:

– *Performances:* to solve, on a mechanical level, part of the problems arising with serial manipulators, one possible solution is to distribute the load among all of the segments, that is, to connect the robot end-effector to the ground via a set of kinematic chains, which consequently only support a fraction of the total load. Let us examine for example Gough's structure. When the structure is centrally positioned, the actuators support approximately only one-sixth of the load placed on the mobile plate. Additionally, the bending stresses on the segments are reduced as the joints are transmitting only traction-compression forces. Both of these factors enable a decrease in the weight of the mobile structure making it possible to use less powerful actuators, as well as smaller size segments. This decrease in mobile weight will also significantly decrease all disruptive effects applied to the mobile platform, such as inertia or the Coriolis force, when moving at high velocity. In turn, this decrease will facilitate control, will yield far greater accelerations than those possible with serial structures and we will then be able to obtain extremely fast motions (we shall see in fact that the world's fastest robot on earth works this way). For the same reasons, it is easier to generate very significant static stresses in the case of certain parallel structures (the Tricept robot is successful in part due to this capacity). Intuitively, it is also hypothesized that the accuracy of position is satisfactory for two reasons: (i) the segment deformations are reduced, and (ii) internal sensor errors (those measuring segment lengths) only slightly impact on the platform position error. For example, when all of the sensors make the same error (which is the worst case), calculation of the platform position will only yield an error on the vertical axis, approximately of the same amplitude as that of the sensor error. Finally, in terms of rigidity, parallel robots often have a design that is advantageous (again because mechanical stresses at work are essentially traction-compression). Thus, considering identical technology, a parallel robot of the "Gough machine" type will be significantly more rigid than an anthropomorphic serial robot. On the contrary, to date we are unable to demonstrate that a parallel structure with 3 axes is more rigid that a serial combination of three linear tables, which is at the basis of machine tools with three axes.

– *User friendliness:* for practical uses, parallel robots are often disadvantaged by their reduced workspace compared to that of serial robots. That is, accessible space for the traveling plate (sometimes called "nacelle") of a parallel robot is at best equal to the intersection of spaces accessible to each of the constituting kinematic chains. Figure 2.4 illustrates this issue in a trivial case of planar robots, the kinematic chains of which have the same stops. The disadvantage of a small workspace is also intensified when considering the ratio between workspace and the footprint required by a parallel robot. That is, certain industries have stringent requirements concerning the use of workshop space. Otherwise, when workspace befits a particular application, maintenance operations may be facilitated via the "systematic" construction of parallel robots (with identical components), and for the case of robots with fixed actuators, via easy access to certain key components such as actuators, coders and connection devices.

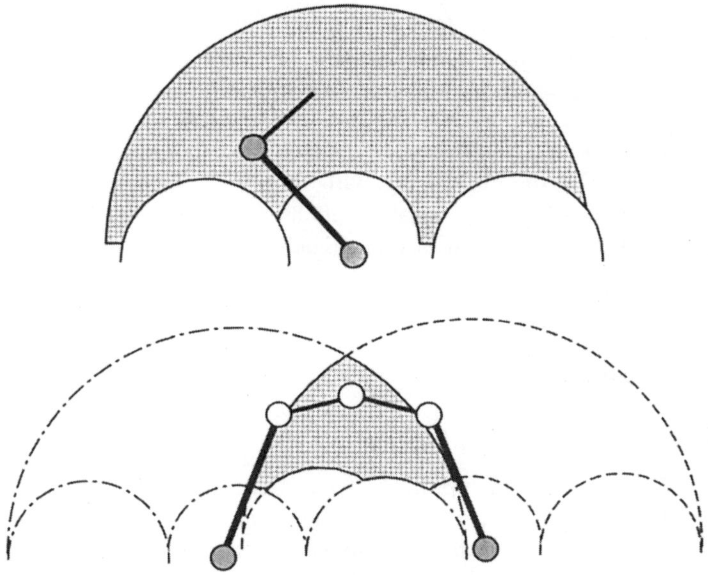

Figure 2.4. *Comparison of the workspace in a serial manipulator and a parallel equivalent*

– *Technology:* one of the general advantages of parallel manipulators is the frequent use of several identical kinematic chains on a single robot: the economic benefits of serial production thus appear sooner, by comparison with serial robots (decrease in the costs of unit analyses and construction, decreases in the volume of spare parts storage, etc.). Another advantage of certain parallel manipulators (with fixed actuators) is the greater freedom in the choice of actuators that they offer the following abilities: revolute, prismatic, with direct or indirect drive, with electric or non-electric power, etc. And finally, we mention an advantage which at this time appears merely potential. If parallel robots become increasingly more popular, we could witness the availability of standard component product lines, which would greatly facilitate the construction of high performance engines. The first example of such a process is the appearance of telescopic actuators in the catalogues of major suppliers (INA, SKF), which are nothing more than "universal-prismatic-revolute" chains ready for use. On the other hand, it becomes essential to possess quality passive joints, often with two to three degrees of freedom. And to build a universal joint, or a compact clearance free spherical joint, while ensuring a decent shelf life in the context of stresses during high amplitude motion, is hardly an easy task.

– *Modeling – singularities:* these points shall be outlined in greater detail in the following sections. However, it appears important to note several relevant issues prior to examining mathematical implications. Contrary to serial robots, inverse models of parallel robots are in general easier to establish than direct models. In particular, it seems impossible to use algebraic methods to describe the direct analytical geometric models of numerous parallel robots (time will tell whether there are any other possible ways). As for serial robots, parallel robots encounter the same singular points issues, where even when they are accessed, the manipulator enters a configuration where it no longer responds correctly to commands, and worst of all, where it is possible that the manipulator will simply be destroyed! The situation is however more complex than for serial robots as there are many different types of singularities, the listing of which is often much more complicated than for serial cases.

2.1.4. *Present day uses*

2.1.4.1. *Simulators and space applications*

The use of this type of manipulator in fact began to expand significantly only with the onset of construction of the first fight simulators. During the 1960s, the development of aerospace industry, the increase in the costs for training pilots, and the necessity to test new aircrafts out of flight, motivated research on manipulators with several degrees of freedom, likely to animate with high dynamics a platform that was heavily loaded with instruments (for example an aircraft cockpit), considering that all of these constraints made it difficult to use serial manipulators.

All kinds of flight simulators use the structure of the Gough platform (see Figure 2.5). And this is also used in many other kinds of simulators, in sometimes quite surprising ways, such as for the equestrian simulator Persival installed at the National School of Horse Riding (see Figure 2.6) or the CINAXE theatre of La Villette (the use of a parallel structure in the entertainment industry is a domain that will likely bear witness to many changes in the future). It is also impossible not to mention automobile simulators, such as those of Daimler-Benz presented by Drosdol [DRO 85], and the driving simulators such as the IDS of the University of Iowa, that of the Swedish Transportation Institute (VTI), and the National Advanced Driving Simulator (NADS).

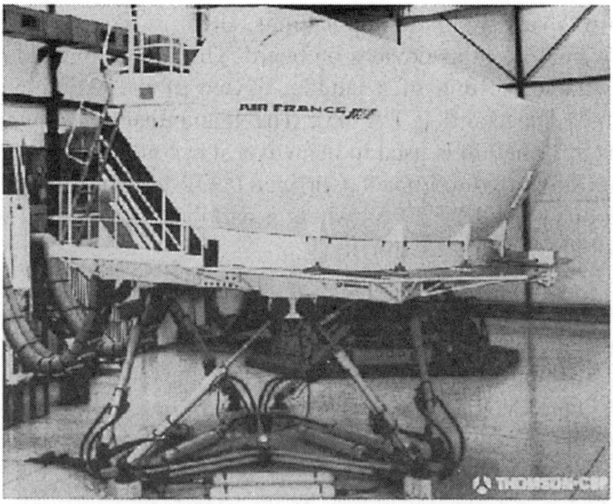

Figure 2.5. *The Airbus A3-40 simulator built by Thomson-CSF (photograph by: P. Palomba)*

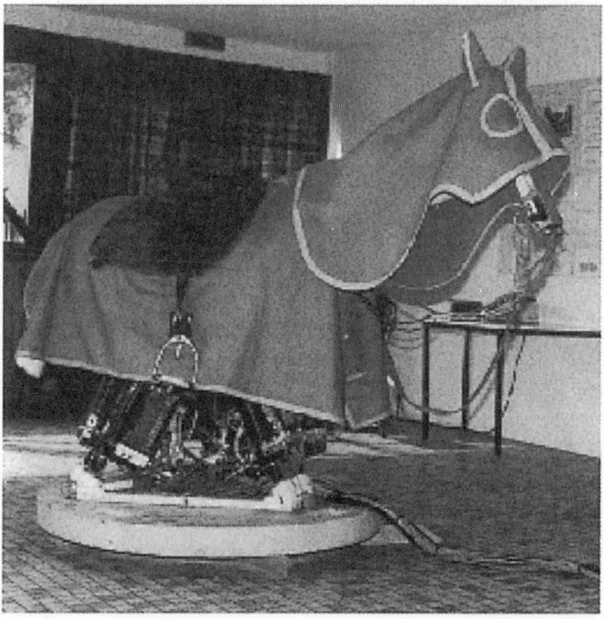

Figure 2.6. *Equestrian simulator "Persival" belonging to the National School of Horse Riding, built in collaboration with ENAC (E.N.E photograph)*

There are also space simulator applications, either as terrestrial devices intended to simulate zero gravity, or as devices on board. The former include, for example, a simulator that enables testing of a landing device of the Columbia space station using the Hermès spacecraft [CLA 85]. The manipulator is equipped with load sensors and this information is used to maneuver it according to a dynamic model of the space station submitted to impact. Corrigan [COR 94] and Dubowsky [DUB 94] developed a simulator called "VES" where a parallel robot is used to simulate the behavior of a serial robot in zero gravity.

Because of the low weight of parallel manipulators, as well as their energy efficiency, they could become prized devices on board. It should also be said that the use of parallel manipulators is old news in the domain of space since such an application was considered for the landing gear of the moon module [ROJ 72]. Moving a bit beyond space activity, there are also uses of parallel manipulators for the positioning of antennae (see Figure 2.7), as anticipated at the University of Canterbury. In fact, a hexapod developed by the Max Planck Institute for Astronomy, in Heidelberg, is used on the UKIRT telescope for all slow focus motion, while a hexapod also controls secondary mirror motion of the Italian Galileo telescope. Using a parallel robot as an active vibration suppression system was also proposed for satellites [WU 92] and for aircrafts, such as the VISS system (Vibration, Isolation, Suppression and Steering System), developed by the US Air Force.

Figure 2.7. *Use of a parallel robot to position antennae. Notice the small size of the actuators in comparison to the parabola*

2.1.4.2. *Industrial applications*

Because of their capacity for precise positioning and their high rigidity, parallel robots are slowly appearing in various domains of industry. Assembly and contour analysis are favorite applications for parallel manipulators and numerous feasibility demonstrations have been set up in laboratories. This domain of expertise, however, has encountered certain difficulties in terms of transfer, even if Marconi appears as the first company to propose a manipulator of this type, called GADFLY, for transportation and assembly of electronic components [POW 821]. Currently, Fanuc is the only company to offer a manipulator of this type, termed the F-100. However, certain special parallel manipulator structures, such as DELTA, which we will discuss later, are used for tasks requiring speedy removal, in particular in the food industry [CLA 94]. Let us note also the *EX 80-0* hexapod, offered by DeltaLab, for teaching purposes. The first drilling machine also deserves to be mentioned, based on the Gough platform principle, offered by the Giddings & Lewis Company, under the name of Variax, thus fulfilling the vision of the Stewart paper reviewers. It was the centerpiece of the machine tool exhibit, in Chicago, in 1994. According to the builder, beyond the fact that the machine possesses six degrees of freedom, it is purported to be five times more rigid than a traditional machine, and to have far greater forward velocity. In any event, the competition was swift to react with Ingersoll also offering a drilling machine, the Octahedra Hexapod HOH6000, and Geodetics launching the G1000. Since then, numerous machines of this type have been built, including (without claiming to be exhaustive): *6X* developed by Mikromat, Hexact developed by the University of Stuttgart and INA, *CMW 300* developed by Constructions Mécaniques des Vosges, HexaM developed by Toyoda (Figure 2.8), *HVS 5000* developed by Honda, MC developed by Okuma and *Eclipse* developed by Sena. These machines all have six degrees of freedom, but there are also machines offered with 3 axes, such as Multicraft *560,* Triaglide developed by Mikron, Tricept developed by Neos Robotics and Urane *SX* (Figure 2.9), developed by Renault Automation-Comau. We shall return to these machines in the chapter dedicated to the typology of parallel robots.

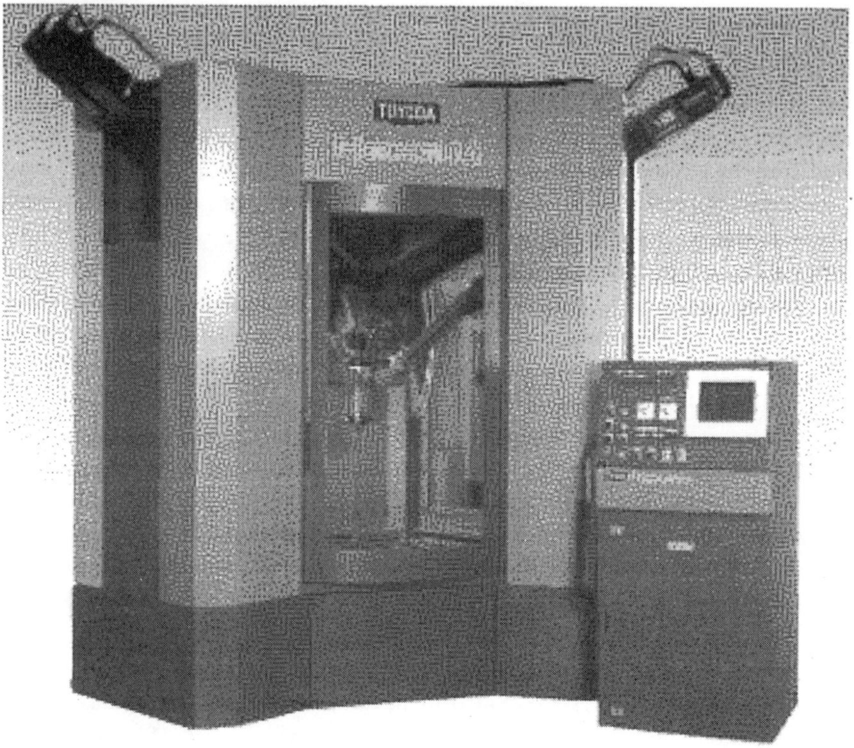

Figure 2.8. *HexaM machine tool developed by Toyoda Machine Works*

Nonetheless, the problems that these companies are facing to make their machines truly operational have had a dampening effect on this development. It is to be regretted, however, that certain developments are occurring outside of the research that was previously conducted by robotics specialists, who had identified and even resolved certain problems now facing industrialists. Nonetheless, it was in 2000 that parallel machine tools were first installed in production sites in the domain of heavy-duty aeronautics machining.

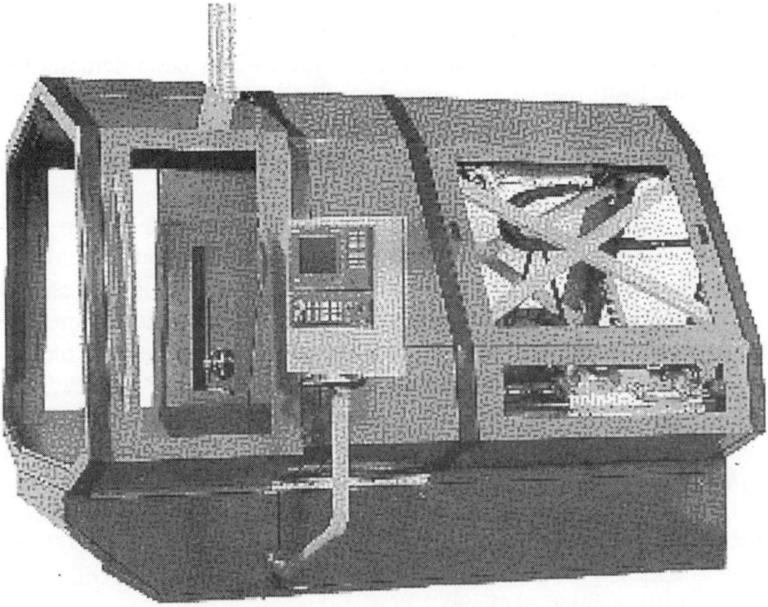

Figure 2.9. *UraneSx machine-tool developed by Renault Automation-Comau*

Another industrial application consists of a parallel robot where the rigid legs of a hexapod are replaced with cables, the length of which may vary using a capstan. In fact, there is nothing that prevents this substitution from occurring, providing that the cables remain under tension, for example using gravity. This is how NIST Albus [ALB 93] created a hexapod crane, the Robocrane, for use in the construction industry. The advantages of this type of machine in terms of rigidity, and accuracy in comparison to traditional cranes, are obvious. A similar system was also offered for unloading containers from a ship under the name of AACT (Automated All-weather Cargo Transfer System).

2.1.4.3. *Medical applications*

The accuracy of parallel robots and the fact that they are more easily "miniaturizable" than serial robots has led to certain research in the medical domain. Thus, active endoscopy heads were proposed [WEN 94], [MER 97]. We are also now seeing parallel robots used in the medical domain for precise positioning, either as permanent devices such as the Delta robot, used for brain surgery to position a microscope at the Necker Hospital, or as a laboratory prototype for orthopedic surgery [BRA 97], for ophthalmic surgery [GRA 93] or for neurosurgery, such as the robot developed by the Fraunhofer Institute in Stuttgart.

2.1.4.4. *Precise positioning*

Precise positioning is a domain where parallel robots have demonstrated their potential. Physik Instrumente, for example, offers the M-850 hexapod, and the F-206 robot (the mechanical structure of which we shall discuss later on), but there are also other companies such as Carl Zeiss, Philips or OHE Hagenbuch, which are equally active in this domain.

The most significant example is undoubtedly the hexapod developed by the European Synchrotron Radiation Facility (ESRF). ESRF is conducting research on the use of X-rays produced by a high power synchrotron, and for certain experiments it is crucial to direct the beam. For this purpose, special mirrors are placed on the path of the beam, and their position and orientation need to be changed at will. The issue is then to move loads in the amount of 500 kg to 2 tons with a degree of accuracy that is equal to less that one micron, a task which is well beyond the capacity of serial robots, and which was accomplished using a hexapod, specially designed for this purpose (Figure 2.10). It is also important to mention the usage of a parallel robot as a master arm, either inside remote controlled systems or for virtual reality [DAN 93], [IWA 90]. An example of a master arm is that developed by V. Hayward, at McGill University in Montreal, and it is presented in Figure 2.11.

Figure 2.10. *The ESRF hexapod. This robot is capable of moving a two-ton load with a degree of accuracy that is greater than the micron*

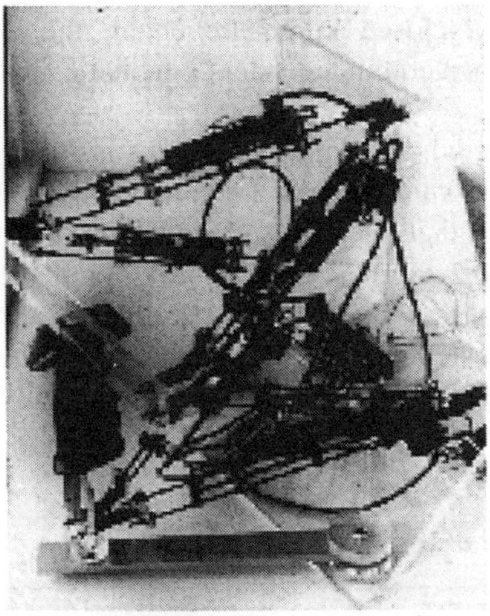

Figure 2.11. *Master arm developed at McGill University in Montreal*

2.2. Machine types

In this chapter we present a few mechanical structures for the parallel robots described in other works[1].

2.2.1. *Introduction*

Due to deficiencies encountered with serial robots, researchers focused on the creation of new robot structures. The forerunners in this domain were Minsky [MIN 72] in 1972, and Hunt [HUN 78] in 1978, who proposed parallel structures. In 1983, Earl [EAR 83] tried to define a systematic method for the construction of various serial robots. More recently, more systematic approaches were proposed by Hervé [HER 95], Danescu [DAN 95] and Yufeng [YUF 95].

1 A more complete list may be consulted on the following web page: http://www-sop.inria.fr/coprin/equipe/merlet/Archi/archi_robot.html

First, let us provide a general definition of parallel robots. A parallel manipulator is a manipulator with a closed kinematic chain, the end-effector of which is connected to the base by several independent kinematic chains.

Parallel manipulators, whose number of chains is strictly equal to the number of degrees of freedom of the end-effector, are called fully parallel manipulators [GOS 88], [PIE 91], considering that the chains in question function to directly connect the base with the end-effector, and that there is only one actuator. A study of mobility serves to demonstrate that there are no fully parallel robots at four or five degrees of freedom, with identical kinematic chains, but that it is nonetheless possible to build robots with four or five degrees of freedom that are not fully parallel.

Joint	**Passive**	**Motorized**	**Mobility**
Prismatic	P	P	1 (translation)
Revolute	R	R	1 (rotation)
Cardan	C	N.A.[2]	2 (rotations)
Spherical	S	N.A.	3 (rotations)

Figure 2.12. *Symbols of the layout graphs*

Using these very general definitions, it is possible to almost infinitely combine the principle of creating several parallel kinematic chains, in view of obtaining a remarkable diversity of manipulators that is often difficult to categorize or analyze in any systematic manner. However, it is possible to distinguish a few large structural families that we illustrate as planar manipulators in diagram format, equivalent in terms of kinematics to a serial robot with three degrees of freedom of Figure 2.13. In this figure, in addition to a kinematic diagram, there is a layout graph, which facilitates the understanding of the sometimes-subtle structure of parallel manipulators. On the layout graph we use symbols explained in Figure 2.12.

2 To date the motorization of more than one degree of freedom in a U or S joint requires technological achievements.

By way of example, the manipulator of Figure 2.14a is fully parallel, but the two following ones are not. Figure 2.14b illustrates a manipulator where a kinematic chain does not act directly on the nacelle, and Figure 2.14c represents a case where a chain contains more than one actuator. In addition to the chains used for actuation, it is sometimes interesting to add a non-motorized chain, and thus, to design manipulators with passive chains (Figure 2.15a). This "additional" chain can be used for example to install special kinematic constraints (the passive chain defines the type of displacement needed, and the active chains ensure motorization), or it may be used for measurement when it is equipped with sensors.

It is also possible to imagine parallel robots that are redundant, and in this sense it is possible to chose between: (i) redundancy in terms of kinematics (Figure 2.15b) where, for each of the nacelle positions, there is an infinite number of positions for the chains, thus optimizing a criteria for mode of function; and (ii) redundancy in terms of actuation (Figure 2.15c), which corresponds to hyperstatic manipulators, and which makes it possible to overcome certain singularities.

Sometimes there are also parallel robots installed as an end-effector on a serial robot, which is then termed an active wrist, and those that are independent which are termed left hand.

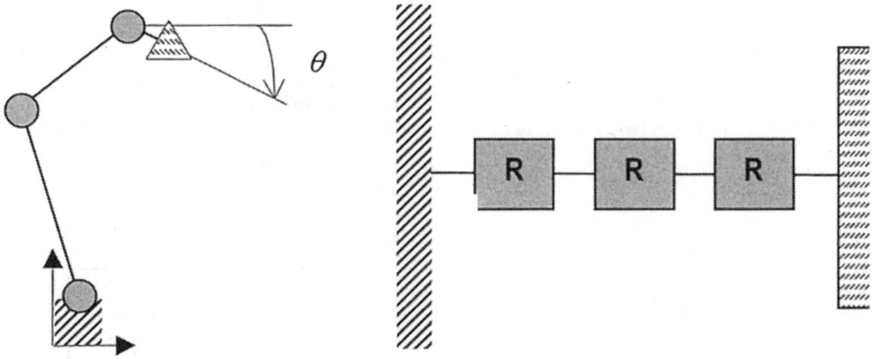

Figure 2.13. *Planar serial robot with three degrees of freedom*

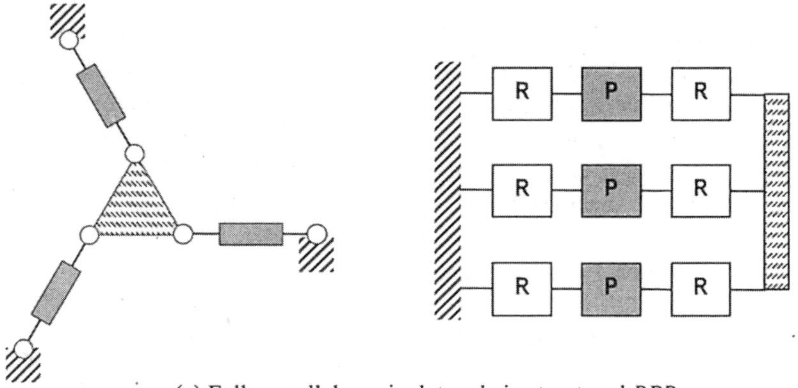

(a) Fully parallel manipulator chain structured *RPR*

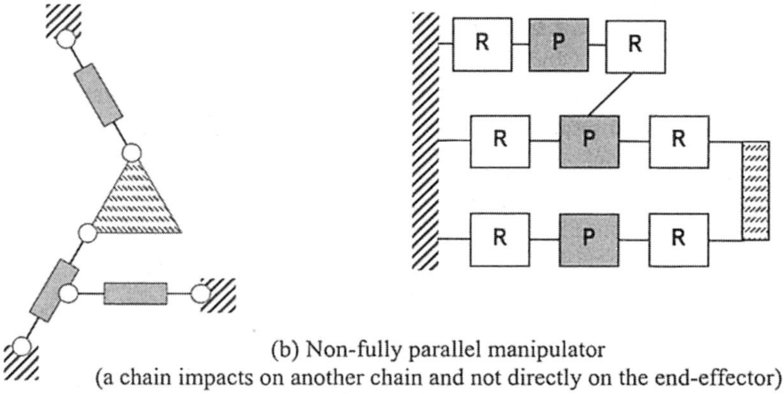

(b) Non-fully parallel manipulator
(a chain impacts on another chain and not directly on the end-effector)

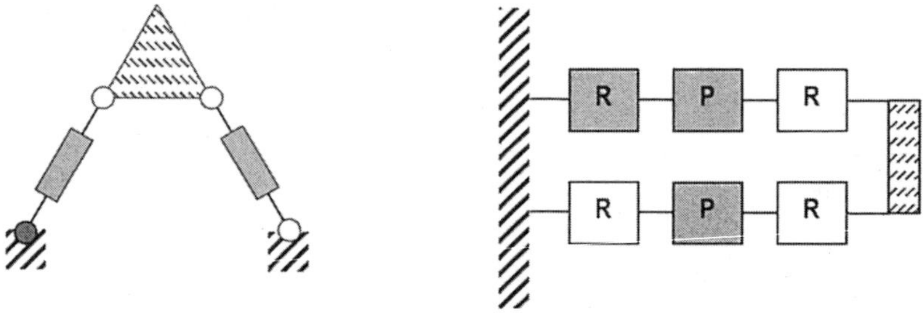

(c) Non-fully parallel manipulator
(a chain has more than one actuator)

Figure 2.14. *Fully and non-fully parallel manipulators*

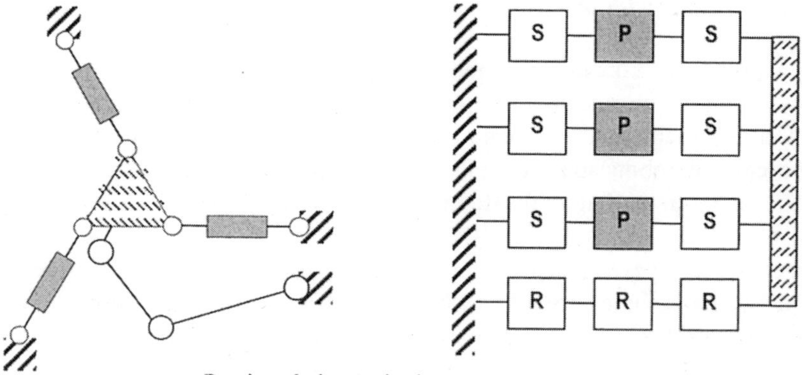

Passive chain manipulator

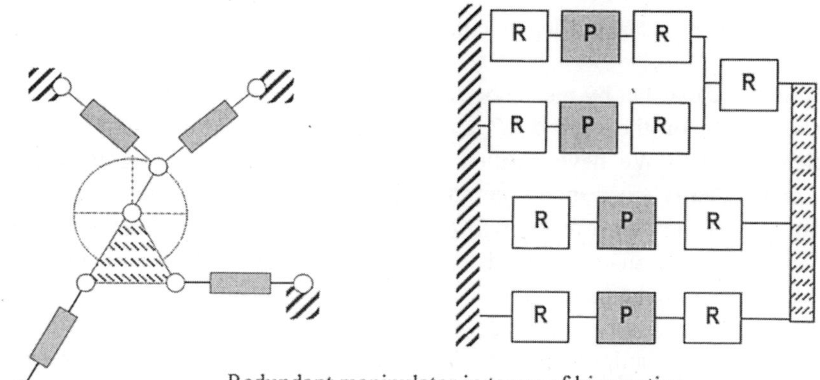

Redundant manipulator in terms of kinematics

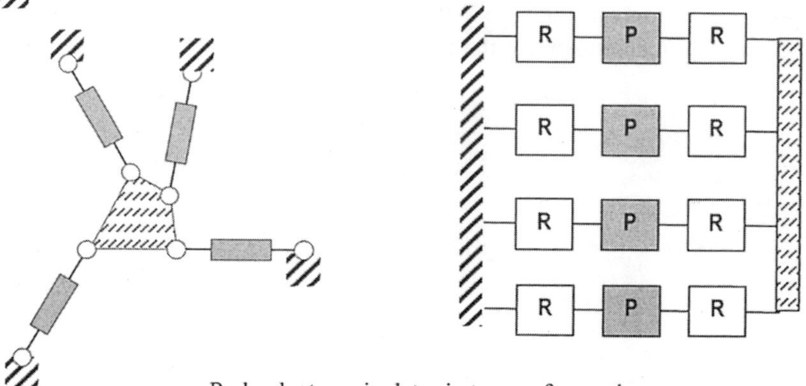

Redundant manipulator in terms of actuation

Figure 2.15. *Passive chain manipulators and redundant manipulators*

2.2.2. *Plane robots with three degrees of freedom*

On a plane, we consider a mobile platform, the three degrees of freedom of which we want to control: both translations according to the x, y axes of a reference frame and the orientation θ around axis z, perpendicular to the plane. We discuss here the case of robots that are fully parallel, thus comprising three independent kinematic chains according to the definition that we specified.

Since each of these chains is connected both to the ground and to the mobile platform, we have three ground and mobile platform attachment points. It is then possible to consider a triangular mobile platform without becoming too specific. And it is possible to describe a chain as a sequence of three joints beginning with the base. The following sequences for the chains become possible: *RRR, RPR, RRP, RPP, PRR, PPR, PRP, PPP* (see Figure 2.16). It is, however, important to remember that the joints must remain independent, thus the sequence *PPP* is excluded.

It becomes clear that by means of a simple exchange of the base with the mobile platform, robots of the type *RRP* are equivalent to *PRR,* and *RPP* robots are equivalent to *PPR*. We have omitted on purpose to specify the motorized joint, which may equally correspond to any one of the three. In general, however, motorization on the end-effector will be avoided to prevent burdening the mobile equipment. It should also be noted that it is quite possible to design robots with completely different chains.

Not all of these manipulators have been studied. 3-*RRR* robots were the focus of an extensive study conducted by Gosselin [GOS 88]. 3-*RPR* robots were studied by [MER 94], while robots of the type 3-*PRR* were mentioned by Hunt [HUN 82], and some of their characteristics were the focus of a study conducted by Gosselin [GOS 96].

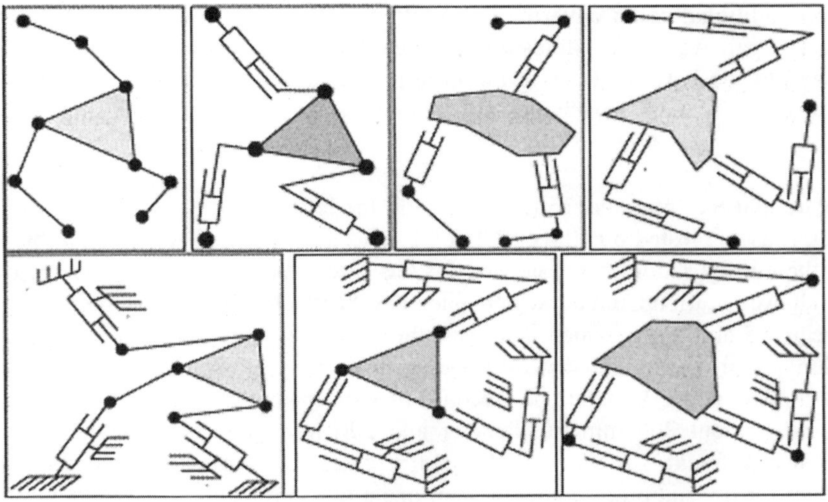

Figure 2.16. *Different fully parallel plane robots with three degrees of freedom and identical chains*

2.2.3. *Robots moving in space*

As pointed out for the Gough platform, parallel robots producing motion in space will require more complex passive joints than those of plane robots, such as revolute or universal joints. In terms of actuators, hydraulic actuators are used almost exclusively for heavy loads (for example, for flight simulators) and electrical actuators are used in most other cases, even if other types have been suggested (pneumatic or piezoelectric, for example).

2.2.3.1. *Manipulators with three degrees of freedom*

2.2.3.1.1. Manipulators for translation motion

Manipulators with three degrees of freedom for translation motion are of significant interest for transfer operations. Consequently, different types of structures have been suggested.

The most famous robot with three degrees of freedom for translation motion is the Delta robot, initially developed at the Polytechnic School of Lausanne, in Switzerland, by a team led by Clavel [CLA 91] (see Figure 2.17). His layout graph (Figure 2.18) indicates that this is somewhat of a particular manipulator, consisting of a "parallel" part comprising three identical chains, covering all three translations, and an additional chain covering the rotation of the tool. Additionally, the strictly "parallel" part was itself interesting since each chain comprises a rotating actuator, attached to the base, and acting on an arm, which is connected to two bars using spherical joints, both bars being also connected to a nacelle also using spherical joints.

Note that the rotational actuator part and the lever could be replaced by a linear actuator, as suggested by Clavel [CLA 94], and then Tsai [TSA 96]. And this is, in fact, the configuration that was used for the Urane Sx machine tool, developed by Renault-Automation/Comau, which manages to reach acceleration points ranging between 3.5 and 5.0 g, which is at least three times more competitive than the best machines with traditional structure. Delta, in its robotic version, was designed for rapid transfer (three transfers per second) of light loads (10 to 30 g) in a workspace measuring about 200 mm in diameter and 200 mm in height. Clavel [CLA 94] discusses different structures that yield a parallel robot with three degrees of freedom for translation motion and mentions a few applications that were built with this robot: manipulation in the food industry and applications in the medical field where Delta is used as a support for microscopes.

Figure 2.17. *The Delta robot using rotating actuators*

It should be noted that Delta's ancestor is a manipulator described in 1942, by Pollard [POL 42], intended for painting automobiles. Another interesting structure was designed in a prototype patented by Neuman [NEU 88]. In this manipulator, the end-effector consists of a rod that is free to move around its axis. This rod is connected to the base using a universal joint and three chains of the type UPS act on the end-effector (see Figure 2.19). The Marconi society has in any event already used the Neumann position device for the Tetrabot (Figure 2.19), a hybrid serial-parallel robot, for the assembly of large parts [DWO 89]. It is important to note that the Tricept family of machine tools, offered by Neos, uses this same principle of design. On the layout graph (Figure 2.20), there is a "parallel" part comprising a passive chain, which constrains motion of the nacelle within a sphere with a variable radius and three identical CPS actuator chains, and another "serial" part which is either a wrist with three degrees of freedom for the robotic version or a manufacturing head with two degrees of freedom for the machine tool version.

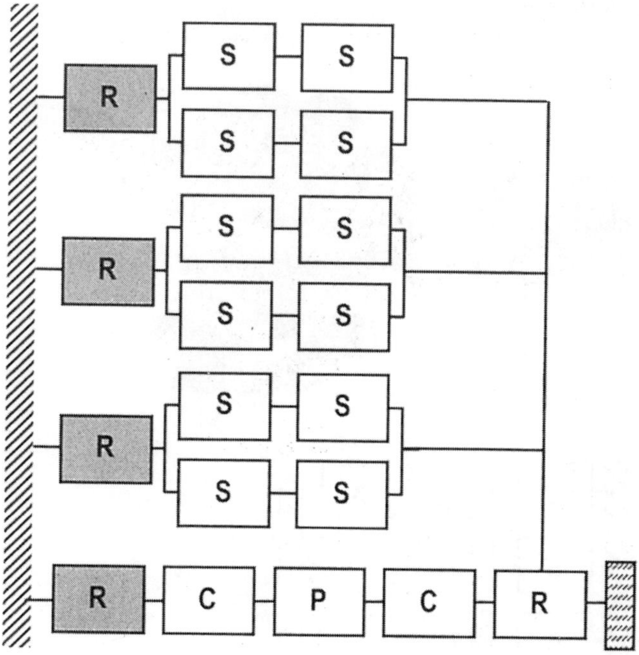

Figure 2.18. *Layout of the Delta robot in its commercial version. The central chain, visible in Figure 2.17, which enables natural rotation of the tool, comprises an engine installed at the base and a transmission via two universal joints and a telescopic rod*

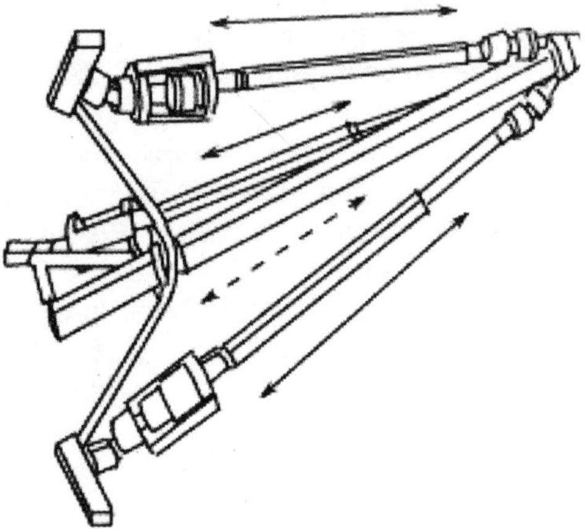

Figure 2.19. *The Neumann prototype. A rod of variable length, mounted on a joint, connects the end-effector to the base (according to Neumann [NEU 88]. Three linear actuators enable motion of the end-effector*

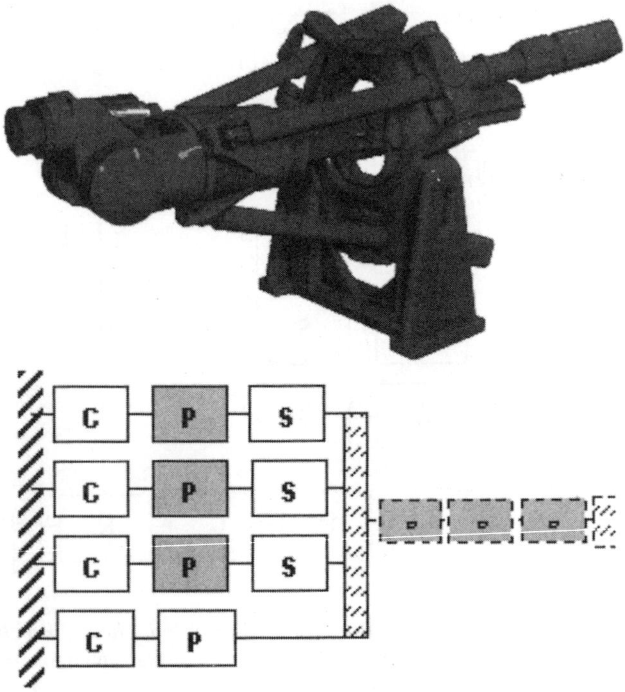

Figure 2.20. *The Tricept machine with layout graph*

2.2.3.1.2. Manipulators for orientation

Manipulators enabling three rotations on a single point offer an interesting alternative to the wrist used traditionally with serial robots (with three revolute joints on converging axes).

An initial possibility to ensure rotational motion exclusively is to place constraints on the mobile platform using a mast with and ball-and-socket joint that forces the platform to rotate around a single point. It is also important to note that it is possible to replace the rigid segments with cables, as Landsberger suggested [LAN 92]. Another possibility for the wrist design is to use chains that generate rotational motion on a single point. Gosselin and his team studied in an exhaustive manner the design of a wrist based on this principle [GOS 88, GOS 94a] for the purpose of creating a control system called the agile eye. This manipulator uses three motorized spherical chains with rotary actuators, the axes of which converge in a single point, which will be the center of rotation (Figure 2.21).

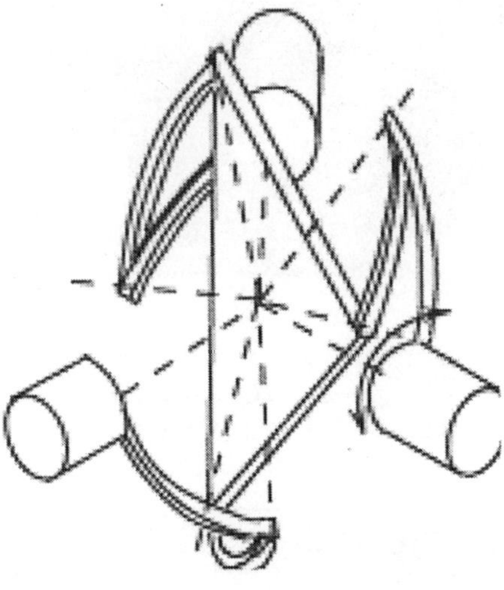

Figure 2.21. *Gosselin's spherical wrist: three spherical chains are used with rotary actuators, the axes of which converge at the center of the mobile plate (according to Gosselin [GOS 88])*

2.2.3.1.3. Manufacturing head unit for machine tools with five axes

The manufacture of complex parts such as molds or structural items in aeronautics require five-axe machine tools, that is, machine tools capable of creating relative displacement between a part and the tool as 3 translations (placing one point in space), and 2 orientations (the axis of the spindle is oriented). In general, common machines have two modules: (i) a set of three linear displacement tables for translations; and (ii) a manufacturing head unit, mounted on the preceding module, for orientation of the spindle. The second module is critical for this kind of machines, and it considerably limits their performance. Several projects have attempted to solve this problem, and a solution based on parallel manipulators is now in service in aeronautics: two tables for translations, according to two displacement axes, and a parallel head unit with three degrees of freedom (see Figure 2.22) enables the third translation motion and the orientation of the spindle.

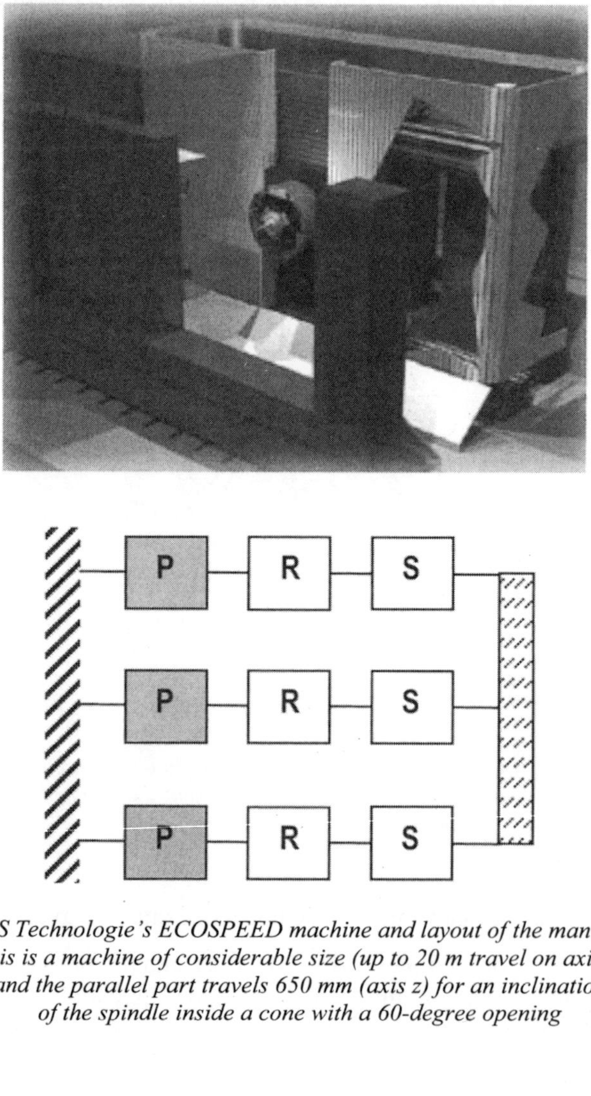

Figure 2.22. *DS Technologie's ECOSPEED machine and layout of the manufacturing head. This is a machine of considerable size (up to 20 m travel on axis x) and the parallel part travels 650 mm (axis z) for an inclination of the spindle inside a cone with a 60-degree opening*

2.2.3.2. *Manipulators with four or five degrees of freedom*

Manipulators with four degrees of freedom were offered quite early in other works but they remain little studied. In 1975, Koevermans [KOE 75] presented a manipulator for a flight simulator using linear actuators with the mobile plate submitted to passive constraints (see Figure 2.23). The degrees of freedom correspond to three rotations and one translation motion.

Zamanov [ZAM 92] suggested a type of structure for parallel manipulators with five degrees of freedom (Figure 2.23). This structure is based on the coupling of two parallel plane manipulators (*A1A2ABA3A4*, *B1B2ABB3B4*), with a shared (*AB*) chain. Such a structure enables the control of the degrees of freedom of the platform, with the exception of the rotation about the normal line of the platform. Moreover, this latter degree of freedom can be controlled with the help of an additional actuator mounted on the platform.

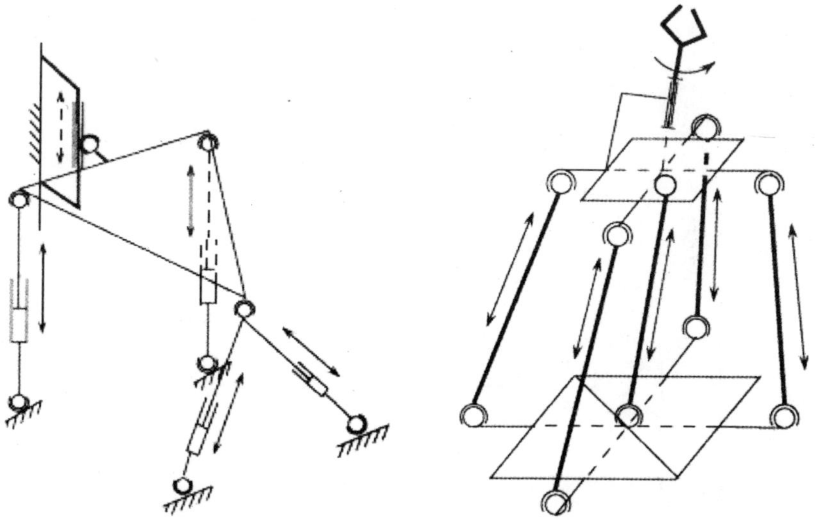

Figure 2.23. *Manipulators with four or five degrees of freedom: on the left a passive constraint ensures that the only degrees of freedom are rotations and translation motion according to axis z (as pointed out by Koevermans [KOE 75]). On the right hand side appears the robot with five degrees of freedom offered by Zamanov: the sixth degree of freedom, rotation on the normal line of the plate, is obtained using an additional manipulator (as pointed out by Zamanov [ZAM 92])*

More recent work has highlighted the possibility of creating manipulators with five degrees of freedom, the structure of which is related to Delta and Hexa kinematics. In the version presented in Figure 2.24, termed H4 [PIE 99], four identical parallel kinematic chains are connected, two by two, on two intermediate transmission elements, also connected to the end-effector, which can be displaced according to 3 translations and one rotation when certain geometric conditions are fulfilled. It is possible to combine this non-fully parallel structure into several versions with varying types of actuators. It is intended for rapid palletization applications, or light manufacturing. Similar proposals are also found under the name of Kanuk or Manta [ROL 99].

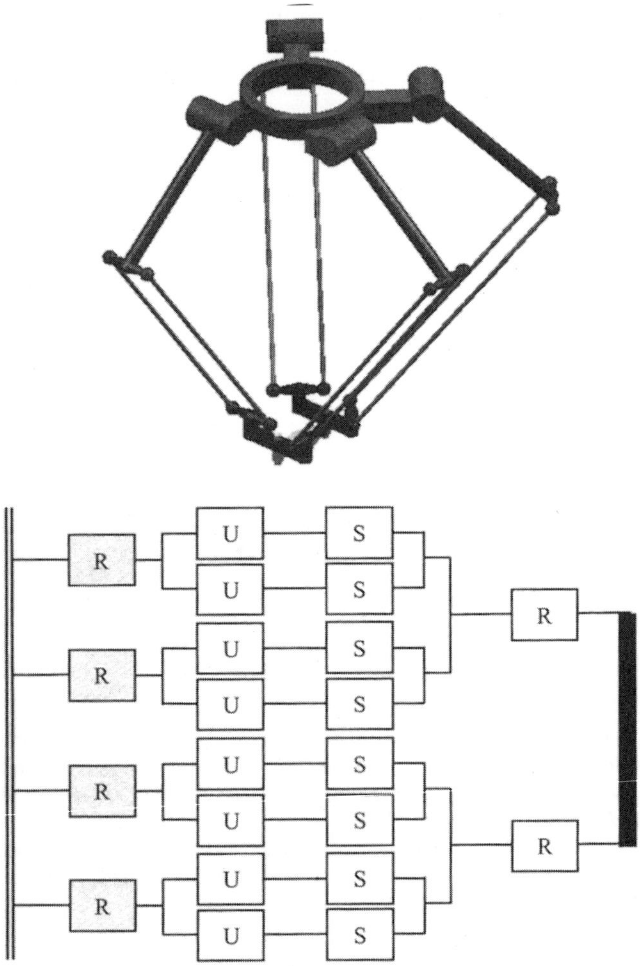

Figure 2.24. *One version of structure H4, with rotary engines*

This family of structure, capable of generating the same displacements as the SCARA robots, was recently the topic of many studies. Figure 2.25 shows a prototype of the Par4 robot [NAB 05] which offers performances that are comparable to the Delta robot, but, without using the kinematic chain **R**UPUR, which is sometime frail, in order to obtain tool rotation. Figure 2.26 shows a prototype of Dual4 robot [COM 05] which makes is possible to group all of the engines onto a single axis, in view of obtaining unlimited rotation of the tool.

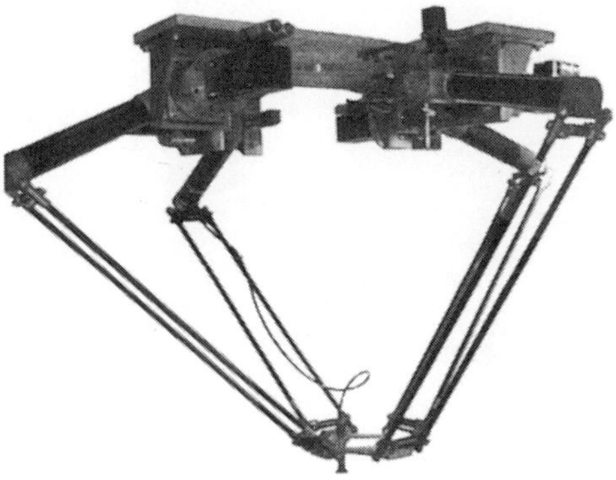

Figure 2.25. *Par4 robot*

Figure 2.26. *Dual4 robot*

2.2.3.3. *Manipulators with six degrees of freedom*

It is in this case that there are the most proposals (even if these are variations based on a single general principle of design) and therefore we will not attempt an exhaustive listing. The designs of fully parallel manipulators, with six degrees of freedom, are based on the use of chains of the type *RRPS, RPRS, PRRS, RRRS*. There are also non-fully parallel manipulators with more complex structures.

2.2.3.3.1. Robots with RRPS chains

This structure, by far the most common, is represented in Figure 2.27, and the Gough platform is a typical example thereof.

Historically, it seems possible to attribute the first robotic design of a manipulator of this type to Professor McCalion, from the University of Christchurch [MCC 79], for a robotized assembly station. As early as the 1980s, Fichter [FIC 80] anticipated a certain number of theoretical problems associated with this type of manipulator and potential applications. C. Reboulet, a pioneer in the construction of this type of manipulator, developed a prototype, as early as 1985, at CERT-DER.A [REB 90]. It is important to also mention the design of a microrobot of this type developed by Arai [ARA 93], where the actuators are piezoelectric elements with a travel distance of 8 micrometers. Variations in length are measured with strain gauges, with accuracy levels of approximately 30 manometers.

The replacement of rigid segments by cables should also be noted, as suggested by Albus and his team [ALB 93] at the National Institute of Standards and Technology (NIST), for the design of a crane element.

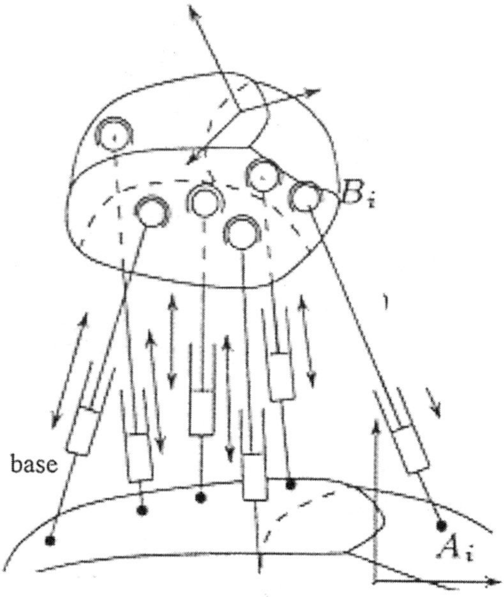

Figure 2.27. *General structure of a parallel manipulator with six degrees of freedom and chains of type RRPS. The platform is connected to the base through six segments. The connection between segments and the base is usually made via a universal joint, and the connection with the mobile platform is ensured via a spherical joint. A prismatic actuator makes it possible to vary the length of the segments*

2.2.3.3.2. Robots with PRRS chains

Parallel manipulators of this type have only recently appeared. For the prototype of the active wrist, patented by INRIA [MER 91], a vertical, motorized, prismatic joint is connected to a segment of fixed length by a universal joint. The other end of the segment is connected to a mobile plate using a spherical joint, and two segments, sharing the same rotation center, use a double spherical joint system. Thus, there are only three joints on the mobile plate.

The advantages of such a structure are the very low center of gravity, the low weight of the mobile equipment and the low collision risks between segments. It is important to also mention that any direction of the prismatic actuators is possible: for example the axes of the prismatic joints of the Hexaglide robot, designed at the Federal Polytechnic School of Zurich, are horizontal and parallel.

Figure 2.28. *Active wrist with six degrees of freedom mounted on the SCARA robot, the joints of which, nearest the base, are displaced vertically using a PRRS chain (INRIA patent). The segments are of fixed length, the engines are on the lower part*

2.2.3.3.3. Robots with RRRS chains

As early as 1983, Hunt [HUN 831] proposed a robot structure using this type of chain (see Figure 2.29). However, the design that is best known is the Hexa robot developed by Pierrot [PIE 91]. This manipulator is different from the Hunt structure in the layout of the axis of the revolute joints of the base, and in the position of the joint centers on the mobile platform. This difference leads to the "Delta" operating mode when the segments of a pair work in an identical manner. It is important to mention that the displacement of the motorized levers on a vertical plane is unnecessary. It is perfectly feasible to build a version where the plane is horizontal.

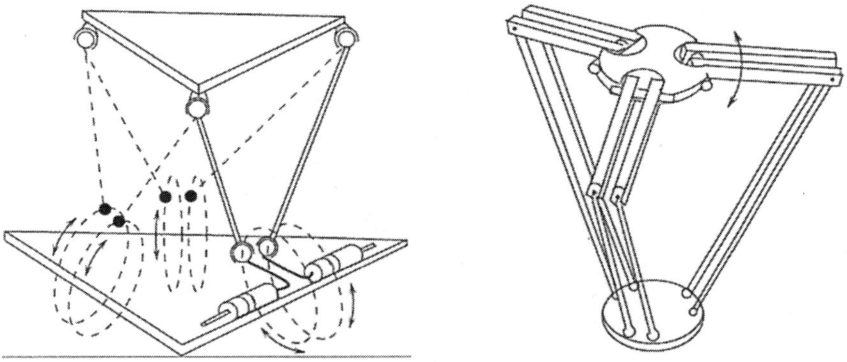

Figure 2.29. *On the left, the robot is using chains of the type RRRS developed by Hunt in 1983 (according to Hunt [HUN 83]). On the right, Pierrot's Hexa robot, a generalization of the "Delta" concept (according to Pierrot [PIE 91])*

We have presented an overview of the most common robot structures, but there are also prototypes using more unusual types of chains. For example, there are the manipulators, designed by Kohli [KOH 88] and Behi [BEH 88], which use double actuators: linear and rotary or linear and linear. It is also worth mentioning the parallel manipulator with six degrees of freedom, called Smartee, designed by Cleary [CLE 93], which unfortunately was unsuccessful when marketed. The end-effector on this robot is linked to the base using three kinematic chains comprising two segments. The segment that is attached to the end-effector is connected to the preceding segment via a passive revolute joint and a differential manipulator makes it possible to control two degrees of freedom of the segment connected to the base.

2.3. Inverse geometric and kinematic models

In this chapter, we describe methods making it possible to obtain relations that supply joint values and velocities, given the pose and velocity of the end-effector (inverse models), and this, for the fully parallel manipulators that are the most common, whether they are equipped with variable length segments (as for example the Gough machine), or with chains using a mobile articulation point controlled by rotary actuators (as with Delta for example), or linear (Hexa for example). Later, we discuss the difficult problem of singularities.

2.3.1. *Inverse geometric model*

For all manipulators discussed here, we consider:

– $\boldsymbol{T}_{n,b} = \begin{bmatrix} s_x & n_x & a_x & x \\ s_y & n_y & a_y & y \\ s_z & n_z & a_z & z \\ 0 & 0 & 0 & 1 \end{bmatrix} = \begin{bmatrix} \boldsymbol{s} & \boldsymbol{n} & \boldsymbol{a} & \boldsymbol{x} \\ 0 & 0 & 0 & 1 \end{bmatrix}$ as the homogenous matrix which describes the nacelle situation compared to the base reference, and which regroups a position vector, $\boldsymbol{x}$, and orientation parameters (represented here as three vectors $\boldsymbol{s}$, $\boldsymbol{n}$, and $\boldsymbol{a}$);

– A_i as the hooking point of the ith chain on the base. This point represents the position of a passive link if the kinematic chain is a segment with a variable length (see Figure 2.30a), or the position of an actuator if the kinematic chain comprises a moving articulation point (see Figures 2.30b and 2.30c);

– C_i as the hooking point of the ith chain of the nacelle;

– q_i as the control parameter. This corresponds to the lengthening of a variable length segment for the case in Figure 2.30a, of an angle for the case in Figure 2.30b, and of a length for the case in Figure 2.30c.

Additionally, for chains comprising a moving joint point, we consider:

– B_i as the passive center of the moving joint for the chains of Figures 2.30b and 2.30c;

– d_i as the length of the segment B_iC_i;

– $\boldsymbol{T}_i$ as the homogenous matrix that describes the transformation from A_i to B_i. This matrix thus depends on constant geometric parameters and on control parameters.

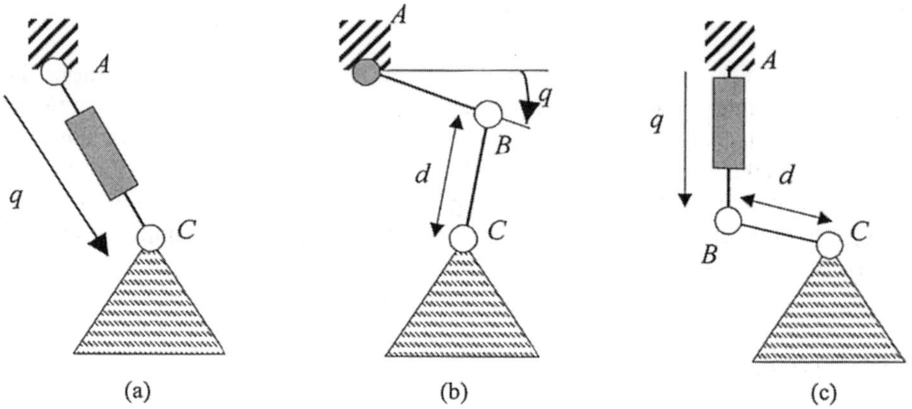

Figure 2.30. *Three typical kinematic chains*

Case a

If $C_{i,n}$ is an expression of the position of C_i according to a frame attached to the nacelle, and $C_{i,b}$ its expression in the base reference frame, then:

$$C_{i,b} = \boldsymbol{T}_{n,b}\ C_{i,n}$$

The case of chains with variable length segment is considered trivial. This is because, when $A_{i,b}$ is an expression of the position of A_i according to the base marker, we have:

$$\left(A_{i,b}C_{i,b}\right)^2 = q_i^2$$

Cases b and c

If the kinematic chain consists of a moving joint point (Figures 2.30a and 2.30b), firstly the coordinates of B_i must be expressed in the base reference frame starting from those of A_i:

$$B_{i,b} = \boldsymbol{T}_i\ A_{i,b}$$

The components of $B_{i,b}$ are thus functions of the geometry and of the control parameter q_i, and thus there is again a relation between the situation of the end-effector and q_i:

$$(B_{i,b}C_{i,b})^2 = d_i^2$$

Polynomial relations thus established make it possible, via traditional resolution, to find the value of the control parameter independently for each of the chains.

2.3.2. *Inverse kinematics*

We consider:

– $\boldsymbol{v}(O_n)$ as the velocity vector for O_n, the center of the reference frame attached to the nacelle;

– $\boldsymbol{\omega}$ as the angular velocity vector of the nacelle;

– $\dot{q}_i$ as the joint velocity of the ith chain.

Additionally, for type a chains:

– $\boldsymbol{s}_i = \dfrac{A_iC_i}{\|A_iC_i\|}$ as the directional vector of the variable length segment.

For type b and c chains:

– $\boldsymbol{b}_i = \dfrac{B_iC_i}{\|B_iC_i\|}$ as the directional vector of the segment with constant length;

– $\boldsymbol{r}_i$ as thc unit vector supplying direction of the control rotation axis when there is a rotary actuator (case b);

– $\boldsymbol{l}_i$ as the vector supplying direction for controlled displacement when the actuator is linear (case c).

For all cases, the velocity of point C_i is given by[3]:

$$\boldsymbol{v}(C_i) = \boldsymbol{v}(O_n) + C_iO_n \times \boldsymbol{\omega}$$

Then the following cases become apparent.

---

3 $\boldsymbol{u} \times \boldsymbol{v}$ designates the vector product of $\boldsymbol{u}$ times $\boldsymbol{v}$.

Case a

If the kinematic chain is a variable length segment, then the joint velocity may be expressed as the projected velocity of C_i on the direction of the segment:

$$\dot{q}_i = \boldsymbol{s_i} \cdot \boldsymbol{v}(C_i)$$

Therefore:

$$\dot{q}_i = \boldsymbol{s_i} \cdot \boldsymbol{v}(O_n) + (\boldsymbol{s_i} \times C_iO_n) \cdot \boldsymbol{\omega}$$

Cases b and c

When applying the equiprojectivity theorem to points B_i and C_i belonging to the fixed length segment, the relation between controlled dimensions and velocity of the nacelle is easily determined:

$$\boldsymbol{v}(B_i) \cdot B_iC_i = \boldsymbol{v}(C_i) \cdot B_iC_i$$

That is, in case b, the velocity of B_i is supplied by:

$$\boldsymbol{v}(B_i) = B_iA_i \times (\dot{q}_i \cdot \boldsymbol{r_i})$$

Therefore we obtain:

$$\dot{q}_i (B_iA_i \times \boldsymbol{r_i}) \cdot B_iC_i = B_iC_i \cdot \boldsymbol{v}(O_n) + (B_iC_i \times C_iO_n) \cdot \boldsymbol{\omega}$$

And in case c:

$$\boldsymbol{v}(B_i) = \dot{q}_i \cdot \boldsymbol{l_i}$$

That is:

$$\dot{q}_i (\boldsymbol{l_i} \cdot B_iA_i) \cdot B_iC_i = B_iC_i \cdot \boldsymbol{v}(O_n) + (B_iC_i \times C_iO_n) \cdot \boldsymbol{\omega}$$

Again, it is possible to see that the kinematic relationships are expressed independently for each of the chains, and that subsequently it is possible to regroup the expressions of a manipulator consisting of k chains in the following matrix format:

$$\boldsymbol{J_q} \cdot \dot{\boldsymbol{q}} = \boldsymbol{J_x} \cdot \dot{\boldsymbol{x}}$$

with:

$$\dot{\boldsymbol{q}} = \begin{bmatrix} \dot{q}_1 \\ \dot{q}_2 \\ \vdots \\ \dot{q}_k \end{bmatrix} \qquad \dot{\boldsymbol{x}} = \begin{bmatrix} \boldsymbol{v}(O_n) \\ \boldsymbol{\omega} \end{bmatrix} \qquad J_x = \begin{bmatrix} \boldsymbol{b_1^t} & (\boldsymbol{b_1} \times C_1O_n)^t \\ \boldsymbol{b_2^t} & (\boldsymbol{b_2} \times C_2O_n)^t \\ \vdots & \\ \boldsymbol{b_k^t} & (\boldsymbol{b_k} \times C_kO_n)^t \end{bmatrix}$$

and for a manipulator consisting exclusively of type a chains:

$$\boldsymbol{J_q} = \boldsymbol{Ident_k}$$

or of type b:

$$J_q = \begin{bmatrix} (B_1A_1 \times \boldsymbol{r_1}) \cdot \boldsymbol{b_1} & 0 & 0 & 0 \\ 0 & (B_2A_2 \times \boldsymbol{r_2}) \cdot \boldsymbol{b_2} & 0 & 0 \\ 0 & 0 & \ddots & 0 \\ 0 & 0 & 0 & (B_kA_k \times \boldsymbol{r_k}) \cdot \boldsymbol{b_k} \end{bmatrix}$$

or of type c:

$$\boldsymbol{J_q} = \begin{bmatrix} \boldsymbol{l_1} . \boldsymbol{b_1} & 0 & 0 & 0 \\ 0 & \boldsymbol{l_2} . \boldsymbol{b_2} & 0 & 0 \\ 0 & 0 & \ddots & 0 \\ 0 & 0 & 0 & \boldsymbol{l_k} . \boldsymbol{b_k} \end{bmatrix}$$

2.3.3. *Singular configurations*

Singular configurations are particular locations of the end-effector where the behavior of parallel manipulators changes drastically. We will explain why such positions are to be avoided in general, and we will show how to characterize such configurations using an inverse Jacobian matrix. Then, we will briefly explain how a geometric method enables the systematic determination of conditions of singularity and how it leads to analytical relations between location parameters of the end-effector, which describe different cases of singularity. We then introduce indices, which make it possible for us to evaluate to what extent we are close to a singular configuration.

The notion of singularity is based on the kinematic equation establishing a linear relation between vectors $\dot{\boldsymbol{q}}$ and $\dot{\boldsymbol{x}}$, which justifies the fact that we are interested in cases where the matrices $\boldsymbol{J}_q$ and $\boldsymbol{J}_x$ degenerate. We will then distinguish among three cases of singularity, when $\boldsymbol{J}_q$ in singular, when $\boldsymbol{J}_x$ is singular, and when $\boldsymbol{J}_q$ and $\boldsymbol{J}_x$ are singular:

– If $\boldsymbol{J}_q$ is singular, then $\dot{\boldsymbol{q}}$ can be different from zero without creating any motion of the platform. This corresponds to a singularity of the serial structure of one of the chains, which we refer to as serial singularity or sub-mobility.

– If $\boldsymbol{J}_x$ is singular, then $\dot{\boldsymbol{x}}$ may be different from zero with no change in the length of the segments. This is then referred to as parallel singularity or over-mobility.

– If $\boldsymbol{J}_q$ and $\boldsymbol{J}_x$ are singular, then there are both serial and parallel singularities.

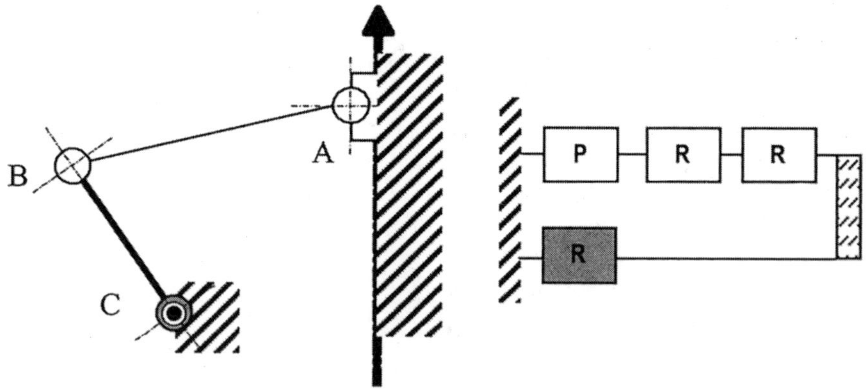

Figure 2.31. *Manipulator with a closed chain and one degree of freedom. A rotary actuator drives a body into translation motion using a crank system*

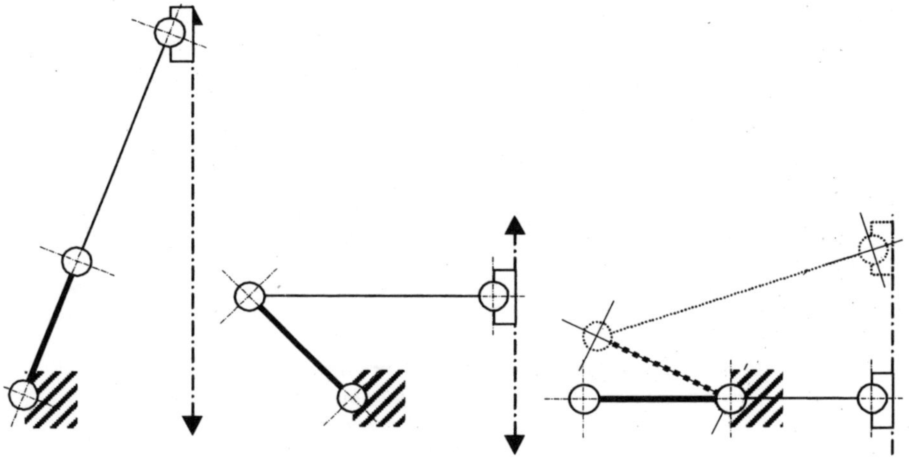

Figure 2.32. *Three types of singularities*

Figure 2.32 illustrates these three cases for the manipulator in Figure 2.31. The first diagram shows a case where $\boldsymbol{J_q}$ is singular, the second is a case where $\boldsymbol{J_x}$ is singular and the third is a case where both matrixes are singular.

It is possible to note that, for the Gough platform, $\boldsymbol{J_q}$ is the identity matrix: thus no serial singularity is possible here, even if parallel singularities can occur. For the Hexa robot, in addition to parallel singularities, there are serial singularities, when for example two segments of a chain are aligned. These singularities will create structural limits in the workspace.

NOTE 1.– serial type singularities are well known. Thus, we are essentially interested in parallel singularities.

NOTE 2.– for certain structures, there may be other types of singularities. Figure 2.33 is an example of such a case. One of the kinematic chains comprises of a parallelogram, which imposes a specific kinematic constraint (the nacelle stays parallel to itself). This constraint is not described in matrices $\boldsymbol{J_q}$ and $\boldsymbol{J_x}$, and thus cannot be detected by an analysis of those matrices. In Figure 2.34, the particular position of the parallelogram causes a singularity termed "constraint" or "internal"; the constraint that is imposed by the parallelogram disappears, and the nacelle gains a degree of freedom. These singularities will not be covered here; anyone who is interested in finding out more should refer to [KRU 03].

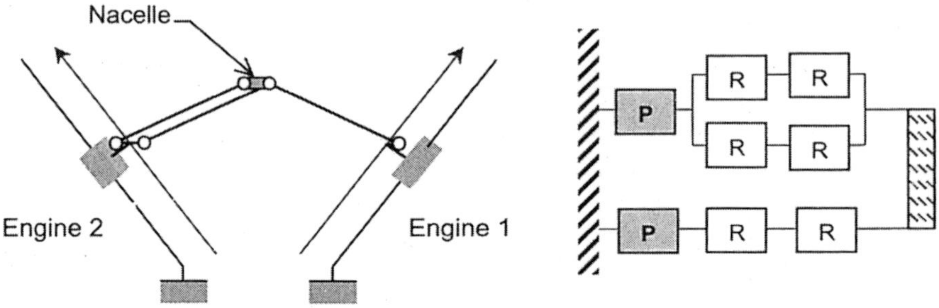

Figure 2.33. *Plane manipulator with two degrees of freedom, where one chain imposes a particular kinematic constraint*

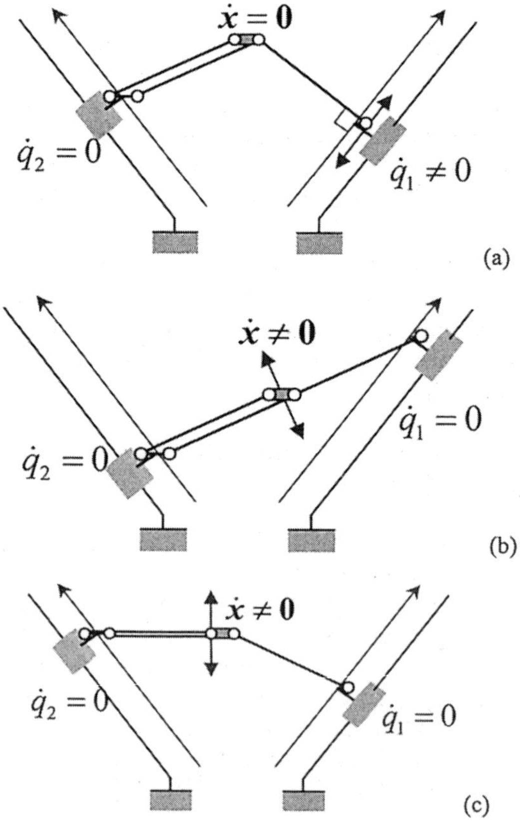

Figure 2.34. *Two common visual singularities of parallel robots ((a) sub-mobility, (b) over-mobility) and "constraint" singularity (c)*

2.3.3.1. *Singularities and statics*

One easy way of introducing singularities is to succinctly discuss the notion of mechanical equilibrium of a parallel robot. For a parallel manipulator, we consider $\boldsymbol{\tau}$ as the vector of the joint forces and $\boldsymbol{f}$ the vector of generalized stresses applied to the end-effector. For a vector $\boldsymbol{f}$ applied to a mobile plate, the mechanical system is in a state of equilibrium when there are joint forces acting on the platform opposite to $\boldsymbol{f}$. When this is not the case, the end-effector of the manipulator will be displaced until a new position of equilibrium is reached. However, there is a well known relationship between $\boldsymbol{\tau}$ and $\boldsymbol{f}$:

$$\boldsymbol{f} = \boldsymbol{J}^{-t} \cdot \boldsymbol{\tau}$$

where $\boldsymbol{J}^{-t}$ is the transpose of an inverse Jacobian matrix (note: $\boldsymbol{J}^{-1} = \left(\boldsymbol{J}_q\right)^{-1} \cdot \boldsymbol{J}_x$). The preceding equation describes a linear system in terms of components of the vector $\boldsymbol{\tau}$, which will in general accept the solution $\boldsymbol{\tau}$ for any $\boldsymbol{f}$ (a solution which thus leads to the mechanical equilibrium of the system) except when the matrix $\boldsymbol{J}^{-1}$ is degenerate. In this case the linear system does not accept any solution and the mechanical system loses its equilibrium. Concretely, this means that the mobile plate will be displaced without motion of the actuators.

An important practical consequence is that in the proximity of a singular configuration, joint forces may become very significant since they are expressed as a quotient where the denominator determines $\boldsymbol{J}^{-1}$. Thus, the risk of deterioration of the manipulator is high and the need to specify singular configurations becomes quite clear.

2.3.3.2. *State of the art*

The search for singular configurations is thus based on the study of the singularity of an inverse kinematic Jacobian matrix. *A priori*, as this matrix is perfectly well known, all that is required is to calculate the roots of the determinant of this matrix in view of obtaining the conditions of singularity. However, we also saw that the matrix $\boldsymbol{J}^{-1}$ in general displayed fairly complicated components. Consequently, even with formal calculation systems, it is quite complicated to obtain the determinant. This method is nonetheless applicable to specific manipulators such as those of spherical robots [SEF 94].

Certain researchers, such as Fichter [FIC 86], have endeavored to analyze in an intuitive manner specific cases of degeneration of the Jacobian inverse matrix, and they obtained a certain number of singularity cases. Another, more numerical approach consists of defining a positive index making it possible to determine whether a configuration is far from a singularity, the index being zero when the configuration is singular, and then to try and find numerically all of the positions that minimize this index. Thus, the index may be based on the absolute eigenvalue of the determinant of the inverse Jacobian matrix or its condition number (the ratio between the lowest value and the highest). However, as Ma and Angeles [MA 91a] have pointed out, this matrix is not homogenous in terms of dimensions. Thus, the index will depend on the choice of dimensional units and hence, does not constitute an intrinsic value.

2.3.3.3. *The geometric method*

In this section, we present a brief overview of the geometric method, which makes it possible to solve the problem in a satisfactory manner for a large number of cases. The basic idea arises from the observation that for many robots the Jacobian matrix consists of Plücker vectors, straight rows related to the segments of the manipulator. As a brief reminder, here is the definition of Plücker coordinates for a straight row: we consider two points M_1 and M_2 on this straight row as well as a reference frame O (see Figure 2.35), and construct a vector of dimension 6 as follows:

$$\boldsymbol{Pr} = [M_1M_2, OM_1 \times OM_2] = [M_1M_2, M_2M_1 \times OM_2]$$

The vector $\boldsymbol{Pr}$ enables us to characterize the straight row crossing through the two points M_1, M_2. Representation of a straight row using its Plücker coordinates is redundant since the dimension of the vector is 6 and 4 parameters are sufficient.

To reduce redundancy, a normalized Plücker vector $\boldsymbol{Prn}$ is introduced, defined by:

$$\boldsymbol{Prn} = \left[\frac{M_1M_2}{\|M_1M_2\|}, \frac{OM_1 \times OM_2}{M_1M_2} \right]$$

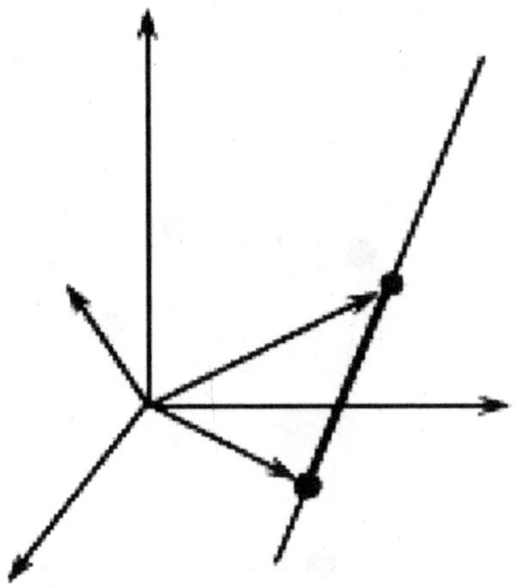

Figure 2.35. *Definition of the Plücker coordinates of a straight row in space*

For example, the inverse Jacobian matrix of the Gough platform consists of 6 normalized Plücker vectors related to the supporting straight rows of the legs of the robot.

Singularity of the inverse kinematic Jacobian matrix thus implies a linear dependence among these vectors. However, Grassmann (1809-1877) demonstrated that the linear dependence of Plücker vectors induces geometric relations with the related straight rows (for a complete introduction to Grassmann geometry, see [VEB 10]). Thus, we know the geometric conditions that are required to be satisfied by sets of 2, 3, 4, 5 and 6 straight rows to obtain a singularity. We will then consider all of the possible pairs of straight rows, and we will calculate the conditions of the pose of the platform so that two straight rows satisfy geometric conditions, which will supply us with a condition of singularity for each pair of straight rows. We then start again with straight-row triplets, which will supply additional conditions of singularity, and so forth, until the straight-row sextuplet. The inverse Jacobian matrix determinant contains all of these conditions, which explains its complexity. We proceed in a certain way to the factorization of this determinant, with the additional bonus of a geometric interpretation of the conditions of singularity.

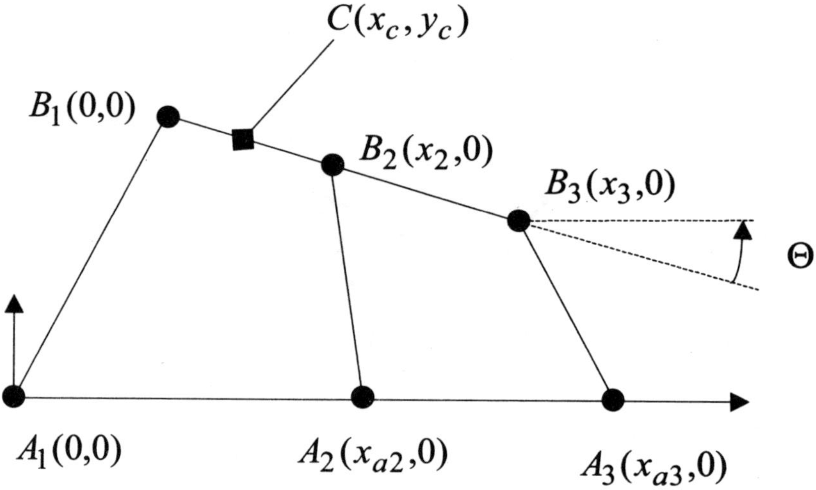

Figure 2.36. *Planar manipulator for which we are seeking singular configurations*

Let us consider the simple example of the planar manipulator presented in Figure 2.36. In this case, it is relatively easy to obtain the determinant of the inverse Jacobian matrix. Using the notation of the figure, we obtain:

$$\begin{aligned}
&\Delta = \cos(\theta)(x_{a2}x_3 - x_2x_{a3})y_c^2 + \\
&((x_{a2} - x_{a3})x_2x_3\cos(\theta) + (x_{a3} - x_c)x_{a2}x_3 + (x_{a3}x_c - x_{a2}x_{a3})x_2)\sin(\theta)y_c \\
&- \sin^2(\theta)x_2x_3x_c(x_{a2} - x_{a3})
\end{aligned}$$

However, the geometric interpretation of the cancellation of this equation is difficult. Thus, a geometric approach is used. The robot here is equipped with three legs, thus three Plücker vectors. Grassmann determined that three of these vectors were dependent only when:

– the three straight rows are in the same plane;

– the three straight rows share a common point.

In our case, the first condition is obviously true. A singularity is thus obtained when the three straight rows share a common point (see Figure 2.37) and when this is expressed according to the position parameters of the platform we find exactly the same condition $\Delta = 0$. Thus, the geometric meaning of the cancellation of the determinant is understood.

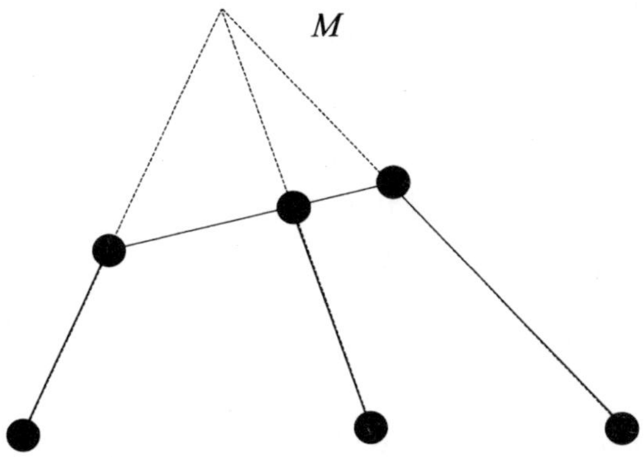

Figure 2.37. *Planar manipulator and one of the singular configurations. With singular configurations, straight rows related to the segments share a point in common (M)*

The example discussed here is simple. There are numerous cases, however, and in particular for robots with six degrees of freedom, where it is impossible to obtain the determinant, whereas the calculation of the conditions of platform position leading to particular positions of the straight rows will, in general, be relatively easy. Examples of processing for the case of space robots are found in [COL 95, MER 89]. We would also like to point out that importing conditions of singularity into an inverse Jacobian matrix makes it possible, through an analysis of vectors, to determine the type of motion of the platform within this singularity, as well as the parameters of this motion.

2.3.3.4. *Maneuverability and condition number*

We read in the introduction about the notion of singular configuration that in the proximity of such a configuration the joint forces could become very significant. Now it is important to quantify what "in the proximity" of a singular configuration means. In order to do this, criteria of performance are defined in view of obtaining interesting information concerning the proximity of a singular configuration, although it can be shown that there is no mathematical distance to measure the "closeness" of a singularity. An index that is commonly used, defined by Yoshikawa [YOS 82], considers the absolute value of the determinant of an inverse Jacobian matrix. This index is generally termed maneuverability. It makes it possible to identify the smallest value of the inverse Jacobian matrix. A low value indicates that the joint forces may become significant given a combination of external forces and moments applied, the ratio of joint forces and external forces being the inverse of the value in question. Then, it is possible to understand the importance of the index in terms of the manipulator's design, although care must be taken when using it.

NOTE.– the crucial importance of the study of parallel-type singularities was previously outlined. However, singularities of the serial type also exert significant importance as they interfere considerably with the analysis. In fact, for manipulators of the type b and c, the $\boldsymbol{J}_q$ matrix does not correspond to the identity. Thus:

$$\det\left(\boldsymbol{J}^{-1}\right)=\det\left(\boldsymbol{J}_q^{-1}\cdot\boldsymbol{J}_x\right)=\frac{\det\left(\boldsymbol{J}_x\right)}{\det\left(\boldsymbol{J}_q\right)}$$

A considerable value of "$\det\left(\boldsymbol{J}^{-1}\right)$" can thus be explained according to two phenomena:

– $\det\left(\boldsymbol{J}_x\right)$ is large, which indicates distance from the parallel type singularity; or

– $\det\left(\boldsymbol{J}_q\right)$ is small, which indicates proximity to a serial type singularity.

This situation leads to recommending great care in the use of certain quality indices during optimization procedures for example.

2.3.3.5. *Singularities in practice*

In the preceding sections we saw that Grassmann geometry made it possible to obtain relations defining the singularities of certain parallel manipulators. In practice, the question is more often posed in terms of finding out whether there are places of singularity within a given workspace.

In certain cases the places of singularity lend themselves well to a joint representation with the workspace. This is the case for example of planar parallel robots for which workspace and places of singularity may be jointly represented [SEF 95]. It is even possible to directly calculate areas of the workspace without singularities [CHA 98]. For space robots the problem is far more complex even if there is a method that makes it possible to solve the problem [MER 98b].

2.4. Direct geometric model

For serial robots, the direct geometric model is in general easy to establish and it makes it possible to go from joint positions ($\boldsymbol{q}$ vector) to the location of the end-effector ($\boldsymbol{x}$ vector) via an analytical analysis. For parallel robots, however, the opposite holds true in general. Let us consider for example a Gough machine. We saw that there was a polynomial relation, which was *a priori* simple, and which made it possible to go independently from the position of the nacelle to the joint position of a chain. However, with the direct geometric model, the point is to solve the opposite problem, and thereby to solve the system composed of all the equations of the inverse model. And if we look at these equations in detail, it turns out that the result is a system of six non-linear equations. Thus, we have:

$$A_{i,b}C_{i,b} = \begin{bmatrix} s_x c_{ix} + n_x c_{iy} + a_x c_{iz} + x - a_{ix} \\ s_y c_{ix} + n_y c_{iy} + a_y c_{iz} + y - a_{iy} \\ s_z c_{ix} + n_z c_{iy} + a_z c_{iz} + z - a_{iz} \end{bmatrix} = \begin{bmatrix} ac_{ix} \\ ac_{iy} \\ ac_{iz} \end{bmatrix}$$

And thus for a Gough machine the following needs to be resolved:

$$ac_{ix}{}^2 + ac_{iy}{}^2 + ac_{iz}{}^2 = q_i^2 \qquad i = 1,\ldots,6$$

Except for certain particular cases (of which the Delta robot is one), getting an analytical form for the solutions is impossible with the current algebraic means. Thus, we need to settle for one of the following two approaches.

2.4.1. *Iterative method*

In general, there are obviously many solutions, real or complex, for the preceding system of six equations, and each real solution corresponds to a possible assembly of the manipulator, if all limits due to end-stops or collisions between two constituting elements of the system are factored out. When we focus on only one of these solutions, it is possible to use an iterative calculation method which, beginning with an estimated solution, for every step of the calculation approaches the value that is required, until it is estimated that sufficient accuracy has been reached. This approach is often used only for robot control because in this case it is possible to supply algorithms with initial high quality estimates.

$$\boldsymbol{x} = mgd(\boldsymbol{q})$$

A first order derivation yields:

$$\boldsymbol{x} = \boldsymbol{x_0} + (\boldsymbol{q} - \boldsymbol{q_0})^t \cdot \frac{\partial \boldsymbol{x}}{\partial \boldsymbol{q}}(\boldsymbol{x_0})$$

with $\boldsymbol{x_0}$ as one estimated solution and $\boldsymbol{q_0} = mgi(\boldsymbol{x_0})$ corresponding to the joint position.

Thus, the following iterative equation may be used:

$$\boldsymbol{x}_{k+1} = \boldsymbol{x}_k + \boldsymbol{H}_k(\boldsymbol{q} - \boldsymbol{q}_k)$$

where $\boldsymbol{H}$ is a matrix which in an "ideal" case will be the Jacobian matrix calculated for point $\boldsymbol{x}_k$, but it is also possible to demonstrate that, using a Jacobian constant (calculated for example at the center of work volume) for all points, is sufficient when the work volume is quite reduced.

NOTE 1.– the end condition for the algorithm may be equal to: $\max(|q_i - q_{i_k}|) < \varepsilon$ where ε is a pre-determined threshold (in general depending on the resolution of the measure of joint positions).

NOTE 2.– in general, this type of algorithm may not converge, or converge to a solution that is not the actual position of the platform, for example if it is within proximity of a singularity. This is one of the basic rationales of algebraic methods.

2.4.2. *Algebraic method*

We saw in the preceding sections that resolution of the direct geometric model amounted to obtaining solutions to systems consisting of equations of the inverse geometric model. First, it is important to note that for most parallel robots the equations of this system consist of algebraic terms (such as x^2 for example), or of terms consisting of sine and cosine of unknowns. It will become clear that their simple transformation enables their conversion into algebraic terms, which in turn makes it possible to consider the system as an algebraic one. The advantage of this type of manipulation is that there is a whole range of tools in algebraic geometry that makes it possible, in certain cases, to solve the system. In order to do this, the next section begins with a small reminder of algebraic geometry.

2.4.2.1. *Reminder concerning algebraic geometry*

2.4.2.1.1. The concept of degree

Given a polynomial P and a variable x:

$$P(x) = \sum_{i=1}^{i=n} a_i x^i$$

where the a_i are coefficients of the polynomial. The unknown x appears in this expression as x, x^2, …, x^n. The degree of the polynomial is n. A polynomial with a degree n has exactly n roots, which are either complex or real, and where complex solutions always come in pairs. Thus, a polynomial with three degrees can have either three real solutions or one real solution and two complex solutions, providing that the roots are counted appropriately; for example the polynomial $(x-1)(x-1)(x-1)$ only has one real root, 1, but it is counted three times because it cancels three factors of the polynomial. In the case of polynomials with several variables, for example x, y, we consider each of the terms $x^i y^j$ of the polynomial. Given the number $m = I + j$: the total degree of the polynomial is the largest of the numbers m. An important theorem in algebraic geometry is that of Bezout, which states that, two curves defined by the total degree algebraic equations m_1 and m_2 intersect at $m_1 m_2$ points.

2.4.2.1.2. The concept of resultant

A traditional method of solving algebraic systems with m equations and m unknowns is to manipulate the equations so they are reduced to a system of $m-1$ equations with $m-1$ unknowns, and then to repeat in view of finally obtaining a single equation with a single unknown. Such an equation is, in fact, relatively easy to solve, either exactly when the degree does not exceed four, or numerically beyond that point.

One possible way of manipulating equations to make an unknown disappear is to use the method of the resultant. Given two polynomials:

$$P_1(x) = \sum_{i=1}^{i=n} a_i x^i = 0 \qquad P_2(x) = \sum_{i=1}^{i=m} b_i x^i = 0$$

A necessary and sufficient condition for both of these equations to have a common root is that the following determinant, termed the resultant of both polynomials, is zero:

$$\begin{bmatrix}
a_n & 0 & 0 & \cdots & 0 & b_m & 0 & 0 & \cdots & 0 \\
a_{n-1} & a_n & 0 & \cdots & 0 & b_{m-1} & b_m & 0 & \cdots & 0 \\
a_{n-2} & a_{n-1} & a_n & \cdots & 0 & b_{m-2} & b_{m-1} & b_m & \cdots & 0 \\
\vdots & & & & \vdots & \vdots & & & & \\
a_1 & a_2 & & & & b_1 & b_2 & & & \\
a_0 & a_1 & & & & b_0 & b_1 & & & \vdots \\
0 & a_0 & a_1 & & & 0 & b_0 & b_1 & & \\
0 & 0 & & & & 0 & & & & \\
\vdots & & & & a_1 & \vdots & & & & b_1 \\
0 & 0 & 0 & & a_0 & 0 & 0 & 0 & \cdots & b_0
\end{bmatrix}$$

Given two polynomials x, y:

$$P(x,y) = \sum_{i=1,j=1}^{i=n,j=m} a_{ij} x^i y^j = 0 \quad Q(x,y) = \sum_{i=1,j=1}^{i=n_1,j=m_1} b_{ij} x^i y^j = 0$$

It is possible to consider that these two polynomials are in fact two x-only polynomials, with coefficients that depend on y. Thus, it is possible to calculate the resultant of these two polynomials, which will yield a y-only polynomial. Consequently, the x variable was simply eliminated. It then becomes possible to solve the y polynomial, and for each of the solutions to carry over the value into the equation, which are now exclusively x polynomials, which makes it possible to calculate that unknown.

Given these basic notions, we now turn to the calculation of the direct geometric model for planar robots.

2.4.2.2. *Planar robots*

This section examines planar manipulators with three degrees of freedom. The equations of the inverse geometric model generate a system of three non-linear equations, which must be solved in order to find the solution of the direct geometric model. First, we will demonstrate that there are several solutions to this system, that is, there are several end-effector locations that respect the fixed values of the joint variables. Thus, the manipulator may be assembled in different ways, and this is why the different configurations are called the modes of assembly of the manipulator.

2.4.2.2.1. Maximum number of solutions

Consider for example the robot 3-RPR described in Figure 2.38.

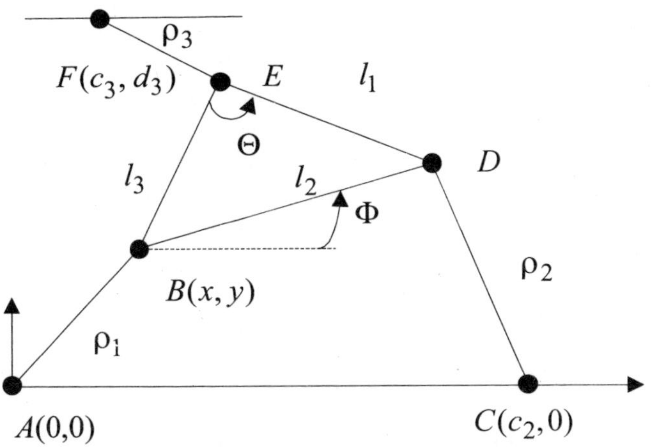

Figure 2.38. *3-RPR robot*

For this manipulator, when the segment *FE* is uncoupled at *E*, there are two manipulators: one is the simple segment *FE,* which revolves around F and the other *(ABEDC)* is a manipulator called a 4-bar manipulator. It is possible to demonstrate that when segment *AB* rotates around *A*, point *E* is then located on an algebraic curve the total degree of which is 6. When *E* belongs to the segment *FE* and rotates around *F*, then *E* is on a circle, and thus, this algebraic curve is of degree 2. For a solution of the direct geometric model, it is necessary that point *E,* considered as belonging to segment *FE*, coincides with point *E* of the 4-segment manipulator. Consequently, the possible positions of e are obtained at the intersection of the circle and the 6-degree curve. Bezout theorem implies that there are at the most 12 (2×6) possible positions for *E*, and thus 12 solutions of the direct geometric model. It is also possible to demonstrate that out of the 12 solutions; at least 6 of them will be complex. Consequently, there are no more than 6 solutions to the problem.

We will now explain how to calculate the solutions.

2.4.2.2.2. Direct polynomial geometric model

The purpose of the study is to obtain a polynomial equation of the direct geometrical model, that is, to reduce the initial problem of solving a system of three equations to the resolution of a polynomial equation with a single variable.

The method suggested by Gosselin [GOS 92] is used with notations according to Figure 2.38.

The origin of the reference frame is chosen here as being the center (*A*) of one of the joints, and the axis *x* is defined as the line that connects *A* to another joint center (*C*). The *y*-axis is then defined as an axis that is perpendicular to *x*. The position of the mobile plate is then defined in reference to the position of the joint center *B*, related to point *A*, the coordinates of which are written (*x*, *y*). The orientation of the mobile plate is determined by the angle Φ between axis *x* and one of the sides of the mobile plate (*BD* here). The mobile plate itself consists of 3 points *B*, *D* and *E* and its geometry is perfectly defined by the length of its 3 sides (l_1, l_2, l_3) and by an angle at the top (that is, angle Θ between sides *EB* and *BD*). The lengths of the 3 segments are written ρ_1, ρ_2, ρ_3. Following these conventions, the coordinates of the 3 joint centers linked to the reference frame are:

$$A:(0,0) \quad C:(c_2,0) \quad F:(c_3,d_3)$$

Following these conditions, the equations of the inverse geometric model are written:

$$\begin{aligned}
\rho_1^2 &= x^2 + y^2 \\
\rho_2^2 &= (x + l_2 \cos\Phi - c_2)^2 + (y + l_2 \sin\Phi)^2 \\
\rho_3^2 &= (x + l_3 \cos(\Phi + \Theta) - c_3)^2 + (y + l_3 \sin(\Phi + \Theta) - d_3)^2
\end{aligned}$$

The preceding system of equations may be simplified in the following manner:

$$\begin{aligned}
\rho_1^2 &= x^2 + y^2 \\
\rho_2^2 - \rho_1^2 &= Rx + Sy + Q \\
\rho_3^2 - \rho_1^2 &= Ux + Vy + W
\end{aligned}$$

The last two equations are linear in *x*, *y*, and this system, the resultant of which is *RV–SU*, is solved in order to obtain:

$$x = -(SA_1 - VA_2)/(RV - SU) \qquad y = (RA_1 - UA_2)/(RV - SU)$$

This result is then carried over to the first equation in order to obtain:

$$(SA_1 - VA_2)^2 + (RA_1 - UA_2)^2 - \rho_1^2 (RV - SU)^2 = 0$$

which only depends on the variable Φ. Then, using the classical Weierstrass substitution:

$$T = \tan(\frac{\Phi}{2}) \quad \Rightarrow \quad \cos(\Phi) = \frac{1-T^2}{1+T^2} \qquad \sin(\Phi) = \frac{2T}{1+T^2}$$

which finally yields a 6-degree polynomial at T:

$$C_0 + C_1 T + C_2 T^2 + ... + C_6 T^6$$

where the coefficients C_i exclusively depend on the geometry of the manipulator. Each of the real solutions of the polynomial makes it possible to determine Φ, which in turn makes it possible to determine x, y.

It is possible to demonstrate that there are configurations where the six solutions are actually all real.

Now, we examine the more difficult case of space robots.

2.4.2.3. *Manipulators with six degrees of freedom*

For these manipulators it is also possible to determine the limits on the number of solutions. However, the methods used are more complex and they are beyond the scope of the present work.

The calculation of the direct geometric model was the object of many publications. It was thus possible to demonstrate that the Stewart platform can have as many as 12 solutions [LAZ 94] and one may refer to [FAU 95] for an exhaustive list of results for other particular cases.

We will limit the focus here on the case of the Gough platform, which offers a good illustration of the complexity of the study. The determination of the maximum number of solutions, then of the solutions for the Gough platform has been one of the biggest challenges facing mechanical engineers in recent years. It is only recently that it was demonstrated that the maximum number of assembly modes for the Gough platform could not exceed 40 [LAZ 92], [RAG 91], [RON 92].

As for the research for solutions, Husty [HUS 96] was the first to show in 1994 that it was possible to manipulate the system of equations of the inverse geometric model in view of reducing it to the resolution of a 40° polynomial with a single variable. And it was only in 1998 that Dietmaier [DIE 98] was in fact able to show an example that actually presented 40 real modes of assembly.

To date, the fastest and most numerically safe method for resolution is a method proposed by Rouiller [ROU 95], which is based on mathematical concepts that are far too complex to be presented here. Alternatively, another safe method is to use interval analysis.

To conclude, it appears that the algebraic approach makes it possible for numerous cases to determine the complete set of solutions, but it sometimes requires heavy manipulations and it cannot, except for certain special cases, be used in a real-time context. It is also important to mention a practical problem. For the calculation of the direct geometric model, we are in general interested in determining the current pose of the robot. As the algebraic approach supplies the full set of solutions, it is thus necessary to proceed with sorting the results in view of determining within the full set of solutions, one of which is actually the pose of the robot. Unfortunately, there are no algorithms known to date that can perform the sorting of solutions.

It is also important to mention that there is another approach that consists of adding sensors to the robot. For example, for a Gough platform, other than those sensors measuring the lengths of the legs, rotation sensors are placed on the base joint, and with a number of sensors that are well distributed, it is then possible to find the solution in a unique manner [TAN 99].

2.5. Bibliography

[ALB 93] J. ALBUS, R. BOSTELMAN and N. DAGALAKIS. The NIST ROBOCRANE *J. of Robotic Systems*, 10(5):709-724, July 1993.

[ARA 93] T. ARAI, R. STOUGHTON and Y.M. JAYA. "Micro hand module using parallel link manipulator", in *Japan/USA Symp. on Flexible Automation*, pp. 163-168, San Francisco, 13-15 July, 1993.

[BEH 88] F. BEHI. "Kinematic analysis for a six-degree-of-freedom 3-PRPS parallel manipulator". *IBEE J. of Robotics and Automation*, 4(5):561-565, October 1988.

[BRA 97] G. BRANDT *et al.* "A compact robot for image guided orthopedic surgery", in *First Joint Conf. of Computer Vision, Virtual Reality and Robotics (CRVMED) II and Medical Robotics and Computer Assisted Surgery (MRCAS) III*, Grenoble, 19-22 March, 1997.

[CHA 98] D. CHABLAT. “Domaine d’unicité et parcourabilité pour les manipulateurs pleinement parallèles”. PhD Thesis, Ecole Centrale, Nantes, 6 November, 1998.

[CLA 85] B. CLAUDINON and J. LIEVRE. Test facility for rendez-vous and docking, in *36th Congress of the IAF*, pp. 1-6, Stockholm, 7-12 October, 1985.

[CLA 91] R. CLAVEL. “Conception d’un robot parallèle rapide à 4 degrés de liberté”. PhD Thesis, EPFL, Lausanne, 1991. No. 925.

[CLA 94] R. CLAVEL. Robots parallèles, 1994. Techniques de l’ingénieur, Traité Mesures et Contrôle.

[CLE 93] K. CLEARY and T. BROOKS. “Kinematic analysis of a novel 6-dof parallel manipulator”, in *IEEE Int. Conf. on Robotics and Automation*, pp. 708-713, Atlanta, 2-6 May, 1993.

[COL 95] C.L. COLLINS and G.L. LONG. “The singularity analysis of an in-parallel hand controller for force-reflected teleoperation”, *IEEE Trans. on Robotics and Automation*, 11(5):661-669, October 1995.

[COM 05] O. COMPANY, F. PIERROT, V. NABAT, M. RODRIGUEZ. “Schoenflies Motion Generator: A New Non-Redundant Parallel Manipulator with Unlimited Rotation Capability”, in *Proc of ICRA ’05*, CD-ROM, 2005.

[COP 94] T.R.J. CORRIGAN and S. DUBOWSKY. “Emulating microgravity in laboratory studies of space robotics”, in *ASME Design Automation Conf.*, pp. 109-116, Minneapolis, 11-14 September, 1994.

[DAN 93] R.W. DANIEL, P.J. FISCHER and B. HUNTFR. “A high performance parallel input device”, in *SPIEE, Telemanipulator Technology and Space Telerobotics*, pp. 272-281, Boston, 3-9 September, 1993.

[DAN 95] G. DANESCU. “Une méthode algébrique de synthèse et conception de mécanismes articulés”. PhD Thesis, Franche-Comté University, Besançon, 22 June 1995.

[DIE 98] P. DIETMAIER. “The Stewart-Gough platform of general geometry can have 40 real postures”, in *ARK*, pp. 7-16, Strobl, 29 June-4 July, 1998.

[DRO 85] J. DROSDOL and F. PANIK. The Daimler-Benz driving simulator. A tool for vehicle development, 25 February-1 March, 1985. SAE Technical Paper Series.

[DUB 94] S. DUBOWSKY *et al.* “The design and implementation of a laboratory testbed for space robotics: the VES mod. II”, in *ASME Design Automation Conf.*, pp. 99-108, Minneapolis, 11-14 September 1994.

[DWO 89] B. DWOLASTZKI and G.S. THORNTON. “The GEC Tetrabot-A serial-parallel topology robot: control design aspects”, in *IEEE Int. Workshop on Robot Control*, pp. 426-431, Oxford, 11-12 April, 1989.

[EAR 83] C.P. EARL and J. ROONEY. “Some kinematics structures for robot manipulator designs”, *J. of Manipulators, Transmissions and Automation in Design*, 105:15-22, March 1983.

[FAU 95] J.C. FAUGÈRE and D. LAZARD. “The combinatorial classes of parallel manipulators”, *Manipulator and Machine Theory*, 30(6):765-776, August 1995.

[FIC 80] E.F. FICUTER and E.D. McDOWELL. "A novel design for a robot arm", in *Proc. Int. Computer Technical Conf.*, pp. 250-255, San Francisco, 1980.

[FIC 86] E.F. FICHTER. "A Stewart platform based manipulator: general theory and practical construction", *Int. J. of Robotics Research*, 5(2):157-181, Summer 1986.

[GOS 88] C. GOSSELIN. "Kinematic analysis optimization and programming of parallel robotic manipulators", PhD Thesis, McGill University, Montreal, 15 June 1988.

[GOS 92] C. GOSSELIN, J. SEFRIOUI and M.J. RICHARD. "Solution polynomiale au problème de la cinématique directe des manipulateurs parallèles plans à 3 degrés de liberté", *Manipulator and Machine Theory*, 27(2):107-119, March 1992.

[GOS 94a] C. GOSSELIN and J.-P. HAMEL. "The Agile Eye: a high performance three-degree-of-freedom camera-orienting device", in *IEEE Int. Conf. on Robotics and Automation*, pp. 781-787, San Diego, 8-13 May 1994.

[GOS 94b] C. GOSSELIN, T. PERREAULT and C. VAILLANCOURT. "Smaps: a computer-aided design package for the analysis and optimization of a spherical parallel manipulators", in *ISRAM*, pp. 115-120, Hawaii, 14-18 August 1994.

[GOS 96] C. GOSSELIN, S. LEMIEUX and J.-P. MERLET. "A new structure of planar three-degree-of-freedom parallel manipulator", in *IEEE Int. Conf. on Robotics and Automation*, pp. 3738-3743, Minneapolis, 24-26 April 1996.

[GOU 57] V.E. GOUGH. "Contribution to discussion of papers on research in auto-mobile stability, control and tire performance, 1956-1957". *Proc. Auto Div. Inst. Mech. Eng.*, pp. 392-394.

[GOU 62] V.E. GOUGH and S.G WHITEHALL. "Universal tire test machine", in *Proceedings 9th Int. Technical Congress FLSLTÀ.*, vol. 117, pp. 117-135, May 1962.

[GRA 93] K.W. GRACE *et al.* "A six-degree of freedom micromanipulator for opthalmic surgery", in *IEEE Int. Conf. on Robotics and Automation*, pp. 630-635, Atlanta, 2-6 May 1993.

[HER 95] J.M. HERVÉ. "Group mathematics and parallel link manipulators", in *9th World Congress on the Theory of Machines and Manipulators*, pp. 2079-2082, Milan, August 30-2 September, 1995.

[HUN 78] K.H. HUNT. *Kinematic Geometry of Manipulators*, Clarendon Press, Oxford, 1978.

[HUN 82] K.H. HUNT. "Geometry of robotics devices", *Mechanical Engineering Transactions*, 7(4):213-220, 1982.

[HUN 83] K.H. HUNT. "Structural kinematics of parallel actuated robot arms". *J. of Manipulators, Transmissions and Automation in Design*, 105:705-712, March 1983.

[HUS 96] M.L. HUSTY. "An algorithm for solving the direct kinematic of Stewart-Gough-type platforms", *Manipulator and Machine Theory*, 31(4):365-380, 1996.

[IWA 90] H. IWARA. "Artificial reality with force-feedback: development of desktop virtual space with compact master manipulator", *Computer Graphics*, 24(4):165-170, August 1990.

[KOE 75] W.P. KOEVERMANS *et al.* "Design and performance of the four d.o.f. motion system of the NLR research flight simulator", in *A GARD Conf. Proc.* No. 198, Flight Simulation, pp. 17-1/17-11, The Hague, 20-23 October 1975.

[KOH 88] D. KOHLI, S.-I. LEE, K.-Y. TSAI and G.N. SANDOR. "Manipulator configurations based on Rotary-Linear (R-L) actuators and their direct and inverse kinematics". *J. of Manipulators, Transmissions and Automation in Design*, 110:397- 404, December 1988.

[KRU 03] S. KRUT. "Contribution à l'étude des robots parallèles légers, 3T-1R et 3T-2R, à forts débattements angulaires", Thesis, Montpellier University, 2, 2003.

[LAN 92] S.E. LANDSBERGER and T.B. SHERIDAN. "A minimal, minimal linkage the tension-compression parallel link manipulator", in *IMACSISICE Int. Symp. on Robotics, Mechatronics, and Manufacturing Systems*, pp. 493-500, Kobe, 16-20 September 1992.

[LAZ 92] D. LAZARD. "Stewart platform and Grôbner basis", in *ARK*, pp. 136-142, Ferrara, 7-9 September 1992.

[LAZ 94] D. LAZARD and J.-P. MEPLET. "The (true) Stewart platform has 12 configurations", in *IEEE Int. Conf. on Robotics and Automation*, pp. 2160-2165, San Diego, 8-13 May 1994.

[MA 91a] O. MA and J. ANGELES. "Structure singularities of platform manipulator", in *IEEE Int. Conf. on Robotics and Automation*, pp. 1542-1547, Sacramento, 11-14 April 1991.

[MA 91b] O. MA and J. ANGELES. "Optimum structure design of platform manipulator", in *ICAR*, pp. 1131-1135, Pisa, 19-22 June 1991.

[MAS 93] O. MASCRY, J. WANG and H. ZHUANC. "On the accuracy of a Stewart platform-part Il: Kinematic calibration and compensation", in *IEEE Int. Conf. on Robotics and Automation*, pp. 725-731, Atlanta, 2-6 May 1993.

[MCC 79] H. MCCALLION and D.T. PHAM. "The analysis of a six degrees of freedom work station for mechanized assembly", in *Proc. 5th World Congress on Theory of Machines and Manipulators*, pp. 611-616, Montreal, July 1979.

[MER 89] J.-P. MERLET. "Singular configurations of parallel manipulators and Grassmann geometry", *Int. J. of Robotics Research*, 8(5):45-56, October 1989.

[MER 91] J.-P. MERLET and C. GOSSELIN. "Nouvelle structure pour un manipulateur parallèle à 6 degrés de liberté", *Manipulator and Machine Theory*, 26(l):77- 90, 1991.

[MER 94] J.-P. MERLET and N. MOULY. Espaces de travail et planification de trajectoire des robots parallèles plans. Technical Report 2291, INRIA, February-October 1994.

[MER 97] J.-P. MERLET. "First experiments with MIPS 1 (Mini In-Parallel Positioning System)", in *ISER*, pp. 372-379, Barcelona, 15-18 June 1997.

[MER 98a] J.-P. MERLET. "Efficient estimation of the external articular forces of a parallel manipulator in a translation workspace", in *IEEE Int. Conf. on Robotics and Automation*, pp. 1982-1987, Louvain, 18-20 May 1998.

[MER 98b] J.-P. MERLET. "Determination of the presence of singularities in 6D workspace of a Gough parallel manipulator", in *ARK*, pp. 39-48, Strobl, 29 June-4 July 1998.

[MIN 72] M. MINSKY. Manipulator design vignettes. Technical Report 267, MIT AI Lab., 1972.

[MUR 96] A.P. MURRAY, P. PIERROT, P. DAUCHEZ and J.M. McCARTHY. "On the design of parallel manipulators for a prescribed workspace: a planar quaternion approach", in V. Parenti-Castelli J. Lenarcic, Eds., *Recent Advances in Robot Kinematics*, pp. 349-357. Kluwer, 1996.

[NAB 05] V. NABAT, O. COMPANY, S. KRUT, M. RODRIGUEZ, F. PIERROT. "Par4: Very High Speed Parallel Robot for Pick-and-Place", in *Proc. of IEEE/RSJ IROS*, CD-ROM, 2005.

[NEU 88] K.E. NEUMANN. Robot, 22 March 1988. United States Patent no. 4,732,525, Neos Product HB Norrtalje Sweden.

[PIE 91] F. PIERROT. "Robots pleinement parallèles légers: conception modélisation et commande". PhD Thesis, Montpellier II University, 24 April 1991.

[PIE 99] F. PIERROT and C. COMPANY. "H4: a new family of 4-dof parallel robots", in *IEEE/ASME International Conference on Advanced Intelligent Mechatronics*, pp. 508-513, 19-22 September 1999.

[POL 42] W.L.V. POLLARD. Position controlling apparatus, 16 June 1942. United States Patent no. 2,286,571.

[POW 82] I.L. POWELL. "The kinematic analysis and simulation of the parallel topology manipulator", *The Marconi Review*, XLV(226):121-138, Third Quarter 1982.

[RAG 91] M. RAGHAVAN. "The Stewart platform of general geometry has 40 configurations", in *ASME Design and Automation Conf.*, volume 32-2, pp. 397-402, Chicago, 22-25 September 1991.

[REB 90] C. REBOULET and R. PICEYRE. "Hybrid control of a 6 dot in parallel actuated micro-macro manipulator mounted on a Scara robot", in *ISRAM*, vol. 3, pp. 293-298, Burnaby, 18-20 July 1990. ASME Press Series.

[ROJ 72] P.J. ROJESKI. "A systems analysis approach to landing gear design". PhD Thesis, Cornell University, May 1972.

[ROL 99] L. ROLLAND. "The manta and the kanuk: Novel 4-dof parallel manipulators for industrial handling", in *IMECE'99 Conference*, volume 67, pp. 831-844, 14-19 November 1999.

[RON 92] F. RONGA and T. VUST. *Stewart platforms without computer?* 1992. Preprint.

[ROU 95] F. ROUILLIER. "Real roots counting for some robotics problems", in B. Ravani J.-P. Merlet, Eds., *Computational Kinematics*, pp. 73-82. Kluwer, 1995.

[SEF 94] J. SEFRIOUI and C. GOSSELIN. "Étude et représentation des lieux de singularités des manipulateurs parallèles sphériques à trois degrés de liberté avec actionneurs prismatiques". *Manipulator and Machine Theory*, 29(4):559-579, 1994.

[SEP 95] J. SEFRIOUI and C.M. GOSSELIN. "On the quadratic nature of the singularity curves of planar three-degree-of-freedom parallel manipulators". *Manipulator and Machine Theory*, 30(4):533-551, May 1995.

[STE 65] D. STEWART. "A platform with 6 degrees of freedom", *Proc. of the Institution of Mechanical Engineers*, 180(Part 1, 15):371-386, 1965.

[TAN 99] L. TANCREDI and M. TEILLAUD. "Application de la géométrie synthétique au problème de modélisation géométrique direète des robots parallèles", *Manipulator and Machine Theory*, 34(2):255-269, February 1999.

[TSA 96] L-W. TSAI. "Kinematics of a three-dof platform with three extensible limbs", in V. Parenti-Castelli J. Lenarcic, Eds., *Recent Advances in Robot Kinematics*, pp. 401-410. Kluwer, 1996.

[VEB 10] O. VEBLEN and J.W. YOUNG. *Projective geometry*. The Athenaeum Press, 1910.

[WEN 94] J.M. WENDLANDT and S.S. SASTRY. "Design and control of a simplified Stewart platform for endoscopy", in *33rd Conf. on Decision and Control*, pp. 357-362, Lake Buena Vista, 14-16 December 1994.

[WU 92] K.C. WU and T.R. SUTTFR. Structural analysis of three space crane articulated truss joint concepts. Technical Report TM-4373, NASA Research Center, Langley, May 1992.

[YOS 82] T. YOSHIKAWA. "Manipulability of robotic manipulators", *Int. J. of Robotics Research*, 1(l), 1982.

[YUF 95] L. YUFFNG and Y. TINGLI. "Structure types and characteristics of six degree-of-freedom closed-chain manipulators with all actuators on base", in *9th World Congress on the Theory of Machines and Manipulators*, pp. 1795-1799, Milan, 30 August-2 September 1995.

[ZAM 92] V.B. ZAMANOV and Z.M. SOTIROV. "Parallel manipulators in robotics", in *IMACSISICE Int. Symp. on Robotics, Mechatronics, and Manufacturing Systems*, pp. 409-418, Kobe, 16-20 September 1992.

Chapter 3

Performance Analysis of Robots

3.1. Introduction

The physical integration of a robot in its environment must guarantee its optimal operation. That is why it is necessary to have relevant tools for assessing the performance of a robot. Consequently, the performance evaluation enables:

– the choice of the best robots for certain tasks;

– their ideal site position;

– the calculation of sure and optimal trajectories with respect to some criteria (minimal cycle time, minimal torque actuators, maximum dexterity, etc.).

We note that these three problems are essentially geometric and kinematic in nature. As such, they must be treated through reliable analyses of the geometric and kinematic performances of robots.

In particular, access to the working points must be guaranteed. A simple solution consists of guaranteeing the inclusion of these points in the robot's work volume envelope. For example, for a planar robot with two revolute joints of parallel axes, the points must be within a disc with a radius L1 + L2, where L1 and L2 represent the two link lengths of the robot. However, this analysis is most often insufficient because it is necessary to consider the effect of the joint limits, as well as the proximity of obstacles. Figure 3.1 shows two different accessibility analyses for the planar robot. The first analysis (Figure 3.1a) shows the envelope of accessible

---

Chapter written by Philippe WENGER.

points, without considering the joint limits. Hence, all points seem accessible. Figure 3.1b takes into account the joint limits ($-120° \leq \theta_1 \leq 100°$, $-120° \leq \theta_2 \leq 150°$, the angles being directed trigonometrically): the workspace is crescent-shaped, and points A and E are not accessible. When there are obstacles in the site, the shape of the workspace becomes very complex.

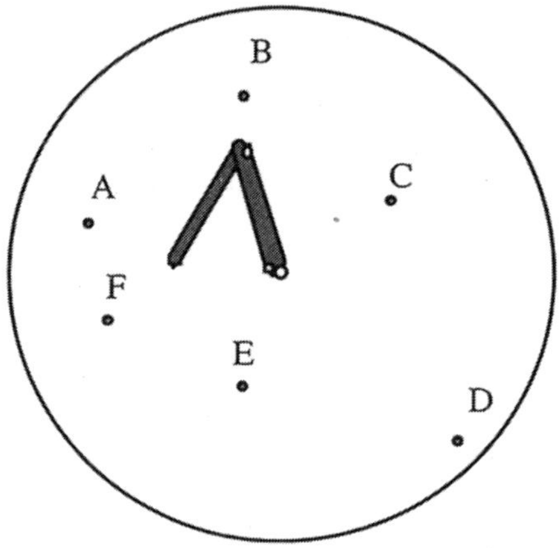

a) without joint limits

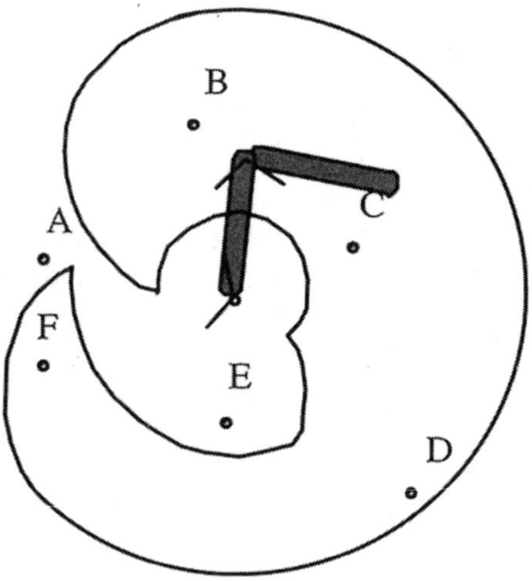

b) with joint limits

Figure 3.1. *Workspaces for a planar robot*

In the following sections, we will show that a simple analysis of accessibility, even considering the joint limits and the obstacles, can prove to be insufficient to assess the movement aptitude of a robot.

Robot performances are initially analyzed through global concepts of accessibility and connectivity. The basic concept of aspect is also pointed out. Then, the local analyses of dexterity, approach lengths and angles are also presented.

The dynamic performances of robots (acceptable accelerations, available torques, etc.) will not be dealt with in this chapter.

3.2. Accessibility

3.2.1. *Various levels of accessibility*

Accessibility is the aptitude of a robot manipulator to partially or fully align a reference frame R_e connected to its end-effector with a target reference frame R_c of the environment.

Generally, a full alignment of two reference frames is possible only for the robots with at least six degrees of freedom (Figure 3.2).

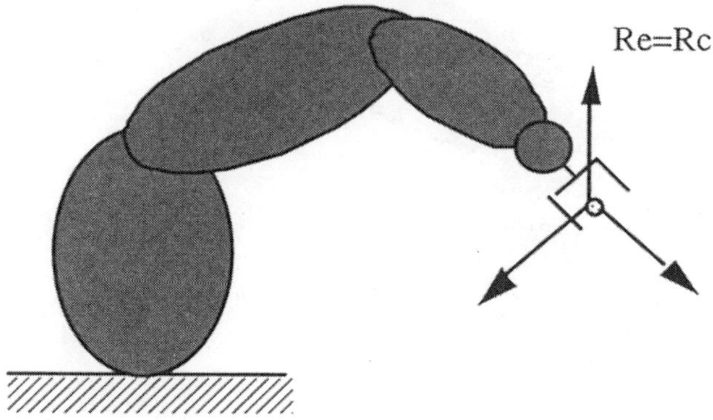

Figure 3.2. *Accessibility to a reference frame with full alignment*

The accessibility in position is defined as the coincidence of the origins of reference frames (see Figure 3.3), the orientation being ignored. In this case, only the origins of reference frames O_e and O_c coincide.

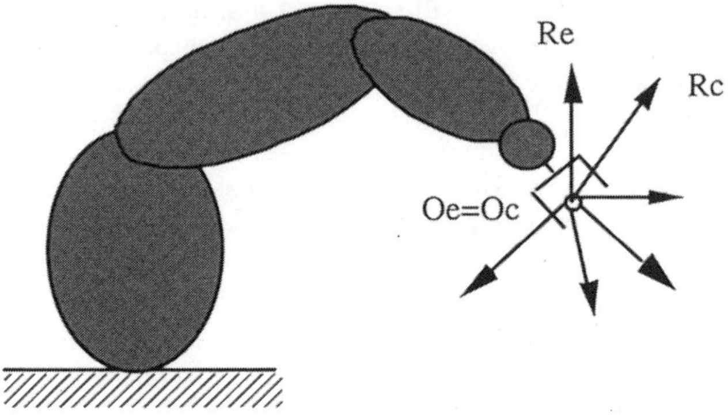

Figure 3.3. *Accessibility in position*

More generally, accessibility can be defined as the coincidence of some position coordinates, and some orientation coordinates. For example, in the absence of a joint tracking sensor, accessibility to a set of reference frames describing an arc welding trajectory will be defined as the coincidence of the center and axis z of these reference frames with the center and axis z of the reference frame connected to the welding torch (see Figure 3.4).

Hence, accessibility depends on the definition of the task.

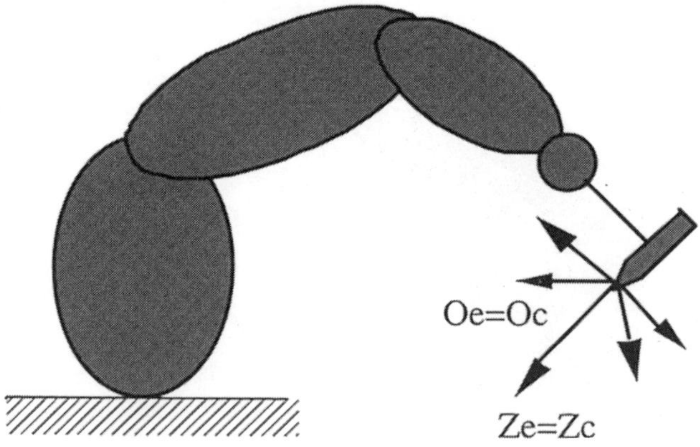

Figure 3.4. *Accessibility in arc-welding tasks*

3.2.2. *Condition of accessibility*

Let **X** be a vector of operational coordinates describing the task. Accessibility to **X** exists when there is a vector **q** of the robot's joint coordinates such that the volume occupied by the robot in the **q** configuration and the volume occupied by all the obstacles of the environment are disjoined. The various causes of non-accessibility are the following (see Figure 3.5):

– vector **X** is of higher dimension than **q** and **X** is badly conditioned: there is no solution to equation **X** = **f(q)**. For example, **X** = $[x,y,z]^t$ and the robot is planar with two joints. If $z \neq 0$ ($z = 0$ being the plane of the robot), **X** is never accessible, irrespective of x and y (see Figure 3.5a). On the other hand, the points **X** = $[x,y,0]^t$ are accessible if (x,y) is within the workspace;

– **X** is too far: there is no solution to the equation **X** = **f(q)** (see Figure 3.5b);

– one or more joint limits are violated: the solutions to equation **X** = **f(q)** are not in the accessible joint domain (see section 3.3.1), which is the case of points A and E in Figure 3.1;

– the robot is in collision with an obstacle of the environment (Figure 3.5c).

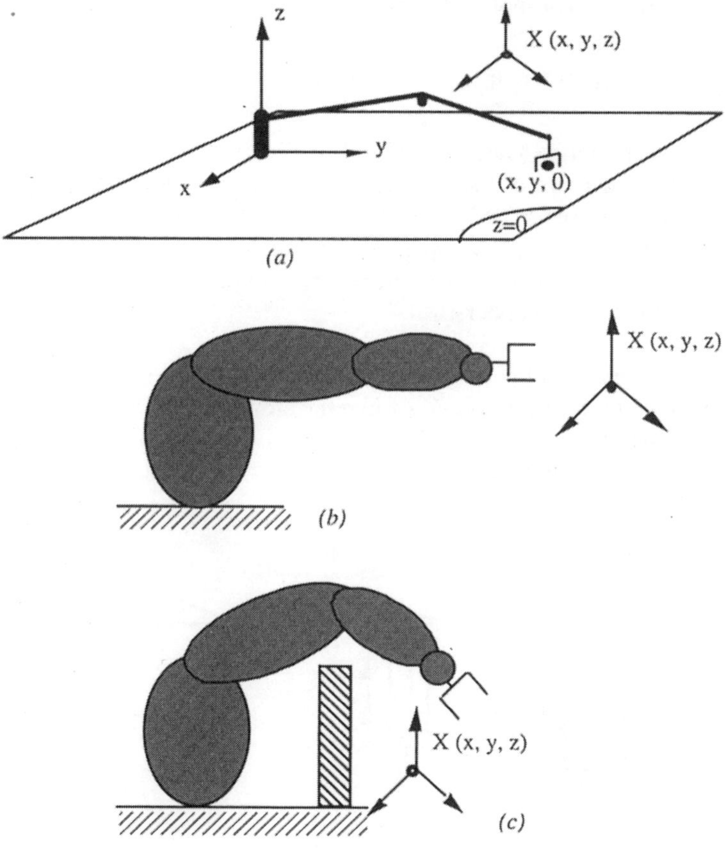

Figure 3.5. *Various causes of non-accessibility*

The accessibility performances of a robot manipulator thus depend on its architecture (the number, the nature and layout of links, link lengths), on its joint limits and on the nature of its environment. When there is no obstacle in the environment of the robot (which is rare), accessibility can be globally defined by the workspace of the robot, or one of its projections. When the collisions between all the robot's parts and the obstacles of its environment must be considered, accessibility can be globally defined by the free workspace of the robot (see section 3.3).

3.3. Workspace of a robot manipulator

3.3.1. *General definition*

The workspace of a robot manipulator is defined as the set of accessible positions and orientations through a particular reference frame connected, in general, to the end-effector of the robot. Let Q be the accessible joint domain:

$$Q = \{ \mathbf{q} \in EA_n, \forall i \leq n, q_{i\,min} \leq q_i \leq q_{i\,max} \} \quad [3.1a]$$

Q represents the set of vectors of the robot's joint coordinates which respect the joint limits. It is a domain of the EA_n joint space (also called configuration space). The structure of Q depends on that of the space EA_n, which itself depends on the architecture of the robot. For example, for a robot with two revolute joints and one prismatic joint, Q is a domain of $T^2 \times R$, where T^2 indicates the classic torus, and R is the straight line of real numbers.

However, for reasons of convenience, we often assimilate Q to a set of R^n. This assimilation does not create ambiguity when the amplitude of the joint limits on the revolute links does not exceed 2π. Thus, in the previous example, we can represent Q as a parallelepiped (Figure 3.6).

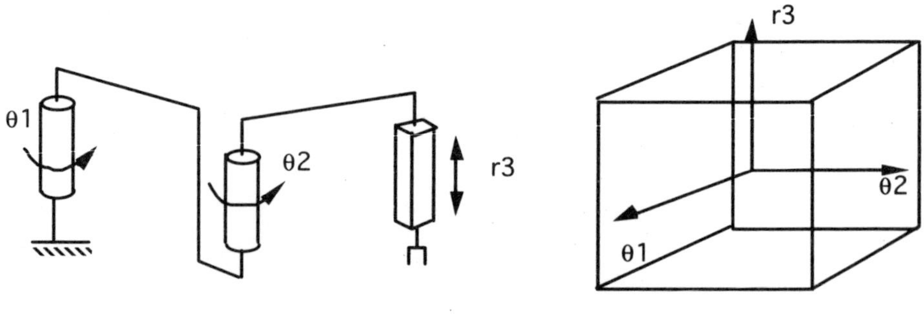

Figure 3.6. *Joint domain for a robot with two revolute joints and one prismatic joint*

The workspace W can be defined by the image of Q by the geometrical mechanical system of the robot:

$$W = f(Q) \quad [3.1b]$$

The workspace is a domain of the operational space. Its structure thus depends on that of the operational space. In the general case, W is a domain of the space $R^3 \times SO(3)$, which represents the product of the space of the position coordinates of R^3 with the group of natural rotations of R^3. W is then 6-dimensional, but its dimensions can be lower when only a part of the positions and orientations of the robot end-effector must be analyzed. For example, if we want to analyze accessibility in position, the workspace is included in R^3 and it is 3-dimensional.

However, it should be noted that the geometry of the workspace depends on the position of the reference frame connected to the robot end-effector. For the current robots with six joints and having a spherical joint, the workspace is defined, generally, by a reference frame connected either to the center of the spherical joint, or to the center of the flask (a generally cylindrical piece which is used as interface between the last segment of the robot and the tool), or in a functional point of the tool (center of the gripper, end of the welding torch, etc.). For the anthropomorphic and spherical robots, the workspace, when defined in position, is located in a sphere, whose radius corresponds to the maximum elongation of the robot. In the case of cylindrical robots, it is located in a cylinder, whose radius and height depend on the limits on the prismatic joints.

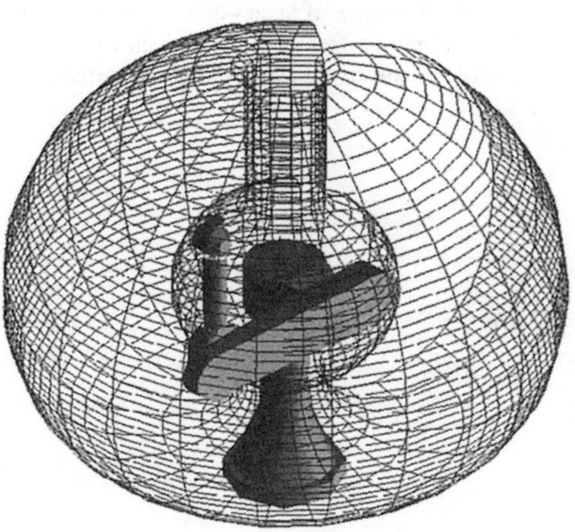

Figure 3.7. *Workspace for an anthropomorphic robot*

Figure 3.7 shows the workspace of an anthropomorphic robot. The workspace is defined here by the set of positions of a reference frame linked to the center of the spherical joint of the robot. It is thus 3-dimensional. The workspace is clearly located in a sphere. The lacking angular sector is due to the angular displacement of the first link which is lower than 360°.

As it is difficult to handle 6-dimensional spaces, the complete evaluation of the workspace is classically carried out by determining its three projections in the Cartesian space [KUM 81]: the space of accessible positions, the primary space and the secondary space. These projections are obtained by decoupling the space of positions from the space of orientations. Apart from these projections, the workspace can be studied for fixed orientations, where we would be talking about sections. These projections and sections are presented hereinafter.

3.3.2. *Space of accessible positions*

The space of accessible positions W_{R^3} ("reachable workspace") is the set of positions which the robot end-effector can reach for at least one orientation [KUM 81]. If $EO_m = R^3 \times SO(3)$, it is defined as the projection of W on R3. Let π_{R^3} be the projection of EO_m on R^3. Let **P** be a point linked to the robot end-effector. Hence, W_{R^3} can be written:

$$W_{R^3}(\mathbf{P}) = \pi_{R^3}\,[\mathbf{f}(Q)] \qquad [3.2]$$

Figure 3.7 represents also the space of accessible positions for the complete robot. In fact, since the selected reference frame is located in the center of the spherical joint, it records all accessible positions by at least one orientation of the wrist.

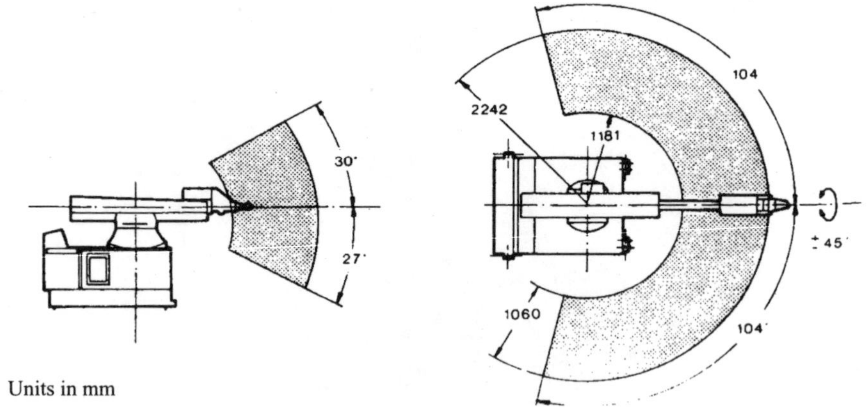

Figure 3.8. *Space of accessible positions represented by sections (spherical robot Unimate 1000)*

The space of accessible positions is the one which is used the most. Manufacturers usually provide it by sections (Figure 3.8).

For a robot which does not have joint limits on its first joint, its space of accessible positions is symmetrical about the first joint axis. It is then possible to represent only one generating section of the space of accessible positions. The overall space can be obtained either by rotating this section around the axis of the first joint of the robot, provided the joint is revolute, or by sliding the section along this axis if the joint is prismatic. For a robot whose first joint is revolute, the generating section is defined in the plane ($\rho = \sqrt{x^2 + y^2}$, z) [YAN 83]. When the first joint is prismatic, the generating section is a parallel projection in the plane (x,y). In practice, industrial robots always have limited range of motion on the first joint and the generating section does not necessarily reflect the structure of the workspace as a system because of boundary losses.

3.3.3. *Primary space and secondary space*

The primary space W_p is the set of positions of W_{R^3} that can be obtained for all possible orientations [KUM 81]. It can be defined by the following formula:

$$W_p(\mathbf{P}) = [\pi_{R^3}[\mathbf{f}(Q)]^c]^c \qquad [3.3a]$$

where exponent "c" indicates the complementary element.

Indeed, we start from the entire workspace $W = \mathbf{f}(Q)$. We project its complementary element $[\mathbf{f}(Q)]^c$ on R^3. We then obtain the set of positions which are not accessible for every orientation, or which are not accessible at all. So, it is enough to consider the complementary element and we obtain the set of accessible positions for all the orientations of the robot end-effector. Figure 3.9 synthesizes this principle.

In [VIJ 86], the authors show that the primary space of a anthropomorphic robot without joint limits is as represented in Figure 3.10. This shows a generating section of the workspace for an anthropomorphic robot, whose two lengths of arm are L1 and L2 respectively (L1 > L2) and whose distance between the control point and the center of the spherical joint is L3 (L3 < L1 – L2).

We show that the primary space (hatched) is limited by the spheres of respective radii L1 – L2 + L3 and L1 + L2 – L3.

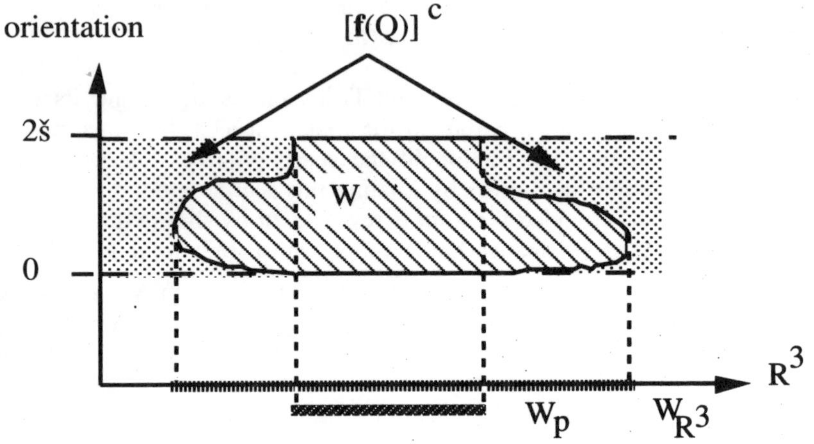

Figure 3.9. *Obtaining the primary space and the space of accessible positions*

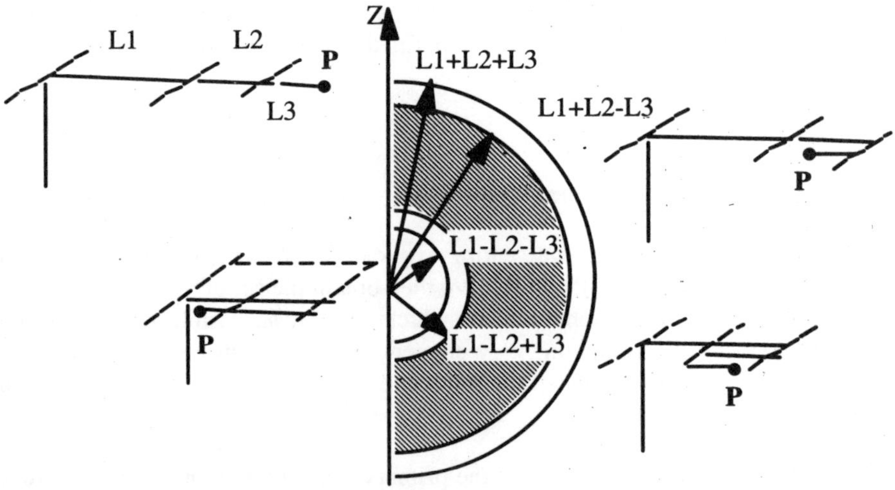

Figure 3.10. *Primary space*

Generally, industrial robots have joint limits and their primary space is often zero. However, in [VIJ 86], the author shows that, under certain conditions, the primary space is unchanged when all joint coordinates are without limits and when some of them have an amplitude of only 180°.

The secondary space W_s is the set of positions of W_{R^3} where orientation is limited. It is the difference between the space of accessible positions and the primary space:

$$W_s(\mathbf{P}) = W_{R^3} - W_p \qquad [3.3b]$$

3.3.4. *Defined orientation workspace*

For certain applications, the tool must preserve the same orientation with respect to a given surface. This is the case, for example, for projection operations (painting, cement, adhesives, cleaning, etc.), arc welding, finishing (fettling, polishing, etc.), or surface inspection (detection of cracks). It is then interesting to study the defined orientation workspace. Contrary to the accessible position space, the defined orientation workspace is not a projection of the workspace W, but rather a section of W. In particular, [YAN 86] is interested in the geometrical and topological characteristics of the defined orientation workspace of a robot with six degrees of freedom on a particular plane when the end-effector is perpendicular to this plane. In this case, the defined orientation workspace is a set of areas of the plane that the authors call work areas. They highlight the existence of sub-areas in these zones: a sub-area is a domain of the work areas, accessible with a given posture of the manipulator (for example elbow up or elbow down; see section 3.4). They implement a procedure to detect these sub-areas, and also their various connected components (Figure 3.11).

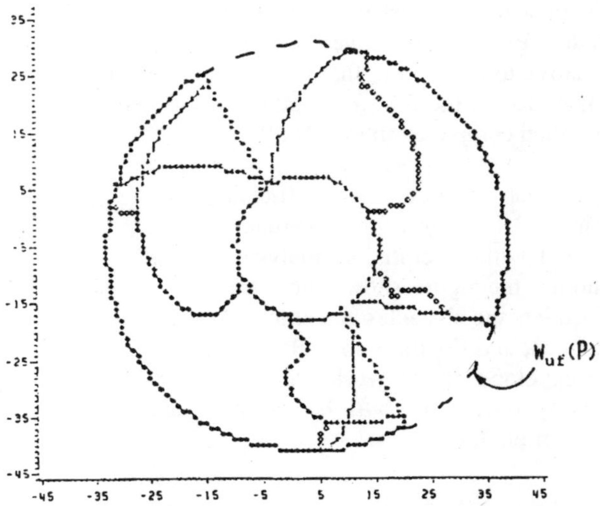

Figure 3.11. *Work areas for a PUMA robot working on the plane X + Y – E = 15 cm with z6 perpendicular to the plane (according to [YAN 86])*

More generally, we can study the defined orientation workspace, which is not, however, necessarily constant. In the case of painting on a general surface, the orientation is given (perpendicular to the surface), but it is not constant if the surface is non-planar.

In the previous examples, the defined orientation workspace was calculated along a surface. For other applications, the defined orientation workspace can be calculated in R^3, for example for handling applications for which the parts must be transported in a certain orientation.

3.3.5. *Free workspace*

When the environment of a robot incorporates obstacles, which is often the case, the concept of workspace is no longer sufficient to characterize the accessibility performances of the robot. It then becomes necessary to consider the collisions between the various parts of the robot and all the obstacles.

Few authors precisely studied the workspace of a robot considering the obstacles of the environment. For a long time, it has been considered sufficient to determine the free areas of the environment by eliminating the obstacles occupying the workspace [HAY 86], [BRO 83a, 83b]. However, this approximation is only valid when the end-effector is the only segment of the robot likely to collide with the obstacles (see example below). Some authors had noticed that the obstacles could generate "shadows". The shadow of an obstacle is defined as the inaccessible zone of the environment where at least one part of the robot collides with this obstacle [LOZ 87]. This definition was improved by [SHI 89] who calculated the shadows of each obstacle relative to a position, that is, to a solution for the inverse geometrical model. To our knowledge, the first formulation of the free workspace was proposed in [CHE 88], and then completed in [WEN 89].

The workspace can be modified significantly by the obstacles occupying the environment. Figure 3.12 shows the workspaces of the planar robot in Figure 3.1, calculated with and without collision analysis. In Figure 3.12a, the effect of the obstacles is ignored: the figure shows the workspace, as presented in section 3.1. Figure 3.12b considers the obstacles and shows the free workspace: the workspace is significantly reduced, and the only obstacle present is of small size and is not even included in the envelope of the work volume. Finally, only points **B** and **F** are accessible (in reality, only B or F will be accessible according to the initial position of the robot (see section 3.5)).

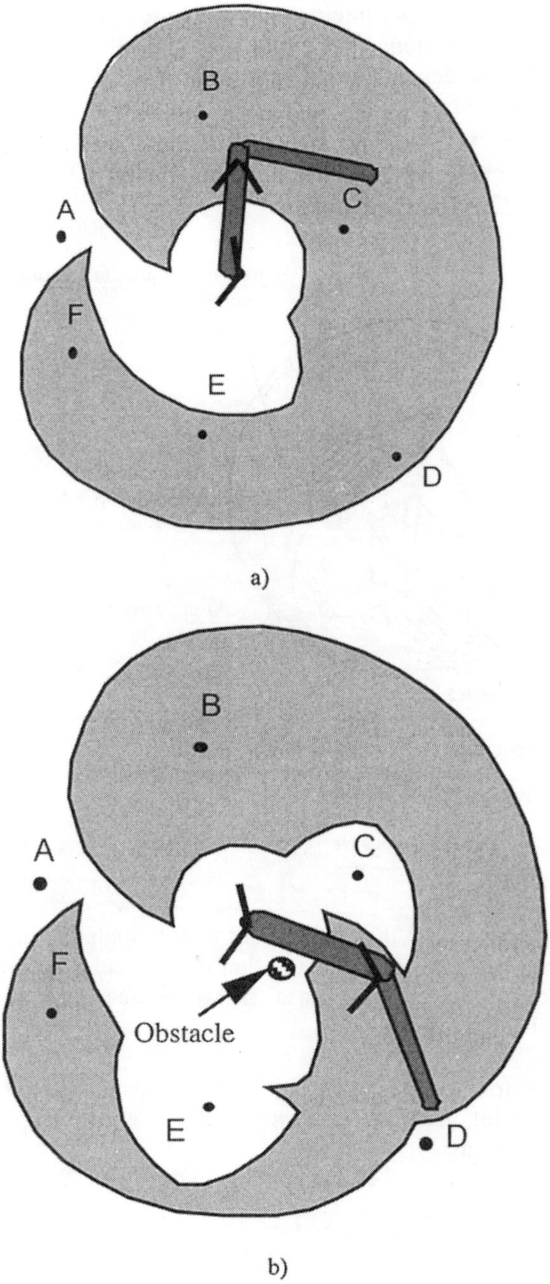

Figure 3.12. *Workspace (a) and free workspace (b)*

In this simple case, we can interpret the influence of obstacles. The inaccessible areas are due to the collisions of the first arm with the obstacle. These collisions generate a joint sector denied for the first joint. This denied sector generates two swept areas corresponding to the two positions "elbow up" and "elbow down" (Figure 3.13). However, a part of each one of these scanning areas is covered by configurations accessible by the other position, finally forming a system of three small inaccessible areas (hatched areas in Figure 3.13).

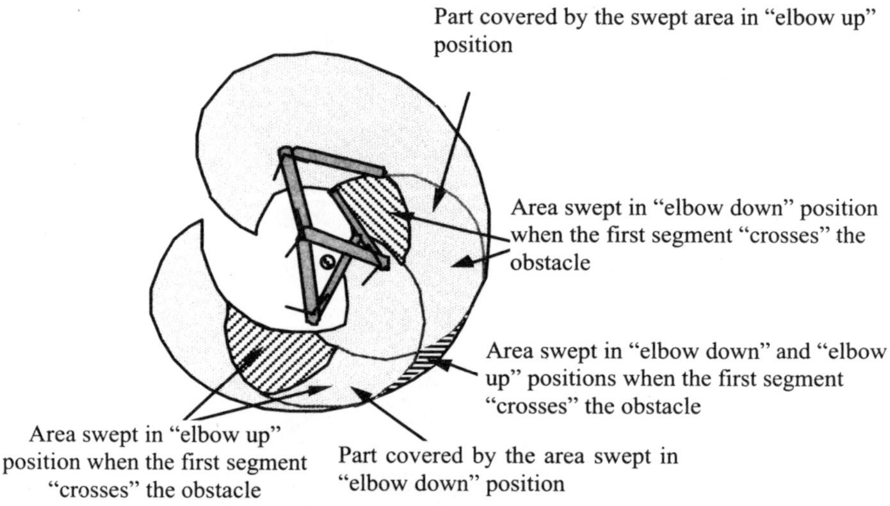

Figure 3.13. *Interpretation of inaccessible zones*

In a more general case, it is very difficult to anticipate the influence of obstacles on the workspace. We can however note that the obstacles that obstruct the robot's links located close to the base decrease the workspace more than the robot's arm located close to the end-effector.

The concept of free workspace, using the collisions between the robot and all the obstacles of its environment, can be formulated in the following way. Let Ql be the set of free configurations:

$$Ql = \{\mathbf{q} \in Q / Rob(\mathbf{q}) \cap OBS = \varnothing\} \quad [3.4]$$

where Rob(**q**) indicates the volume of the robot in the configuration **q** and OBS the volume occupied by all the obstacles. The free workspace Wl is defined by the image, through the geometric operator **f** of the robot, of the set of free configurations:

$$Wl = \mathbf{f}(Ql) \qquad [3.5]$$

The free workspace can thus be obtained on the basis of the set of free configurations.

NOTE.– the software of CAO-Robotics do not enable a visualization of the workspace that considers the obstacles. A general method of calculation of the free workspace is given in [CHE 98].

3.3.6. *Calculation of the workspace*

It is not possible to represent spaces of dimension higher than three. The workspace, most of the time, is thus determined by the space of the accessible positions (see section 3.3.2) or by the workspace generated by the first three joints of the robot.

If the current systems of CAO-Robotics do not make it possible to represent the workspace of any robot, this is undoubtedly because there are no methods which are at the same time reliable, simple and sufficiently fast. In addition, we are always confronted with the double problem of the calculation time (especially for the calculation of the free workspace) and of the data storage necessary for a sufficiently fine modeling of the workspace. The methods of workspace calculation can be gathered into two main categories: those which consist of calculating the limits of the workspace (the workspace is then represented by its envelope) and those which proceed from scanning (workspace is then generally represented by its volume). Table 3.1 synthesizes the various existing methods. They are detailed in [CHE 98].

Method	References	Applications	Limit obstacles considered		Observations
Limits					
sing. esp. joint	[KOH 87], [BUR 95]	robots RRR	no	yes	can be applied to > 3R based on [BUR 95]
	[ELO 96]	robots XXX	axes 2, 3	no	
sing. Cartesian space	[RAN 94]	robots RRR	no	no	less heavy than [TSA 93]
	[BAI 04], [TSA 93]	robots XXX	no	no	
toroidal surfaces	[CEC 89], [CEC 95]	robots 3-6R	no	no	does not apply to the robots with axes //
Scanning					
in the Cartesian space	[DOM 82]	usual robots	yes	no	by cutting plan
	[SHI 89]	planar robots	yes	yes	regular grid
	[LAL 94]	planar robots	yes	yes	adapted scanning
	[LUE 92]	spatial robots	yes	yes	regular grid high memory storage
in the joint space	[WEN 89]		yes	yes	scanning optimized mod. quadtree/octree
by aspects	[BOR 86], [GOR 94]	usual robots	yes	no	

Table 3.1. *Synthesis of the principal methods for workspace calculation*

3.4. Concept of aspect

3.4.1. *Definition*

The concept of aspect was introduced by Paul Borrel in 1986 [BOR 86]. Let m be the number of independent operational coordinates used to describe the position of the end-effector. An aspect A is a connected part of the accessible joint domain Q on which none of the minors of order m from the Jacobian matrix is zero, except if this minor is zero everywhere on Q. For non-redundant robots, the aspects are the highest open domains (connected) of Q not having singular configurations.

These aspects create a partition of Q. For conventional robot architectures, which account for the majority of the industrial robots, the aspects are the domains of solution uniqueness for the inverse geometric model. An aspect is thus associated with a posture of the robot. However, for less ordinary robots, there may be several solutions in the same aspect (such robots are referred to as "cuspidal robots"; the interested reader may refer to [WEN 04] for more details on this subject).

Figure 3.14 shows the two aspects of a planar robot with two revolute joints with joint limits (the latter was already presented in Figures 3.1 and 3.12). The first aspect is associated with posture "elbow up", whereas the second corresponds to the configurations of posture "elbow down". The figure shows the two aspects in the joint space (a) and their images in the Cartesian space (b) (the hatchings show the correspondence). We notice an overlapping of the two images thus defining an area where the number of postures is 2.

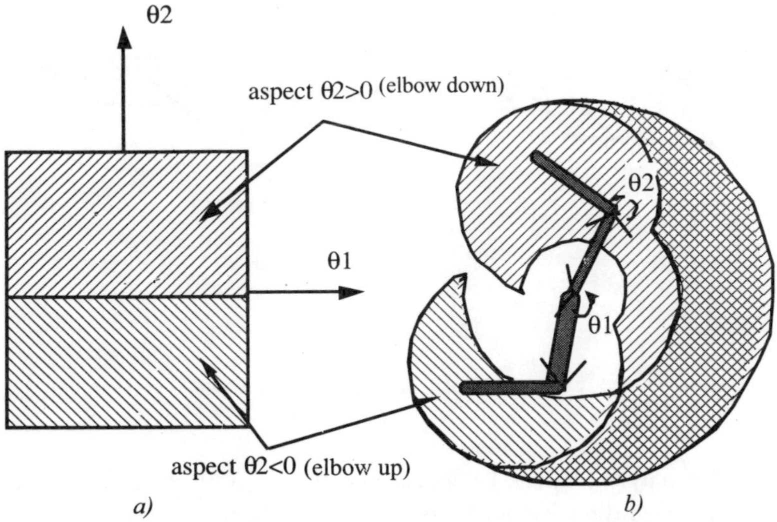

Figure 3.14. *Aspects and postures for a planar robot*

For a given robot, the number of aspects is difficult to anticipate. It depends not only on the shape of the singularities in the joint domain, but also on the values of joint limits.

A concept close to that of aspect was introduced later by Burdick [BUR 88] and [PAR 88], as the "c-sheets" and the "regions" respectively. The "c-sheets" and the "regions" are defined like the aspects, but only for the non-redundant robots with revolute joints and without limits. Contrary to the aspects whose form and number depend also on the joint limits, the "c-sheets" and the "regions" are related only to the architecture of the robot.

3.4.2. *Mode of aspects calculation*

For current robots, the determinant of the Jacobian matrix can be written in a factored form. The aspects are then identified by the combination of signs of the various factors. They are separated by the singularity hypersurfaces defined by the factors and are bordered by the joint limits. For example, for anthropomorphic robots, the determinant can be written in the following way:

$$\det({}^{3}\mathbf{J}_{6}) = -\, C3\ D3\ RL4\ S5\ (S23\ RL4 - C2\ D3) \qquad [3.6]$$

where D3 and RL4 are the parameters of DHM. In this case, D3 is the length of the first arm and RL4 is that of the second arm (Figure 3.15).

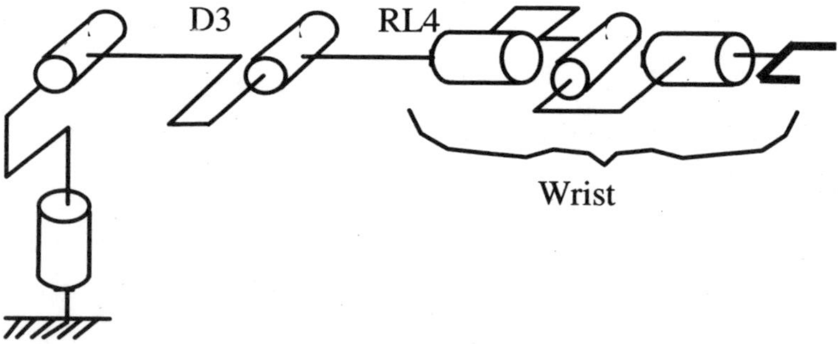

Figure 3.15. *An anthropomorphic robot*

We see that only θ_2, θ_3 and θ_5 appear in equation [3.6]; the aspects can thus be represented in three dimensions. Note that this is always the case as soon as the robot has a spherical wrist [KHA 98].

Figure 3.16 shows that the joint domain is decomposed into 12 aspects[1], separated by the singularity hypersurfaces defined respectively by:

$$C3 = 0 \qquad [3.7]$$

$$S23\ RL4 - C2\ D3 = 0 \qquad [3.8]$$

$$S5 = 0 \qquad [3.9]$$

These equations correspond to the three singularities of the anthropomorphic robot: equation [3.7] describes the arm singularity, equation [3.8] represents the shoulder singularity (represented by a plane in Figure 3.16), and finally [3.9] defines the wrist singularity.

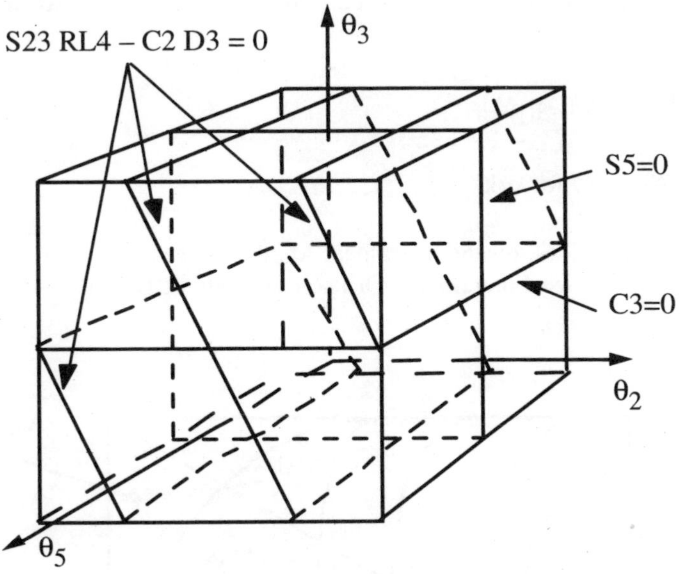

Figure 3.16. *The aspects of an anthropomorphic robot*

1 In the presence of joint limits, the joint domain has the structure of a parallelepiped. In the opposite case, the joint domain has the structure of a torus, and the opposite sides of the parallelepiped in Figure 3.16 will have to be identified and there will be only eight aspects.

3.4.3. *Free aspects*

The concept of aspect introduced by Borrel does not consider the presence of obstacles. The free aspects generalize the aspects for robots moving in the presence of obstacles [CHE 87a], [WEN 89]. Their formalization is necessary for the feasibility study of continuous trajectories in the presence of obstacles.

The free aspects Al_j are defined on the basis of the set of free configurations Ql (see section 3.3.5):

$$\forall j \in J,\ Al_j = A_j \cap Ql \tag{3.10}$$

where the set J describes the indices of the various aspects.

Contrary to the aspects defined by Borrel [BOR 86], the free aspects are not necessarily connected. We define the partition of a free aspect Al_j in connected components Al_{jk}:

$$\forall j \in J,\ Al_j = \cup_{k \in I(j)} Al_{jk} \tag{3.11}$$

with $Al_{jk} \cap Al_{jk'} = \varnothing$ for $k \neq k'$

The set I(j) describes, for the free aspect Al_j, the indices of the various connected components of Al_j (see Figure 3.17; Ql itself consists of two connected parts Ql_1 and Ql_2).

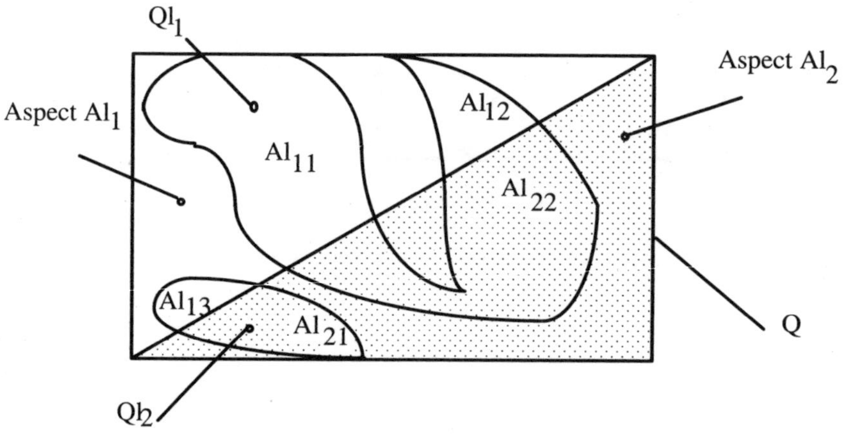

Figure 3.17. *Representation of the connected components of the free aspects*

To calculate the free aspects, it is necessary to begin with the aspects and to remove the domains which generate collisions. For that, we can use for example the method of Faverjon [FAV 84]. This method makes it possible to model the set of free configurations with an octree. Instead of enriching only one octree that represents the set of free configurations, we enrich as many octrees as there are free aspects. The identification of the free aspects is done using the signs of the determinant factors of the Jacobian matrix, as explained in section 3.4.2. The calculation of the connected components of the free aspects is done easily with the help of octrees.

EXAMPLE 3.1.– Figure 3.18 shows the free aspects and their image in the Cartesian space (the correspondence is defined by the fill patterns) for the planar robot moving in the presence of a small obstacle, already presented in Figure 3.12 (section 3.3.5). Here, the free aspects can be obtained easily since their limits consist of segments of straight lines only. We see that the presence of the obstacle generates a "cutting" in the joint domain, which divides each free aspect into two connected components. The free aspect Al_1, associated with the posture "elbow down" (respectively Al_2, associated with the posture "elbow up") consists of two connected parts Al_{11} and Al_{12} (respectively Al_{21} and Al_{22}).

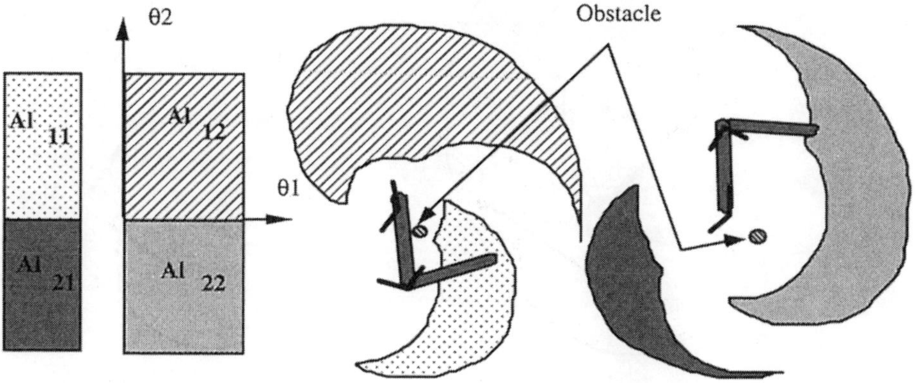

Figure 3.18. *Free aspects for a planar robot moving in the presence of an obstacle*

3.4.4. *Application of the aspects*

The main application of the aspects is the analysis of continuous trajectories in the workspace. Indeed, Borrel shows that the decomposition into aspects makes it possible to anticipate joint locking on a joint limit during motion, which can be avoided by a judicious choice of the initial posture on the continuous trajectory. Figure 3.19 illustrates this result for a planar robot with two revolute joints. This robot has two aspects, corresponding to the configurations $\theta_2 > 0$ (posture elbow down) and $\theta_2 < 0$ (posture elbow up) respectively. The hatched area shows the set of points reachable in the aspect $\theta_2 > 0$. The non-hatched area represents the set of points only reachable in the other aspect. The trajectory **AB** to follow starts in region reachable in the two aspects and stops in a region accessible only in the aspect corresponding to the posture "elbow up". The trajectory can thus be entirely followed only if the initial configuration is chosen in the aspect associated with the posture "elbow up" (on the right). On the left, the robot starts from the posture "elbow down", and locks on a joint limit before reaching the final point of the straight trajectory. The joint can be unlocked by changing the posture, which means changing the aspect. However, this makes the robot leave the trajectory because the transition between the two aspects is the fully "outstretched" singular configuration, accessible only on the external limit of the workspace, and this is impossible for continuous trajectories like cutting or arc welding. The solution here consists of starting from the initial posture "elbow up".

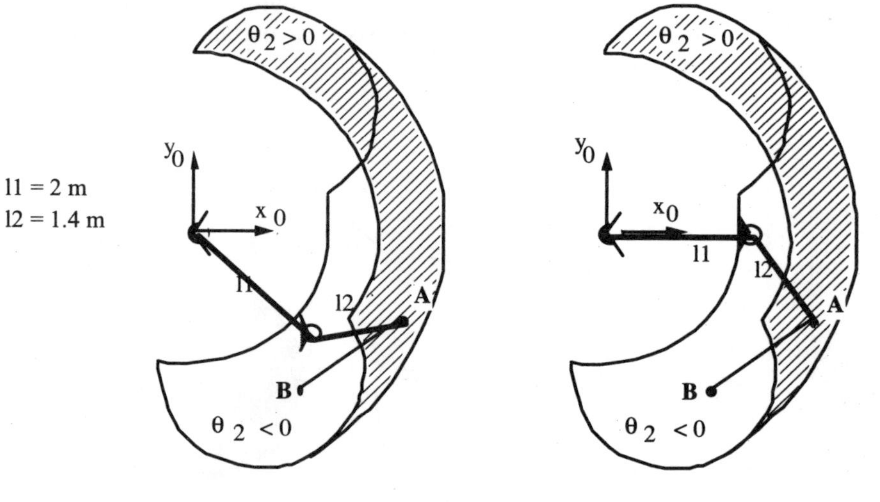

Figure 3.19. *Analysis of continuous trajectories (from [BOR 86])*

The decomposition into aspects also enables the reconstruction of the workspace of industrial robot manipulators The workspace is then obtained by combining, in the operational space, the images of each aspect. An algorithm was proposed by Borrel [BOR 86] based on the software CATIA-Robotics. This algorithm was used later by [GOR 94], who implemented it in the CFAO EUCLID software. The algorithm is generalized to eight architecture classes of XXXRRR type with wrist spherical joint and for which it is possible to calculate singularity surfaces analytically. Knowing the analytical equations of the limits of the aspects then makes it possible to find out the equations of the boundary surfaces of the images of the aspects in the workspace. A topological combination of the aspects' images, carried out using a CAD software, makes it possible to generate the entire workspace.

Figure 3.20 shows the workspace of a RPRRRR type robot calculated starting from the aspects and modeled with a CAD software.

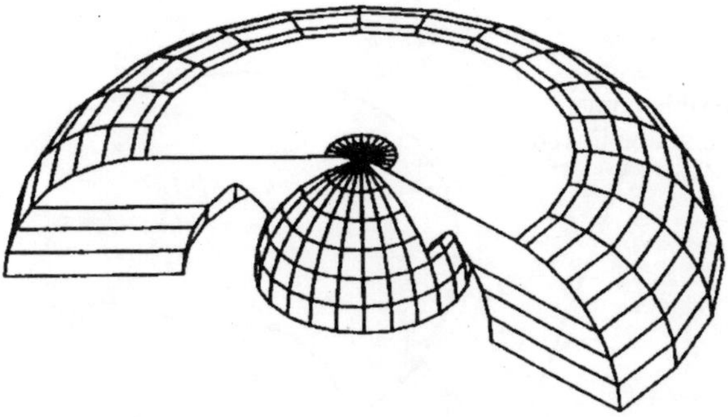

Figure 3.20. *Workspace of a RPRRRR type robot obtained through the aspects' image (according to [GOR 94])*

It should be noted that this method is not appropriate for robot manipulators of general architecture, because the analytical equations of the aspects can no longer be obtained [ELO 96].

Finally, the aspects, and more particularly the free aspects, were at the origin of the definition of the t-connected regions, i.e. the domains in the operational space where any continuous trajectory is feasible (see section 3.5.3).

3.5. Concept of connectivity

3.5.1. *Introduction*

The workspace makes it possible to analyze the total performances of accessibility of a robot manipulator. However, a simple analysis of accessibility can prove to be insufficient, even when all constraints (limits, collisions) seem to be considered, like in Figure 3.12b.

Indeed, the points **B** and **F** are actually accessible, but a displacement of the end-effector between these two points is impossible since the first arm of the robot will get locked on the obstacle (Figure 3.21). The robot can reach all the points of its workspace, but, contrary to appearance (the workspace is connected i.e. all one piece), we see that the robot cannot move freely in this space, in the sense here that it cannot move between every set of points. This academic example was selected on purpose in order to better show the impossibility of certain displacements, but generally, a simple visual analysis does not make it possible to draw a conclusion regarding the consequences of an obstruction.

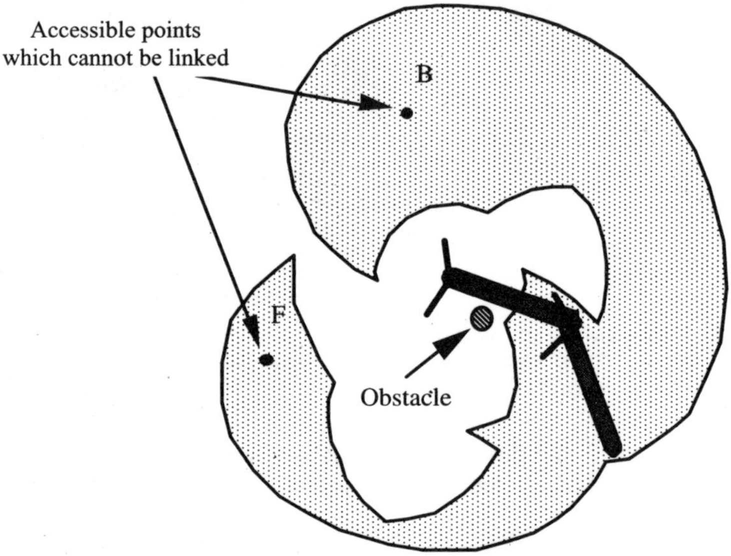

Figure 3.21. *Impossible displacement between points B and F*

If the end-effector can move between two prescribed points, this means that there exists a continuous trajectory between them. However, this trajectory is not arbitrary. The example in Figure 3.22 shows that the robot can move between points $\mathbf{X}_1$ and $\mathbf{X}_2$, but not in a straight line. Indeed, point $\mathbf{X}_1$ is accessible only in the posture "elbow down", because of the joint limits; while point $\mathbf{X}_2$ is accessible only in the posture "elbow up", this time because of the obstruction caused by the obstacle. A change of posture during motion is thus necessary, which forces the robot end-effector to go through a fully-outstretched arm configuration, and thus through a point located on the workspace boundary: the trajectory must be left. This problem, already discussed by Borrel [BOR 86] (see Figure 3.19) and which can be analyzed using the aspects, must be generalized in order to consider the obstacles.

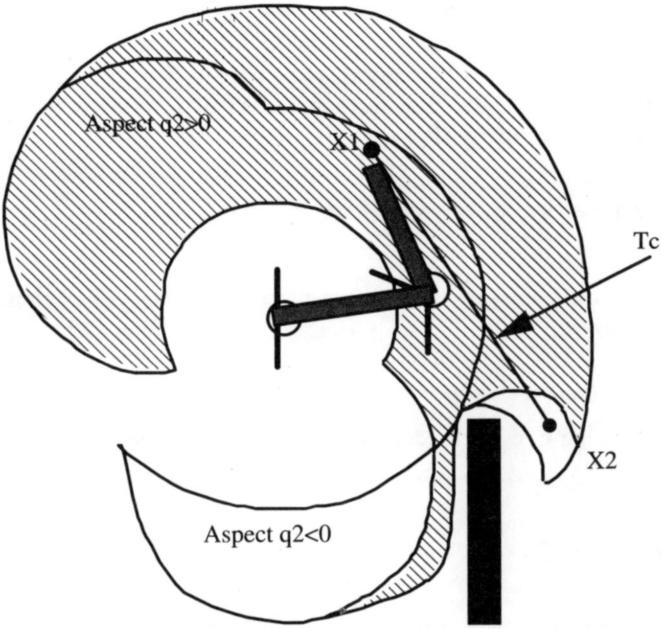

Figure 3.22. *Non-feasible continuous trajectory*

A concept stronger than accessibility must be introduced: connectivity. This concept was gradually formulated in [CHE 87b], [CHE 87c], [CHE 88], [WEN 88], [WEN 89] and [WEN 91]. As implied in the two previous examples, there are two main connectivity levels, according to whether the trajectory to be followed between the points is specified or not. For each level of connectivity it is very useful, for the design of robotic sites and for robots' offline programming, to determine the workspace regions where the robot can move freely.

3.5.2. *Characterization of n-connectivity*

From now on, we will call work domains the workspace regions that are obtained as images through the robot's geometrical operator, of the connected components Ql_i of the set of free configurations:

$$\forall\ i \in I:\ Wl_i = \mathbf{f}(Ql_i) \qquad [3.12]$$

where I describes the number, which is assumed to be finite, of the connected components Ql_i.

We note that the work domains are in general not disjoined, but they overlap.

N-connectivity guarantees that any discrete trajectory of the free workspace Wl is feasible. By definition, the free workspace Wl of a robot is n-connected if it verifies the following property (P_n):

$$(P_n): \forall \mathbf{T}_d = \{\mathbf{X}_1, \mathbf{X}_2, \ldots, \mathbf{X}_p\} \subset Wl, \exists\, i \in I / \mathbf{T}_d \subset Wl_i \qquad [3.13]$$

i.e. if the set of the points of an arbitrary discrete trajectory $\mathbf{T}_d$ belong to a work domain.

Necessary and sufficient condition

We show that the free workspace of a robot is n-connected in the sense of (P_n) if and only if:

$$\exists\, i \in I / Wl = Wl_i \qquad [3.14]$$

This means that there is a work domain that contains all others and that covers the free workspace as a whole. We notice that in the absence of obstacles, the workspace is always n-connected. Indeed, in this case the accessible joint domain is connected (in the mathematical sense). There is thus only one work domain which represents the entire workspace.

EXAMPLE 3.2.– Figure 3.23 shows the case of an n-connected workspace in the presence of obstacles. Although the set of free configurations is not connected (it has two connected components), the robot can connect any set of points within its workspace because one of the connected components (Wl_1) generates the entire free workspace Wl (on top left) and the image of the other (Wl_2) is included in Wl (on the right). The latter work domain, Wl_2, is generated when the elbow of the robot is blocked in the notch of the obstacle.

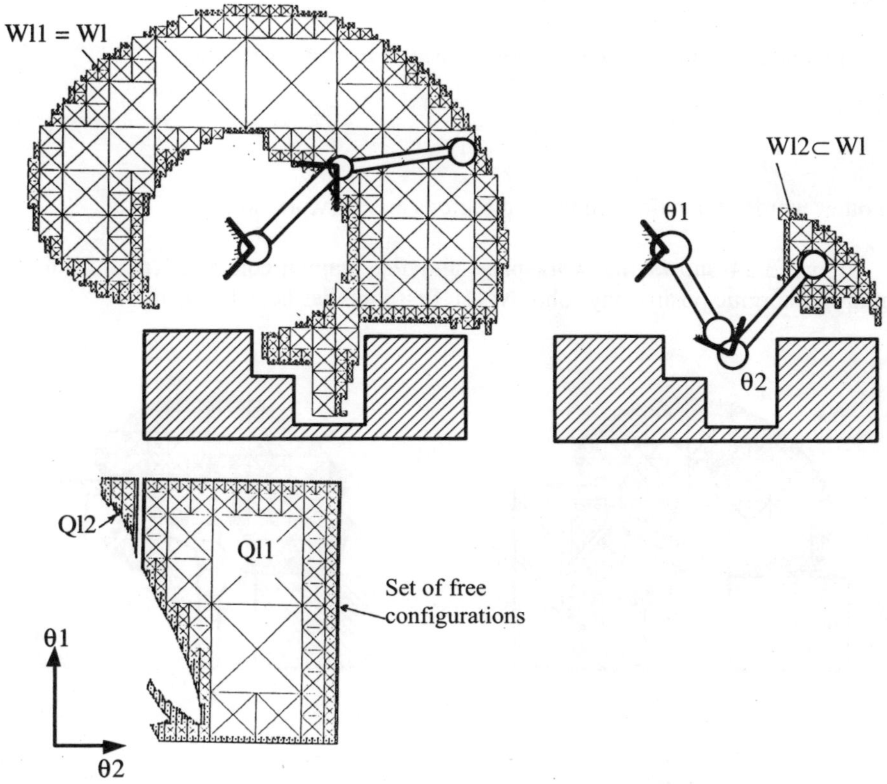

Figure 3.23. *Case of an n-connected workspace*

Overview of n-connected regions

When the free workspace is not n-connected, it is appropriate to search for the n-connected regions of Wl, meaning the areas where the end-effector can really move between any set of work points. A region Wlp of Wl will be said n-connected if it verifies the generalized property (P_{ng}):

$$(P_{ng}) \colon \forall\, \mathbf{T}_d = \{\mathbf{X}_1, \mathbf{X}_2, \ldots, \mathbf{X}_p\} \subset Wlp,\ \exists\, i \in I / \mathbf{T}_d \subset Wl_i \tag{3.15}$$

We show that Wlp is n-connected if and only if:

$$\exists\, i \in I / Wlp \subset Wl_i \tag{3.16}$$

in other words, any area belonging to a work domain is n-connected.

The main n-connected regions are defined in the following way:

$$\exists\ i \in I / Wlp = Wl_i \quad [3.17]$$

in other words, the main n-connected regions are the work domains.

Figure 3.24 shows the workspace and the main n-connected regions for an anthropomorphic positioning robot whose first joint has been blocked.

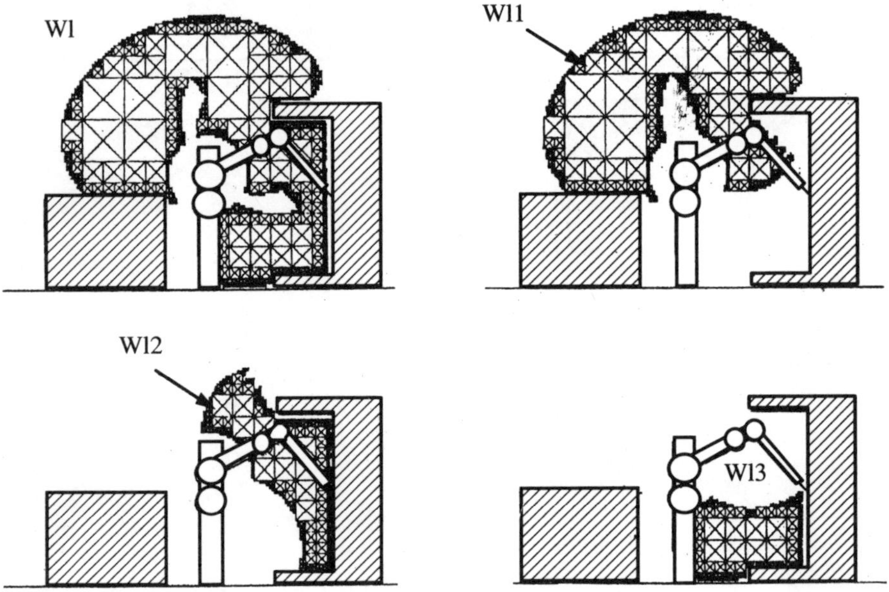

Figure 3.24. *Main n-connected regions*

3.5.3. *Characterization of t-connectivity*

Feasibility of continuous trajectories

The n-connectivity properties do not guarantee the feasibility of continuous trajectories. Nothing in fact proves that the robot end-effector is able to follow, completely and without ever leaving it, any path in the free workspace. However, for many tasks, following continuous trajectories is necessary. It is the case in arc-welding, cutting (lazer or by water jet) or painting. It is thus necessary to introduce a stronger concept which ensures following continuous trajectories: t-connectivity.

We already noted that the joint limits or the presence of obstacles reduce the capacities of the robot to follow continuous trajectories. Before evaluating the connectivity of the workspace in the sense of following continuous trajectories, it is necessary to examine the feasibility problem of a continuous trajectory defined in the free workspace.

The results shown by Borrel in obstacle-free environments show that, for a standard robot architecture, a continuous trajectory $\mathbf{T}_c$, whose aspect does not change can be completely followed by the robot end-effector if and only if all the points of $\mathbf{T}_c$ are accessible in the same aspect, i.e. if $\mathbf{T}_c$ is within the image of the aspect.

The generalization of this result to obstructed environments requires distinguishing the case of non-redundant robots from the case of redundant robots.

In both cases, and according to what precedes, it is necessary that $\mathbf{T}_c$ belongs to the image of a connected component of a free aspect.

If the robot considered is non-redundant, we know that any point $\mathbf{T}_c$ is accessible with only one configuration per aspect (for a standard robot architecture). If $\mathbf{T}_c$ entirely belongs to the image of a connected component Al_{jk} of a free aspect, it is then certain that the pre-image of $\mathbf{T}_c$ in the joint domain forms a path in Al_{jk} and that $\mathbf{T}_c$ is feasible. The analysis in the case of non-redundant robots is thus rather simple.

On the other hand, the analysis becomes complicated for a redundant robot because any point $\mathbf{T}_c$ has infinity of reciprocal images in Al_{jk} and $\mathbf{f}^{-1}(\mathbf{T}_c)$ is not necessarily connected. The trajectory will be feasible if and only if there is a connected component of $\mathbf{f}^{-1}(\mathbf{T}_c)$ whose image covers $\mathbf{T}_c$. The example in Figure 3.25 shows a redundant planar robot with three rotations for which only the two coordinates of position are specified. The continuous trajectory $\mathbf{T}_c$, with the origin in $\mathbf{X}_1$ and extremity in $\mathbf{X}_2$, which it must follow is entirely within the image of an aspect that proves to be connected (Figure 3.25: $\mathbf{T}_c$ is within the image of the aspect defined by $L_2\ S2 + L_3\ S23 < 0$; $L_2\ S3 + L_1\ S23 < 0$; $S3 < 0$). Nevertheless, the robot cannot follow this trajectory in a continuous way, being "rolled up" around the small obstacle. Indeed, Figure 3.26, which synthesizes the reciprocal image $\mathbf{f}^{-1}(\mathbf{T}_c)$ of this trajectory in the joint space, shows that if all the points of the trajectory have at least an antecedent, $\mathbf{f}^{-1}(\mathbf{T}_c)$ is not connected and there is no connected component of $\mathbf{f}^{-1}(\mathbf{T}_c)$ whose image covers $\mathbf{T}_c$. This example highlights that for a redundant robot, the image of a connected component of a free aspect is not a t-connected domain. However, we cannot state that the t-connectivity performances of redundant robots are worse than those of non-redundant robots (it is generally the opposite) but only that the aspects are not adapted to the t-connectivity analysis of redundant robots because they are not uniqueness domains.

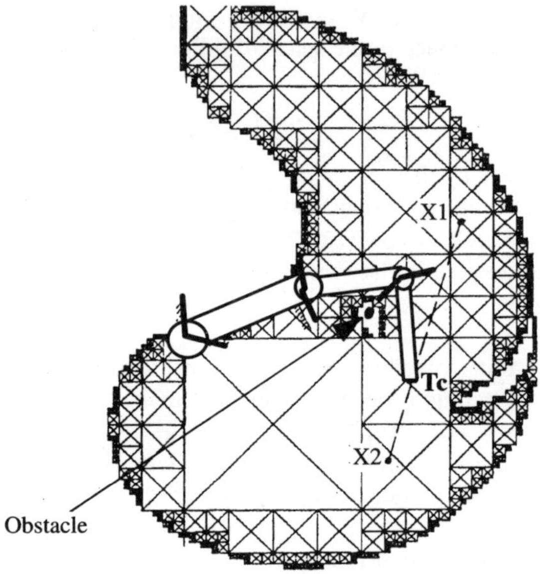

Figure 3.25. *Image of the aspect in EO_2 defined by (x,y)*

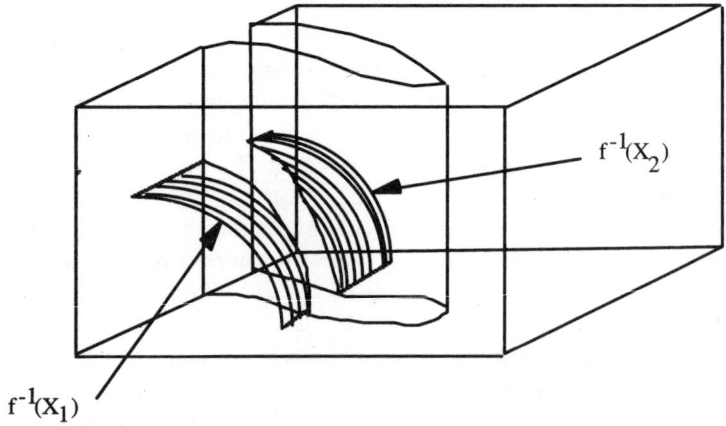

Figure 3.26. *Diagram of reciprocal image of the continuous trajectory (X_1, X_2). Representation of reciprocal images of a certain number of points on the trajectory*

In summary, let $\mathbf{T}_c$ be a continuous trajectory of the free workspace; for $\mathbf{T}_c$ to be feasible by a non-redundant robot, it is necessary and sufficient that the trajectory is entirely within the image of a connected component of a free aspect, or that it passes through an aspect changing domain which, as we pointed out, is a domain of dimension m-1. For $\mathbf{T}_c$ to be feasible by a redundant robot, it is necessary and sufficient that there is a connected component of $\mathbf{f}^{-1}(\mathbf{T}_c)$ whose image covers $\mathbf{T}_c$. However, this condition does not make it possible to analyze the t-connectivity of the workspace.

The results on t-connectivity of the workspace areas that will be discussed on in this section apply only to non-redundant robots. Indeed, to our knowledge, there is no result enabling a global t-connectivity analysis of the free workspace for the redundant robots. However, a necessary and sufficient, practical feasibility condition of a continuous trajectory was formulated in [WEN 93] and implementation tools are presented there. Finally, we note that, if n-connectivity analyses are general, the t-connectivity analyses described hereinafter are valid only for standard robot architectures, for which there is only one solution of the inverse geometric model in each aspect [ELO 96].

Definition of t-connectivity

The free workspace Wl of a non-redundant robot will said t-connected if any continuous trajectory of Wl is feasible, i.e. if for any continuous trajectory $\mathbf{T}_c$ of Wl there is a connected component k of a free aspect j, Al_{jk}, so that its image $WA_{jk} = \mathbf{f}(Al_{jk})$ contains $\mathbf{T}_c$:

$$(P_t): \forall\ \mathbf{T}_c \subset Wl,\ \exists\ j \in J \text{ and } \exists k \in J_j / \mathbf{T}_c \subset WA_{jk} \qquad [3.18]$$

Necessary and sufficient condition

We show that the free workspace Wl of a non-redundant robot is t-connected if and only if:

$$\exists\ j \in J \text{ and } \exists k \in J_j / Wl = WA_{jk} \qquad [3.19]$$

which means that the free workspace is entirely generated by the image in the operational space of only one connected component Al_{jk} of a free aspect Al_j.

EXAMPLE 3.3.– Figure 3.27 shows the case of a t-connected workspace: this planar robot is able to follow any continuous trajectory of its workspace. Indeed, one of the two aspects ($\theta_2 > 0$) has as an image the entire free workspace (Figure 3.27a). The image of the other aspect ($\theta_2 < 0$) is included in the previous one (Figure 3.27b).

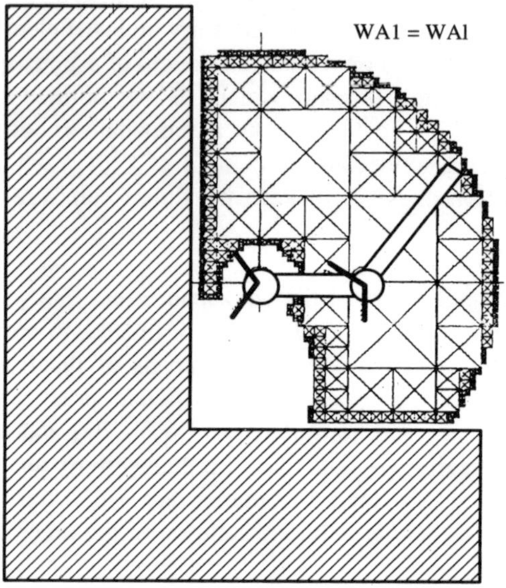

Figure 3.27a. *Image of the aspect* $\theta_2 > 0$

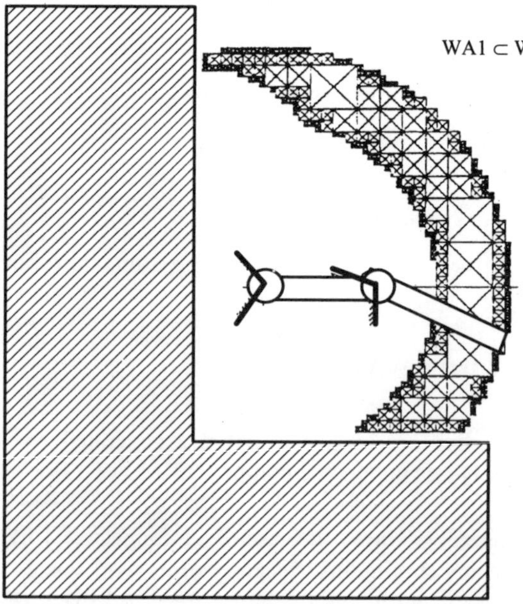

Figure 3.27b. *Image of the aspect* $\theta_2 < 0$

Overview of t-connected regions

The t-connectivity concept can be extended to the regions of Wl. We show that a region Wlp of Wl is t-connected i.e. it verifies the following generalized property:

$$(P_{tg})\ \forall\ \mathbf{T}_c \subset Wlp\ \exists\ j \in J \text{ and } \exists k \in I(j) / \mathbf{T}_c \subset WA_{jk} \cap Wlp \qquad [3.20]$$

if and only if:

$$\exists\ j \in J \text{ and } \exists k \in I(j) / Wlp \subset WA_{jk} \qquad [3.21]$$

The main t-connected regions in the direction of (P_{tg}) are WA_{jk}, i.e. the images of the connected components of the free aspects.

EXAMPLE 3.4.– Figure 3.28 shows the main t-connected regions for the location of Figure 3.24.

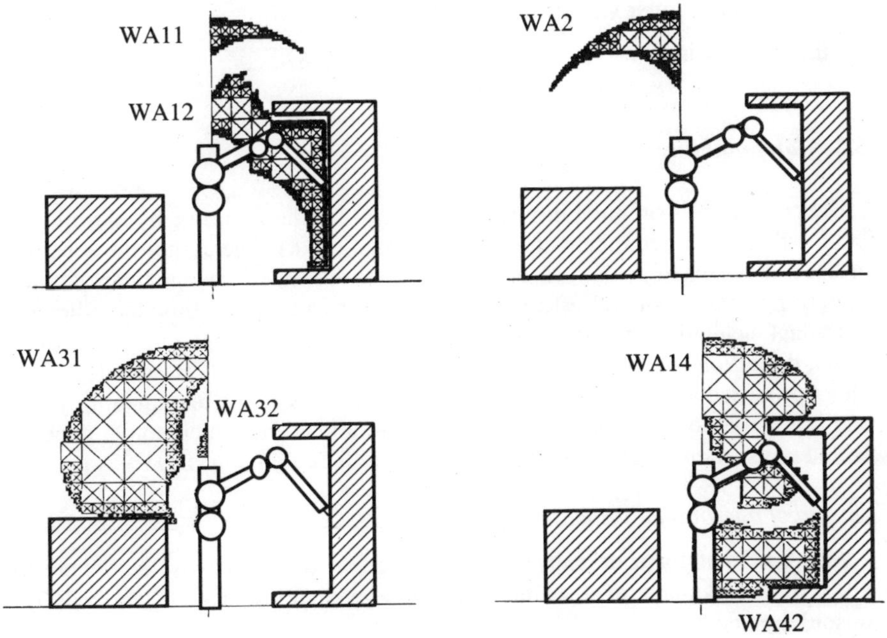

Figure 3.28. *Main t-connected regions*

3.6. Local performances

3.6.1. *Definition of dexterity*

Dexterity means "agility of the hand". It can be intuitively defined as measuring the aptitude of the robot end-effector to "easily" perform small and arbitrary displacements about a given working point.

From a mathematical point of view, dexterity can be defined in various ways. In all the cases, it is related to a given joint configuration and uses the expression of the robot's Jacobian matrix. Dexterity is thus closely related to the relation between operational velocities and joint rates (or, to the relation between the forces and moments acting on the robot end-effector and the torques measured on the actuators). Dexterity can be also useful for the measurement of the proximity of a singular configuration.

There are three main measurements for dexterity [ANG 92]:

– manipulability;

– the lowest singular value;

– the isotropy index.

3.6.2. *Manipulability*

The concept of manipulability was introduced for the first time by Yoshikawa in 1985 [YOS 85], for redundant and non-redundant robots. The author introduces this concept as a means of evaluation of "the ability to arbitrarily change the position and orientation of the robot end-effector". He defines two tools for this: the ellipsoid of operational velocities and the manipulability index. We point out here only the original definitions of Yoshikawa, which are the most frequently used today. Certain authors proposed afterward more exhaustive definitions of the manipulability which make it possible to consider the singular configurations and which do not depend on selected units. A complete work on this subject will be found in [DOT 95].

Ellipsoid of operational velocities

We show that for a domain of joint velocities $\dot{\mathbf{q}}$ defined by the sphere unit: $\|\dot{\mathbf{q}}\| \leq 1$, the domain of feasible operational velocities is characterized by an ellipsoid defined by:

$$\dot{\mathbf{X}}^{\mathrm{T}} (\mathbf{J}\mathbf{J}^{\mathrm{T}})^{-1} \dot{\mathbf{X}} \leq 1 \qquad [3.22]$$

where $\dot{\mathbf{X}}$ is an operational speed.

Moreover, we show that the main axes of the ellipsoid of operational velocities are defined by the eigenvectors $\mathbf{e}_i$ of $(\mathbf{JJ}^T)^{-1}$, and the lengths of the semi-axes are the singular values σi of the Jacobian matrix $\mathbf{JJ}^T$, i.e. the square roots of the eigenvalues λ_i of the matrix $\mathbf{JJ}^T$. The eigenvalues of matrix $\mathbf{JJ}^T$ can be obtained by finding out the solutions of the equation:

$$\det(\mathbf{JJ}^T - \lambda \mathbf{I}) = 0 \qquad [3.23]$$

The singular values are the square roots of the solutions of [3.23].

Figure 3.29 illustrates the situation for a planar robot with three revolute joints, whose task is defined by the two position coordinates x and y in the plane.

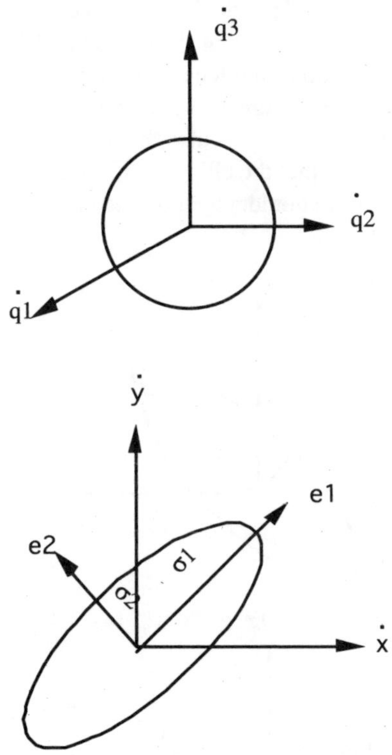

Figure 3.29. *Definition of ellipsoid of operational velocities*

The ellipsoid defines either the domain of operational velocities or that of operational infinitesimal displacements.

We note that the direction of the longest axis of the ellipsoid is that where:
– the operational velocity could be the highest;
– the smoothness of movement (resolution) will be the worst.

Inversely, it is along the shortest axis that:
– the operational velocity will be the lowest;
– the smoothness of movement will be the best.

NOTE.– according to the principle of virtual works, there is a duality between the ellipsoid of velocities or displacements, and the ellipsoid of forces defined based on the matrix $\mathbf{JJ}^T$ by $\mathbf{f}^T\mathbf{JJ}^T\mathbf{f} \leq 1$ (where $\mathbf{f}$ is the wrench of forces and torques exerted by the end-effector on the environment). This duality is created through the energy operator. Thus, it is shown that the main directions of the ellipsoid of forces are identical to those of the ellipsoid of velocities, but that the lengths of the semi-axes of the former are the inverse of those of the latter (see Figure 3.30). This shows that the direction where the operational velocity capacity is maximal is also the direction where the force capacity is the lowest. Finally, we must note that the ellipsoid of forces can be balanced by the limits on the joint torques when they are not identical on each joint [CHI 96].

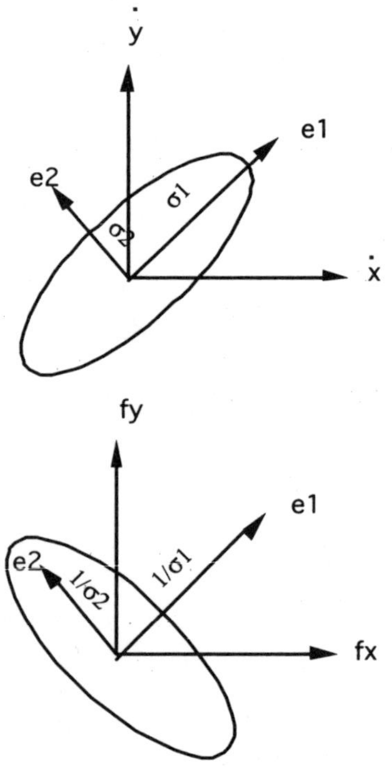

Figure 3.30. *Ellipsoids of manipulability in speed and stress*

In general, these values must be calculated numerically using specialized software like Mathematica [MAT 91]. For simple cases such as a planar robot with two revolute joints, these values can be calculated analytically (see Example 3.5).

Eigenvectors $\mathbf{e}_i$ of $(\mathbf{JJ}^T)^{-1}$ are obtained from the preceding eigenvalues λ_i through the following relation:

$$(\mathbf{JJ}^T)^{-1}\mathbf{e}_i = \frac{1}{\lambda_i^2}\mathbf{e}_i \qquad [3.24]$$

As for the eigenvalues, the eigenvectors will be generally obtained using specialized software.

Manipulability index

The manipulability index is defined by the scalar:

$$w = \sqrt{\det(\mathbf{JJ}^T)} \qquad [3.25]$$

We show that this index is proportional to the volume of the ellipsoid of velocities and that it can be also calculated as the product of the singular values of $\mathbf{JJ}^T$, i.e. as the product of the square roots of the eigenvalues of $\mathbf{JJ}^T$. In the particular case of a non-redundant robot, we show that this index is written:

$$w = |\det(\mathbf{J})| \qquad [3.26]$$

Hence, the manipulability index constitutes, in a defined point and for a given joint configuration, a measurement of the aptitude of the end-effector to move starting from this point. By definition, the manipulability index is zero in singular configurations. The manipulability index is to be carefully considered. Indeed, a value of this index considered to be "correct" can hide a strong disproportion between the lengths of the main axes of the ellipsoid of operational velocities and thus, a relatively low value of the lowest singular value. Moreover, the manipulability index does not give any indication regarding the privileged direction.

NOTE.– the force manipulability index can also be defined, for the non-singular configurations, by:

$$w_f = \frac{1}{\sqrt{\det\left(\mathbf{JJ}^T\right)}} \qquad [3.27]$$

EXAMPLE 3.5 – let us consider a non-redundant planar robot with two revolute joints. With the usual notations, the Jacobian matrix of this robot is written:

$$\mathbf{J} = \begin{bmatrix} -S1L1 - S12L2 & -S12L2 \\ C1L1 + C12L2 & C12L2 \end{bmatrix} \quad [3.28]$$

To make things simpler, this matrix is rather expressed in the reference frame of the second joint. The following matrix ${}^2\mathbf{J}$ is thus obtained:

$${}^2\mathbf{J} = \begin{bmatrix} S2L1 & 0 \\ L2 + C2L1 & L2 \end{bmatrix} \quad [3.29]$$

The eigenvalues, the eigenvectors and the determinant of matrix $\mathbf{JJ}^T$ are the same as those of ${}^2\mathbf{J}{}^2\mathbf{J}^T$. We have:

$$\det(\mathbf{JJ}^T - \lambda\mathbf{I}) = \lambda^2 - \lambda(L2^2 + L^2) + L1^2L2^2s2^2 \quad [3.30]$$

where:

$$L^2 = L1^2 + L2^2 + 2L1L2c2 \quad [3.31]$$

The singular values σ_i are thus defined by:

$$\begin{cases} \sigma 1 = \sqrt{\dfrac{L^2 + L2^2 + \sqrt{(L^2 + L2^2)^2 4L1^2 L2^2 S2^2}}{2}} \\ \sigma 2 = \sqrt{\dfrac{L^2 + L2^2 \sqrt{(L^2 + L2^2)^2 4L1^2 L2^2 S2^2}}{2}} \end{cases} \quad [3.32]$$

The long axis of the ellipsoid of manipulability has thus $\sigma 1$ as half-length, and the half-length of the shortest axis is $\sigma 2$.

Hence, we see that for $\theta_2 = 0$, $\sigma 2$ vanishes while $\sigma 1$ reached its maximum value of $\sqrt{L1^2 + 2L2^2 + 2L1L2}$.

Starting from formula [3.24] and standardizing the obtained vectors, we will obtain the vectors **v1** and **v2** of the main axes of the ellipsoid of velocities by:

$$\mathbf{v1} = \begin{bmatrix} \text{sign}(L2 + L1C2)\sqrt{\dfrac{\sigma 1^2 - L2^2}{\sigma 1^2 - \sigma 2^2}} \\ \sqrt{\dfrac{L2^2 - \sigma 2^2}{\sigma 1^2 - \sigma 2^2}} \end{bmatrix}$$

$$\mathbf{v2} = \begin{bmatrix} -\text{sign}(L2 + L1C2)\sqrt{\dfrac{L2^2 - \sigma 1^2}{\sigma 1^2 - \sigma 2^2}} \\ \sqrt{\dfrac{\sigma 1^2 - L2^2}{\sigma 1^2 - \sigma 2^2}} \end{bmatrix} \quad [3.33]$$

Figure 3.31 shows various ellipsoids of manipulability along a radial direction when L1 = L2 = 1.

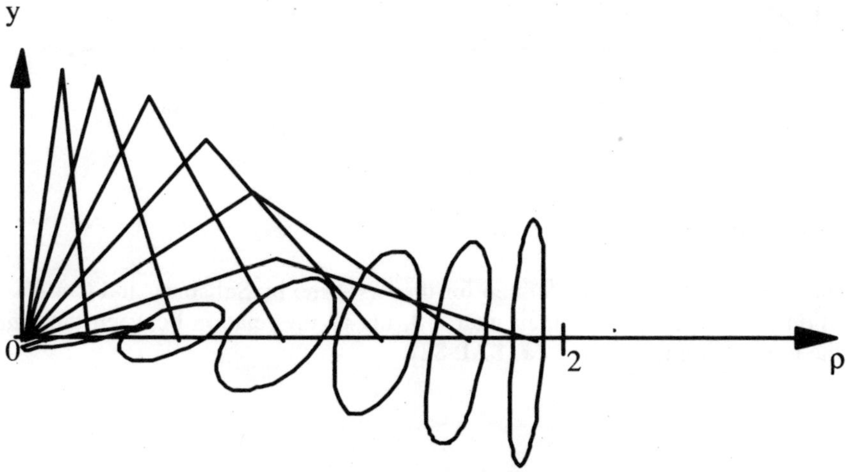

Figure 3.31. *Ellipsoids of manipulability based on the radial distance (according to [YOS 85])*

The determinant is written: $\det(^2\mathbf{J}) = L1L2S2$ and the manipulability index is thus:

$$w = L1L2|S2| \quad [3.34]$$

This index depends only on θ_2. We see that for this robot, the manipulability index is maximal for all joint configurations such as $\theta_2 = \pm 90°$. In the Cartesian space, the place of maximal manipulability index is thus the circle with the radius $\sqrt{L1^2 + L2^2}$. We also see that, for a maximum given elongation L1 + L2, the manipulability index is optimal when L1 = L2.

Figure 3.32 shows, when L1 = L2 = 1, the evolution of w according to the radial distance between the end of the last arm and the center O of the robot's base reference frame.

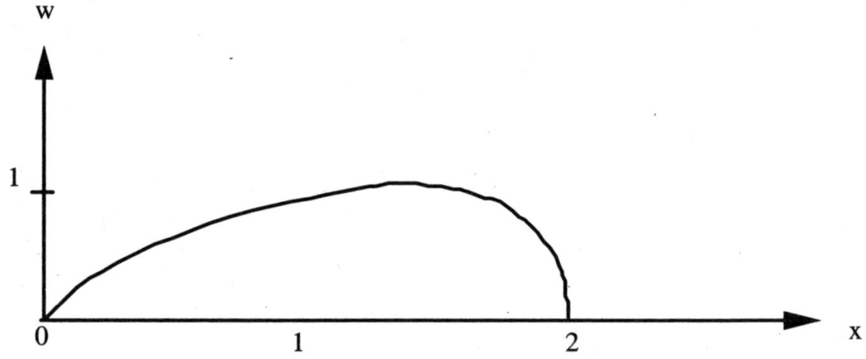

Figure 3.32. *Plot of the manipulability index versus the radial distance (according to [YOS 85])*

3.6.3. *Isotropy index*

The isotropy index k was defined for the first time by Salisbury and Craig as the condition number of the Jacobian matrix **J**, i.e. as the relation between the highest and the lowest singular value of **J** [SAL 82]:

$$k = \frac{\sigma_{max}}{\sigma_{min}} \quad [3.35]$$

It should be specified that this definition is valid only when the elements of **J** have the same units. In the opposite case (i.e. when the task position and orientation coordinates), it is possible to define a homogenous matrix by dividing the lines of **J** corresponding to the positions by "characteristic length" of the studied robot [ANG 97], [CHA 02]. A Jacobian matrix unit is thus obtained. The characteristic length will be calculated as that which minimizes the conditioning index of the new dimensionless Jacobian matrix.

The condition number is well-known in linear algebra. It defines the "inversibility" or the conditioning of a matrix: in other words, it makes it possible to predict the errors obtained while solving a linear system according to the entry errors. The higher this index is, the worse the conditioning of the linear system, and the highest the sensitivity to entry errors will be.

Applied to robotics, this index makes it possible to evaluate the precision with which the prescribed operational velocities could be controlled using the joint rates calculated by the inverse of **J.** It constitutes a measure of uniformity of the ellipsoid of velocities, from where we get the term of isotropy. It reaches its ideal minimal value of 1 when the ellipsoid is a sphere. Physically, when the robot has an isotropic configuration (i.e. when k = 1), the robot end-effector moves with the same ease in all directions. For Example 3.5, we can show, based on the formulae [3.41], that the robot has an isotropic configuration in $\theta_2 = 135°$ when $L1 = (\sqrt{2}/2)L2$. The manipulability index is then twice smaller than the maximum obtained for L1 = L2 in $\theta_2 = \pm 90°$.

3.6.4. *Lowest singular value*

Another way of measuring the dexterity of a robot is the lowest singular value of **J**, i.e. the half-length of the shortest axis of the ellipsoid of velocities [KLE 87a]. Whereas the manipulability and isotropy indices refer to the robot's capacity to move its robot end-effector in all directions about a given point, the lowest singular value measures the ability of displacement in the most unfavorable direction. The lowest singular value is a more reliable way of measuring the proximity of a singularity than the manipulability index. Hence, it is particularly recommended in the applications where we want to move away from singular configurations.

3.6.5. *Approach lengths and angles*

Approach angles

In the secondary space, the orientation of the robot end-effector is limited (see section 3.33). The possible orientations are called approach angles and their limit values are the limit angles. The set of possible orientations in a given point **P** of the workspace forms a solid angle and can be represented by a diagram (see Figures 3.37a and 3.37b).

A service factor in point **P** can be determined [ROT 76]. This factor is written:

$$\theta\,(\mathbf{P}) = \frac{(\text{solid accessible angle in } \mathbf{P})}{4\pi} \qquad [3.36]$$

The concept of approach angles is interesting for applications where the orientation of the tool in a localized zone plays an important role. For example, it is the case for screwing or complex assembly operations.

Approach distances

The approach distances were introduced by [HAN 83] and measure the ability of the tool to slide, from a point **P** of its workspace, along the axis defined by a given approach angle. The envelope of the limit approach distances about point **P** can be represented in various sections by diagrams. These diagrams also make it possible to visualize the approach angles. Figure 3.33a illustrates the graphs in a section XY for a primary space point: point **P** can thus be approached from all directions. In Figure 3.33b, the point is chosen in the secondary space and the approach angles are thus limited.

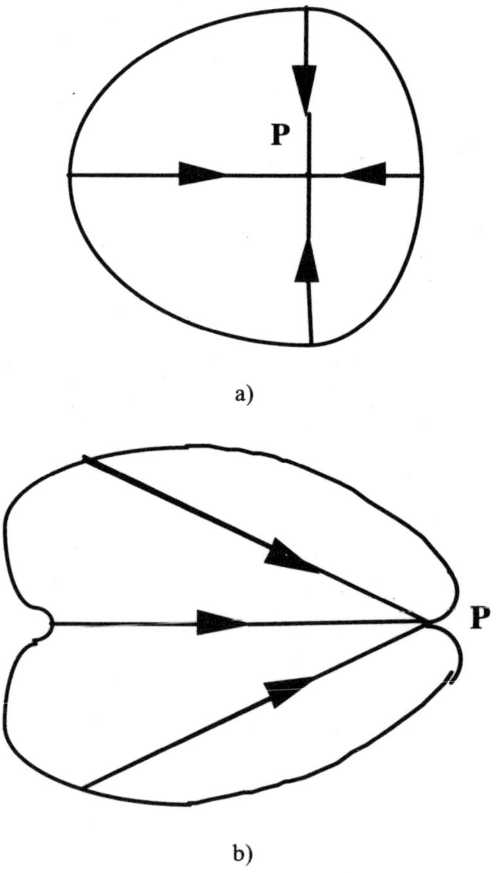

Figure 3.33. *Approach angle and distance*

The concept of approach distances is interesting for certain applications like assembling (insertions) and machining (drilling, milling, etc.).

3.7. Conclusion

This chapter presented the main tools for assessing the geometrical and kinematic performances of robots. The workspace, if it is correctly defined, is certainly the most natural tool to assess the total geometric performances. The robot joint limits and the obstacles in the environment often result in limiting the possibilities of displacement within the workspace, whether it is a point-to-point movement or a movement following continuous trajectories. The connectivity analyses seem then an essential complement to those of accessibility. These tools for performance assessment were implemented as a criterion in robotic site design software POSOPT, developed at IRCCyN [CHE 98] and were validated through industrial applications like the aircraft robotized cleaning [CHE 95] and the inspection of surfaces in hostile environment [WEN 97]. To better assess the robot/task adaptation, it is necessary to evaluate the abilities of local displacement of the robot end-effector. The dexterity analyses represent thus an ideal tool for this local performance analysis.

We must note that this chapter dealt only with the case of current robot architectures. For non-standard robots, other properties appear (like the "cuspidality", i.e. the possibility of changing posture without crossing a singularity) which call into question the concepts of aspect and connectivity. The interested reader may refer to [ELO 96], [WEN 04] or [BAI 04] for more details on this subject.

3.8. Bibliography

[ANG 92] ANGELES J., RANJBARAN F., PATEL R.V., "On the design of the kinematic structure of seven-axes redundant manipulators for maximum conditioning", *Proc. IEEE Int. Conf. on Robotics and Automation*, Nice, May 1992, p. 494-499.

[ANG 97] ANGELES J. "Fundamentals of robotic mechanical systems", *Theory, Methods and Algorithms*, Mechanical Engineering series, Springer-Verlag, 1997.

[BAI 04] BAILI M., "Analyse et classification des manipulateurs 3R à axes orthogonaux", PhD Thesis, Nantes University and Ecole Centrale of Nantes, December 2004.

[BOR 86] BORREL P., "Contribution à la modélisation géométrique des robots manipulateurs; application à la conception assistée par ordinateur", PhD Thesis, USTL Montpellier, July 1986.

[BRO 83a] BROOKS R.A., "Planning collision free motions for pick-and-place operations", *The Int. J. of Robotics Research*, Vol. 2(4), Winter 1983, p. 19-44.

[BRO 83b] BROOKS R.A., "Solving the path-finding problem by good representation of free space", *EEE Trans. on Systems, Man, and Cybernetics*, Vol. SMC-13, p. 190-197.

[BUR 88] BURDICK J.W., "Kinematic analysis and design of redundant manipulators", PhD Thesis, Stanford, 1988.

[BUR 95] BURDICK J.W., "A recursive method for finding revolute-jointed manipulator singularities", *ASME J. of Mechanical Design*, Vol. 117, 1995, p. 55-63.

[CEC 89] CECCARELLI M., "On the workspace of 3R robot arms", *Proc. 5th IFToMM Int. Symp. on Linkages and Computer Aided Design Methods*, Bucharest, July 1989, p. 37-46.

[CEC 95] CECCARELLI M., "On the workspace of general 4R manipulators", *The Int. J. of Robotics Research*, Vol. 14(2), April 1995, p. 152-160.

[CHA 02] CHABLAT D., ANGELES J., "On the kinetostatic optimisation of revolute-coupled planar manipulators", *Mechanisms an Machine Theory*, 2002, Vol. 37, p. 351-374.

[CHE 87a] CHEDMAIL P., WENGER P., "Domaine atteignable par un robot: généralisation de la notion d'aspect à un environnement avec obstacles, application aux robots articulés plans", *Proc. AUM'87*, Nantes, September 1987, p. 312-314.

[CHE 87b] CHEDMAIL P., WENGER P., "Condition nécessaire et suffisante de 'parcourabilité' de l'espace opérationnel accessible par un robot dans un environnement avec obstacles", internal memo of LAN, no. 06-87, November 1987.

[CHE 87c] CHEDMAIL P., WENGER P., "Parcourabilité de l'espace opérationnel accessible par un robot dans un environnement avec obstacles", internal memo of LAN, no. 09-87, November 1987.

[CHE 88] CHEDMAIL P., WENGER P., "Ability of a robot to move between two points within cartesian free workspace with an encumbered environment", *Proc. IFAC Symp. on Robot Control, SYROCO'88*, 1988, p. 73.1-73.6.

[CHE 95] CHEDMAIL P., WENGER P., "Conception optimale d'un site robotisé de nettoyage d'avion", *Proc. Congrès International de Génie Industriel*, Montreal, October 1995, p. 367-376.

[CHE 98] CHEDMAIL P., DOMBRE E., WENGER P., *La CAO en Robotique, outils et méthodologies*, Hermès, Paris, 1998.

[CHI 96] CHIACCHIO P., BOUFFARD-VERCELLI Y., PIERROT F., "Evaluation of force capabilities for redundant manipulators", *Proc. IEEE Int. Conf. on Robotics and Automation*, Minneapolis, April 1996, p. 3520-3525.

[DOM 82] DOMBRE E., BORREL P., LIÉGEOIS A., "A rate control algorithm for obtaining the workspaces and the strategies of an underwater manipulator", *Proc. Conf. IASTED-MECO*, Tunis, September 1982, p. 59.1-59.3.

[DOM 91] DOMBRE E., FOURNIER A., GEHORS D., VERGELY L., "Intégration de modèles de capteurs dans un système de CAO Robotique", *Proc Conf. Int. CFAO et Infographie, MICAD'91*, Paris, February 1991, p. 83-97.

[DOT 95] DOTY K.L. *et al.*, "Robot manipulability", *IEEE Trans. on Robotics and Automation*, Vol. RA-11(3), June 1995, p. 462-468.

[ELO 96] EL OMRI J., "Analyse géométrique et cinématique des mécanismes de type manipulateur", PhD Thesis, Nantes University and Ecole Centrale of Nantes, February 1996.

[FAV 84] FAVERJON B., "Obstacle avoidance using an octree in the configuration space of a manipulator", *Proc. IEEE Int. Conf. on Robotics and Automation*, Atlanta, March 1984, p. 504-510.

[FRE 84] FREUDENSTEIN F., PRIMROSE E.J.F., "On the analysis and synthesis of the workspace of a three-link, turning-pair connected robot arm", *ASME J. Mechanisms, Transmissions, and Automation in Design*, Vol. 106, 1984, p. 365-370.

[GAR 89] GARCIA G., "Contribution à la modélisation d'objets et à la détection de collision en robotique à l'aide d'arbres octaux", PhD Thesis, Nantes University, ENSM, September 1989.

[GOR 94] GORLA B., AZKUE URGATE J., "Une méthode de génération du domaine de travail opérationnel d'un robot manipulateur", *Revue Internationale de CFAO et d'Infographie*, Vol. 9(5), 1994, p. 695-707.

[HAN 83] HANSEN J.A., GUPTA K.C., KAZEROUNIAN S.M.K., "Generation and evaluation of the workspace of a manipulator", *The Int. J. of Robotics Research*, Vol. 2(3), 1983, p. 22-31.

[HAY 86] HAYWARD V., "Fast collision detection scheme by recursive decomposition of a manipulator workspace", *Proc. IEEE Int. Conf. on Robotics and Automation*, San Francisco, April 1986, p. 1044-1049.

[KHA 98] KHALIL W, DOMBRE E., *Modélisation et commande des robots*, 2nd edition, Hermès, Paris, 1998.

[KLE 87a] KLEIN C.A., BLAHO B.E., "Dexterity measures for the design and control of kinematically redundant manipulators", *The Int. J. of Robotic Research*, Vol. 6(2), 1987, p. 72-83.

[KOH 85] KOHLI D., SPANOS J., "Workspace analysis of mechanical manipulators using polynomial discriminant", *ASME J. Mechanisms, Transmission and Automation in Design*, Vol. 107, June 1985, p. 209-215.

[KOH 87] KOHLI D., HSU M.S., "The Jacobian analysis of workspaces of mechanical manipulators", *Mechanisms and Machine Theory*, Vol. 22(3), 1987, p. 265-275.

[KUM 81] KUMAR A., WALDRON K., "The workspaces of mechanical manipulators", *Trans. of ASME, J. of Mechanical Design*, Vol. 103, 1981, p. 665-672.

[KUM 86] KUMAR A., PATEL S., "Mapping the manipulator workspace using interactive computer graphics", *The Int. J. of Robotics Research*, Vol. 5(2), 1986, p. 112-121.

[LAL 94] LALLEMAND J.-P., ZEGHLOUL S., *Robotique; aspects fondamentaux*, Masson editions, Paris, 1994.

[LEE 83] LEE T.W., YANG D.C.H., "On the evaluation of manipulator workspace", *Trans. of the ASME, J. of Mechanisms, Transmissions and Automation in Design*, Vol. 105, March 1983, p. 70-77.

[LOZ 87] LOZANO-PÉREZ T., TOURNASSOUD P., LANUSSE A., "Handley: a robot system that recognises, planes and manipulates", *Proc. IEEE Int. Conf. on Robotics and Automation*, Raleigh, March-April 1987, p. 843-849.

[LUE 92] LUETH T.C., "Automatic planning of robot workcell layouts", *Proc. IEEE Int. Conf. on Robotics and Automation*, Nice, May 1992, p. 1103-1108.

[MAT 91] WOLFRAM S., *MATHEMATICA: a system for doing mathematics by computer*, 2nd edition, Addison Wesley, 1991.

[PAR 88] PARENTI C.V., INNOCENTI C., "Position analysis of robot manipulator: regions and subregions", *Proc. Advances in Robot Kinematics, ARK'88*, Ljubjana, Kluwer Academic Publishing, 1988, p. 150-158.

[RAG 89] RAGHAVAN M., ROTH B., "Kinematic analysis of the 6R manipulator of general geometry", *Proc. 5th Int. Symp. on Robotics Research*, Tokyo, August 1989, p. 314-320.

[RAN 94] RANJBARAN F., Angeles J., "On positioning singularities of 3-revolute robotic manipulators", *Proc. 12th Symp. on Engineering Applications in Mechanics*, Montreal, June 1994, p. 273-282.

[RAS 87] RASTEGAR J., DERAVI P., "Methods to determine workspace with different numbers of configurations and all the possible configurations of a manipulator", *Mechanism and Machine Theory*, 1987, Vol. 22(4), p. 343-350.

[ROT 76] ROTH B., "Performance evaluation of manipulators from a kinematic viewpoint", Lecture on Robotics at IRIA, Toulouse, 1976, p. 233-263.

[SAL 82] SALISBURY J.K., CRAIG J.J., "Articulated hands: force and kinematic issues", *The Int. J. of Robotic Research*, Vol. 1(1), 1982, p. 4-17.

[SHI 89] SHILLER Z., "Interactive time optimal robot motion planning and work-cell layout design", *Proc. IEEE Int. Conf. on Robotics and Automation*, Scottdale, May 1989, p. 964-969.

[TSA 93a] TSAI K.Y., KHOLI D., "Trajectory planning in task space for general manipulators", *ASME J. of Mech. Design*, Vol. 115, December 1993, p. 915-921.

[TSA 93b] TSAI K.Y., KOHLI D., ARNOLD J., "Trajectory planning in joint space for mechanical manipulators", *ASME J. of Mech. Design*, Vol. 115, December. 1993, p. 909-914.

[VIJ 86] VIJAYKUMAR R.J., WALDRON J., TSAI M., "Geometric optimization of serial chain manipulators structures for working volume and dexterity", *The Int. J. of Robotics Research*, Vol. 5(2), 1986, p. 91-103.

[WEN 88] WENGER P., CHEDMAIL P., "Ability of a robot to travel through its workspace with an encumbered environment: five characterizations", *Proc. Int. Symp. on Robotics and Manufacturing, ISRAM'88*, New Mexico, November 1988, p. 1011-1018.

[WEN 89] WENGER P., "Aptitude d'un robot manipulateur à parcourir son espace de travail en présence d'obstacle", PhD Thesis, Nantes University, ENSM, January 1989.

[WEN 91] WENGER P., CHEDMAIL P., "Ability of a robot to travel through its free workspace", *The Int. J. of Robotics Research*, Vol. 10(3), June 1991, p. 214-227.

[WEN 93] WENGER P., CHEDMAIL P., REYNIER F., "A global analysis of following trajectories by redundant manipulators in the presence of obstacles", *Proc. IEEE Int. Conf. on Robotics and Automation*, Atlanta, May 1993, p. 3.901-3.906.

[WEN 97] WENGER P., "Design of robotic workcells: synthesis and application", *in Integrated Design and Manufacturing in Mechanical Engineering,* Kluwer Academic Publishing, 1997.

[WEN 98] WENGER P., "Classification of 3R positioning manipulators", *ASME J. of Mechanical Design*, Vol. 120(2), June 1998, p. 327-332.

[WEN 04] WENGER P., "Uniqueness domains and regions of feasible paths for cuspidal manipulators", *IEEE Transactions on Robotics*, Vol. 20(4), pp. 745-750, August 2004.

[YAN 83] YANG D.C.H., LEE T.W., "On the workspace of mechanical manipulators", *Trans. of the ASME, J. of Mechanisms, Transmissions and Automation in Design*, Vol. 105, March 1983, p. 62-70.

[YAN 86] YANG D.C.H., CHIUEH T.S., "Work area of six-joint robots with fixed hand orientation", *Int. J. of Robotics and Automation*, Vol. 1(1), 1986, p. 23-32.

[YOS 85] YOSHIKAWA T., "Manipulability and redundant control of mechanisms", *Proc. IEEE Int. Conf. on Robotics and Automation*, St Louis, March 1985, p. 1004-1009.

[YU 87] YU Z.Q., "Détection de collision en robotique", PhD Thesis, Nantes University, ENSM, November 1987.

[ZEG 91] ZEGHLOUL S., "Développement d'un système de CAO Robotique intégrant la planification de tâches et la synthèse de sites robotisés", Thesis, Poitiers University, February 1991.

Chapter 4

Trajectory Generation

4.1. Introduction

A robotic motion task is specified by defining a path along which the robot must move. A path may be defined by a geometric curve or a sequence of points defined either in operational coordinates (end-effector coordinates) or in joint coordinates. The issue of trajectory generation is to calculate, for the control system, the desired reference joint or end-effector variables as a function of time, such that the robot tracks the desired path. Thus, a trajectory refers to a path and a time history along this path.

The trajectories of a robot can be classified as follows:

– trajectory between two points with free path between them;

– trajectory between two points via a sequence of desired intermediate points, also called via points, with free paths between these via points;

– trajectory between two points with constrained path between them (straight line segment for instance);

– trajectory between two points via intermediate points with constrained paths between these via points.

For the first two classes, the trajectory is generally created in the joint space. For the last two classes, it is usually better to generate the trajectory in the operational space.

Chapter written by Moussa HADDAD, Taha CHETTIBI, Wisama KHALIL and Halim LEHTIHET.

Trajectory generation in the joint space presents several advantages:

– it requires fewer online calculations, since there is no need to calculate the inverse geometric or kinematic models;

– the trajectory is not affected by crossing singular configurations;

– maximum velocities and torques are determined from actuator data sheets.

The drawback of joint-space trajectory generation is that the corresponding end-effector path in the operational space is not predictable, although it is repetitive, which increases the risk of undesirable collisions when the robot works in a cluttered environment. Consequently, the joint-space scheme is more appropriate to achieve fast motions in a free space.

Trajectory generation in the operational space enables the prediction of the geometry of the path. It has, however, a number of disadvantages:

– it may fail when the calculated trajectory crosses a singular configuration, when the generated points are out of the joint limits;

– velocities and accelerations of the operational coordinates depend on the robot configuration. If the trajectory generation is based on kinematic constraints, lower bounds are generally used so that joint velocity and torque limits are satisfied. Consequently, the robot may work under its nominal performance. This problem is solved by using trajectory optimization techniques based on the dynamic model.

The following two sections of this chapter cover the problem of trajectory generation under kinematic constraints between two points in the joint space and in the operational space, respectively. The last sections extend the problem to trajectory generation under kinodynamic constraints, with and without via points.

4.2. Point-to-point trajectory in the joint space under kinematic constraints

We consider a robot with n degrees of freedom (dof). Let $\boldsymbol{q}^{ini}$ and $\boldsymbol{q}^{fin}$ be the joint coordinate vectors corresponding, respectively, to the initial and the final configuration. Let $\mathbf{k}_v$ and $\mathbf{k}_a$ be the vectors of maximum joint velocities and maximum joint accelerations, respectively. The value of k_{vj} can be exactly calculated from the actuator specifications and transmission ratios, while the value of k_{aj} is approximated by the ratio of the maximum actuator torque to the maximum joint inertia (upper bound of the diagonal element A_{jj} of the inertial matrix defined in Chapter 1). Once the trajectory has been computed with these kinematic constraints, we can proceed to a time scaling [HOL 83] in order to better match the maximum joint torques using the dynamic model. In section 4.4, we will consider the problem of using optimization techniques to deal with dynamic constraints.

The trajectory between $\boldsymbol{q}^{ini}$ and $\boldsymbol{q}^{fin}$ is determined by the following equations:

$$\boldsymbol{q}(t) = \boldsymbol{q}^{ini} + r(t) \cdot \boldsymbol{D} \quad \text{for} \quad 0 \leq t \leq T \tag{4.1}$$

$$\dot{\boldsymbol{q}}(t) = \dot{r}(t) \cdot \boldsymbol{D} \tag{4.2}$$

where $\boldsymbol{D} = \boldsymbol{q}^{fin} - \boldsymbol{q}^{ini}$, $r(t)$ is an interpolation function and T is the duration of the motion (or transfer time).

The boundary conditions for $r(t)$ are given by:

$$\begin{cases} r(0) = 0 \\ r(T) = 1 \end{cases}$$

Several interpolation functions can provide a trajectory such that $\boldsymbol{q}(0) = \boldsymbol{q}^{ini}$ and $\boldsymbol{q}(T) = \boldsymbol{q}^{fin}$. We will study successively the fifth-order polynomial model and the trapezoidal velocity model.

4.2.1. *Fifth-order polynomial model*

The fifth-degree polynomial interpolation ensures a smooth trajectory that is continuous in positions, velocities and accelerations. The polynomial is obtained by using the following boundary conditions:

$$\boldsymbol{q}(0) = \boldsymbol{q}^{ini}, \ \boldsymbol{q}(T) = \boldsymbol{q}^{fin}, \ \dot{\boldsymbol{q}}(0) = 0, \ \dot{\boldsymbol{q}}(T) = 0, \ \ddot{\boldsymbol{q}}(0) = 0, \ \ddot{\boldsymbol{q}}(T) = 0$$

Solving the six constraints yields the following interpolation function:

$$r(t) = 10 \cdot \left(\frac{t}{T}\right)^3 - 15 \cdot \left(\frac{t}{T}\right)^4 + 6 \cdot \left(\frac{t}{T}\right)^5 \tag{4.3}$$

The position, velocity and acceleration with respect to time for joint j are plotted in Figure 4.1. Maximum velocity and acceleration are given by:

$$\left|\dot{q}_{j\max}\right| = \frac{15 \cdot \left|D_j\right|}{8 \cdot T} \qquad [4.4]$$

$$\left|\ddot{q}_{j\max}\right| = \frac{10 \cdot \left|D_j\right|}{\sqrt{3} \cdot T^2} \qquad [4.5]$$

Generally, the transfer time T from $\mathbf{q}^{ini}$ to $\mathbf{q}^{fin}$ is not specified. The goal is to minimize T while satisfying velocity and acceleration constraints. The approach is to calculate the minimum time separately for each joint, and then to synchronize all the joints at the largest minimum time between them.

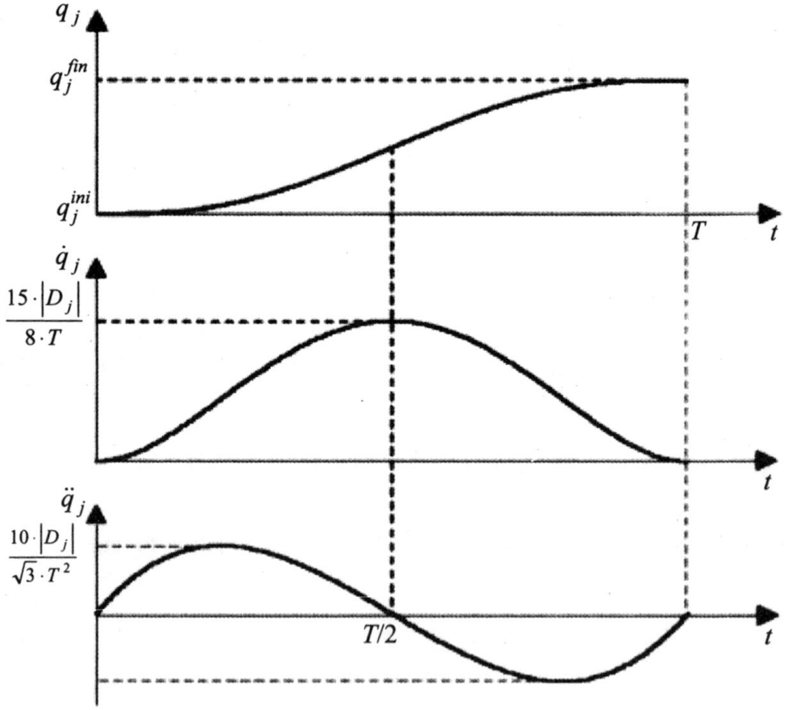

Figure 4.1. *Position, velocity and acceleration profiles for a quintic polynomial*

The minimum traveling time T_j for joint j occurs when either the velocity or the acceleration is saturated during the motion. It is calculated by:

$$T_j = \max\left[\frac{15 \cdot |D_j|}{8 \cdot \mathrm{k}_{vj}}, \sqrt{\frac{10 \cdot |D_j|}{\sqrt{3} \cdot \mathrm{k}_{aj}}}\right]$$

Thus, the global minimum traveling time T is equal to:

$$T = \max(T_1, \ldots, T_n) \qquad [4.6]$$

4.2.2. *Trapezoidal velocity model*

A trajectory generated using a trapezoidal velocity model consists of a constant acceleration phase followed by a constant velocity phase and terminated by a constant deceleration phase (see Figure 4.2). The initial and final velocities are zero. Thus, the trajectory is continuous in position and velocity, but discontinuous in acceleration. For kinematic constraints, we consider the symmetric case where the duration of acceleration and deceleration phases are equal and denoted by τ (in section 4.4 we will treat the problem with asymmetric phases). In this case, it is possible to saturate both the acceleration and the velocity in order to minimize the traveling time (see Figure 4.3) if the following condition is verified:

$$|D_j| > \frac{\mathrm{k}_{vj}^2}{\mathrm{k}_{aj}} \qquad [4.7]$$

If condition [4.7] is not verified, the maximum velocity that can be achieved for joint j will be obtained by:

$$\mathrm{k}_{vj} = \sqrt{|\mathrm{D}_j| \mathrm{k}_{aj}} \qquad [4.8]$$

Consequently, in the following, for the joints not verifying [4.7], the maximum velocity will be obtained from [4.8].

The trapezoidal velocity trajectory results in the minimum traveling time among those providing the continuity of velocity. The joint j trajectory (Figure 4.3) is represented by the following equations:

$$\begin{cases} q_j(t) = q_j^{ini} + \frac{1}{2} \cdot t^2 \cdot \mathrm{k}_{aj} \cdot \mathrm{sign}(D_j) & \text{for } 0 \le t \le \tau_j \\ q_j(t) = q_j^{ini} + (t - \frac{\tau_j}{2}) \cdot \mathrm{k}_{vj} \cdot \mathrm{sign}(D_j) & \text{for } \tau_j \le t \le T_j - \tau_j \\ q_j(t) = q_j^{fin} - \frac{1}{2} \cdot (T_j - t)^2 \cdot \mathrm{k}_{aj} \cdot \mathrm{sign}(D_j) & \text{for } T_j - \tau_j \le t \le T_j \end{cases} \quad [4.9]$$

with: $\tau_j = \frac{\mathrm{k}_{vj}}{\mathrm{k}_{aj}}$

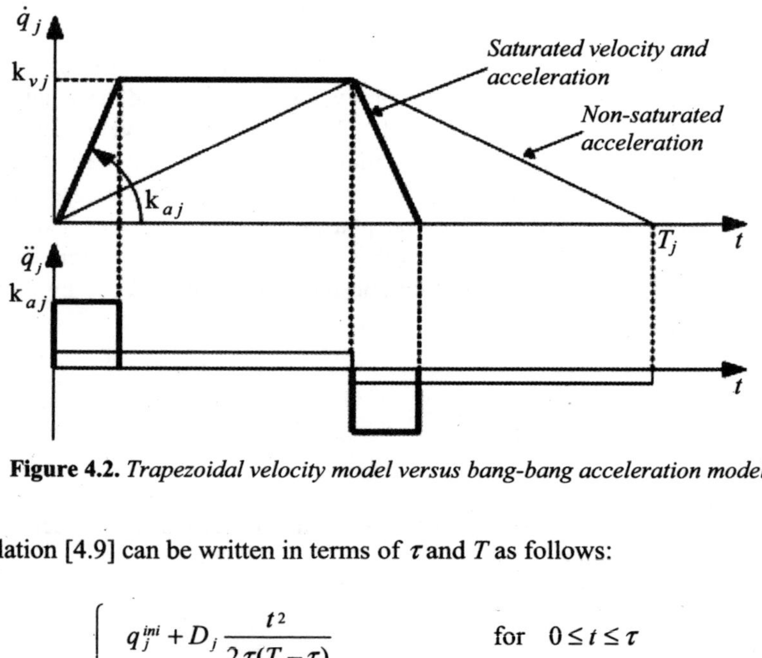

Figure 4.2. *Trapezoidal velocity model versus bang-bang acceleration model*

Relation [4.9] can be written in terms of τ and T as follows:

$$q_j(t) = \begin{cases} q_j^{ini} + D_j \dfrac{t^2}{2\tau(T-\tau)} & \text{for } 0 \le t \le \tau \\ q_j^{ini} + D_j \dfrac{(2t-\tau)}{2(T-\tau)} & \text{for } \tau \le t \le T - \tau \\ q_j^{ini} + D_j \left(1 - \dfrac{(T-t)^2}{2\tau(T-\tau)}\right) & \text{for } T - \tau \le t \le T \end{cases} \quad [4.10]$$

The area of the trapeze under the velocity curve is equal to the distance traveled in the interval [0, T_j], which can be written as:

$$\left|D_j\right| = \mathrm{k}_{vj} \cdot T_j - \frac{\mathrm{k}_{vj}^2}{\mathrm{k}_{aj}} \qquad [4.11]$$

Hence, the minimum time for joint j is given by:

$$T_j = \frac{\mathrm{k}_{vj}}{\mathrm{k}_{aj}} + \frac{\left|D_j\right|}{\mathrm{k}_{vj}} = \tau_j + \frac{\left|D_j\right|}{\mathrm{k}_{vj}} \qquad [4.12]$$

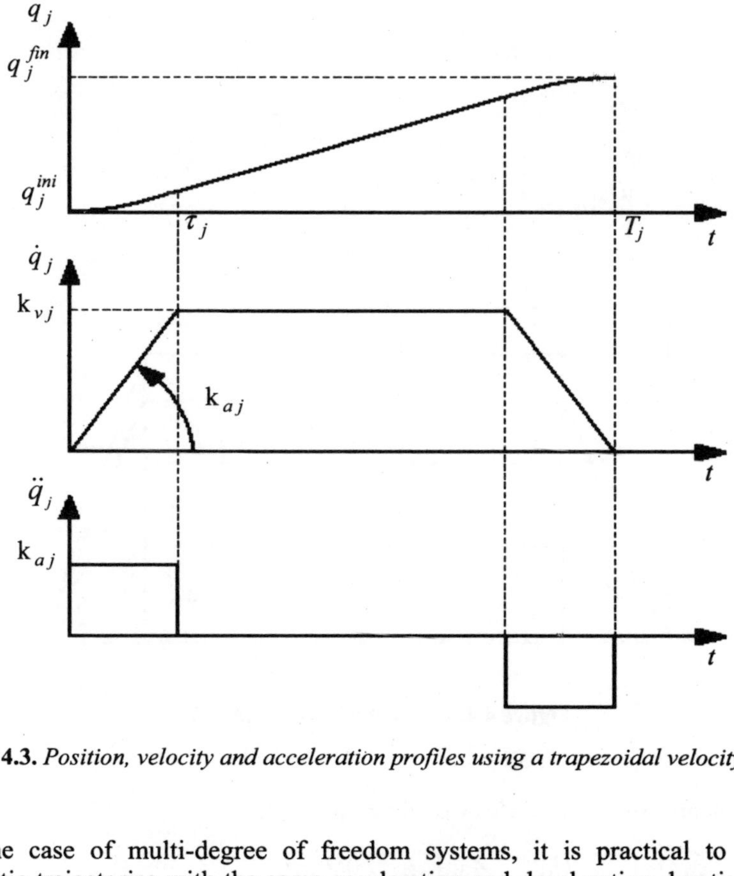

Figure 4.3. *Position, velocity and acceleration profiles using a trapezoidal velocity model*

In the case of multi-degree of freedom systems, it is practical to produce homothetic trajectories with the same acceleration and deceleration duration for all joints. Consequently, the velocity profile of joint j can be obtained in terms of the velocity of an arbitrary joint k using the following relation (see Figure 4.4):

$$\dot{q}_j(t) = \alpha_j \cdot \dot{q}_k(t) \quad \text{for } j = 1, \ldots, n \tag{4.13}$$

where α_j is constant for a given motion and $D_k \neq 0$ (see equation [4.10]).

In the following, we detail how to obtain a common minimum traveling time T under the condition that all the joints have a common τ. By doing so, the optimum τ and T are *a priori* not equal to any optimum τ_j and T_j, respectively, calculated for each joint separately (see equation [4.12]).

To calculate the optimal τ, resulting in a minimum time T, let us first solve the problem for two joints. Let $\lambda_j \, \mathrm{k}_{vj}$ be the maximum velocity of the synchronized trajectory for joint j and let $\upsilon_j \, \mathrm{k}_{aj}$ be the corresponding maximum acceleration. According to equation [4.11], the minimum traveling time for each joint, if calculated separately, would be:

$$\begin{cases} T_1 = \tau_1 + \dfrac{|D_1|}{\mathrm{k}_{v1}} \\ T_2 = \tau_2 + \dfrac{|D_2|}{\mathrm{k}_{v2}} \end{cases} \tag{4.14}$$

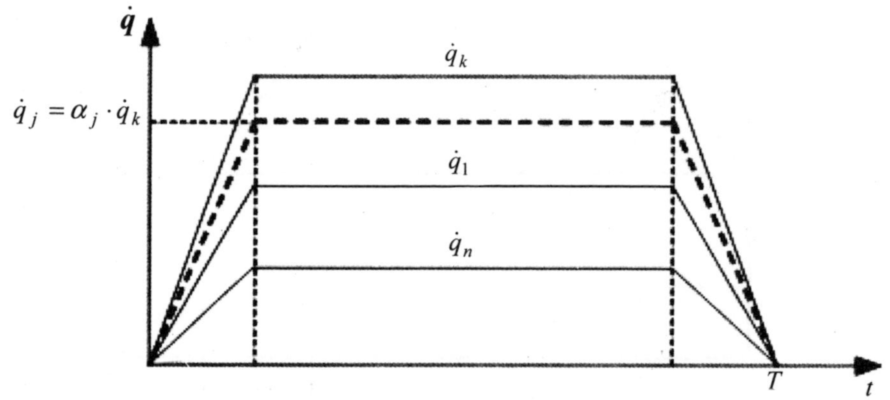

Figure 4.4. *Homothetic velocity profiles*

The synchronized trajectories would be such that:

$$T = \frac{\lambda_1 \cdot \mathrm{k}_{v1}}{\upsilon_1 \cdot \mathrm{k}_{a1}} + \frac{|D_1|}{\lambda_1 \cdot \mathrm{k}_{v1}} = \frac{\lambda_2 \cdot \mathrm{k}_{v2}}{\upsilon_2 \cdot \mathrm{k}_{a2}} + \frac{|D_2|}{\lambda_2 \cdot \mathrm{k}_{v2}} \tag{4.15}$$

with $T \geq \max(T_1, T_2)$.

From equation [4.15], it is straightforward to obtain:

$$\lambda_2 = \lambda_1 \cdot \frac{k_{v1} \cdot |D_2|}{k_{v2} \cdot |D_1|} \tag{4.16}$$

$$\upsilon_2 = \upsilon_1 \cdot \frac{k_{a1} \cdot |D_2|}{k_{a2} \cdot |D_1|} \tag{4.17}$$

Using [4.16] and taking into account that $0 \leq \lambda_j \leq 1$ yields:

$$0 \leq \lambda_1 \leq \frac{k_{v2} \cdot |D_1|}{k_{v1} \cdot |D_2|} \text{ and } \lambda_1 \leq 1 \tag{4.18}$$

Likewise, from the acceleration constraints, we obtain:

$$0 \leq \upsilon_1 \leq \frac{k_{a2} \cdot |D_1|}{k_{a1} \cdot |D_2|} \text{ and } \upsilon_1 \leq 1 \tag{4.19}$$

The minimum time T is obtained when the parameters λ_1 and υ_1 are at their largest value and satisfy simultaneously the above constraints, which result in:

$$\begin{cases} \lambda_1 = \min\left[1, \dfrac{k_{v2} \cdot |D_1|}{k_{v1} \cdot |D_2|}\right] \\ \upsilon_1 = \min\left[1, \dfrac{k_{a2} \cdot |D_1|}{k_{a1} \cdot |D_2|}\right] \end{cases} \tag{4.20}$$

and the corresponding duration of the acceleration phase is:

$$\tau = \frac{\lambda_1 \cdot k_{v1}}{\upsilon_1 \cdot k_{a1}} \tag{4.21}$$

These equations are easily generalized for n joints (assuming that $D_1 \neq 0$ and $D_j \neq 0$):

$$\begin{cases} \lambda_1 = \min\left[1\,,\ \dfrac{\mathrm{k}_{vj} \cdot |D_1|}{\mathrm{k}_{v1} \cdot |D_j|}\right] \\ \upsilon_1 = \min\left[1\,,\ \dfrac{\mathrm{k}_{aj} \cdot |D_1|}{\mathrm{k}_{a1} \cdot |D_j|}\right] \end{cases} \text{for } j = 2, \ldots, n \qquad [4.22]$$

We note that if $D_j = 0$, the joint j does not move; $q_{j(t)} = q_j^{ini}$

4.2.3. *Smoothed trapezoidal velocity model*

We can modify the previous method in order to have a trajectory that is continuous in acceleration, by replacing the acceleration and deceleration phases either by a second-degree polynomial (Figure 4.5a) or by a trapezoidal acceleration profile (Figure 4.5b) [CAS 84]. In this section, we detail the first approach, which is simpler to implement.

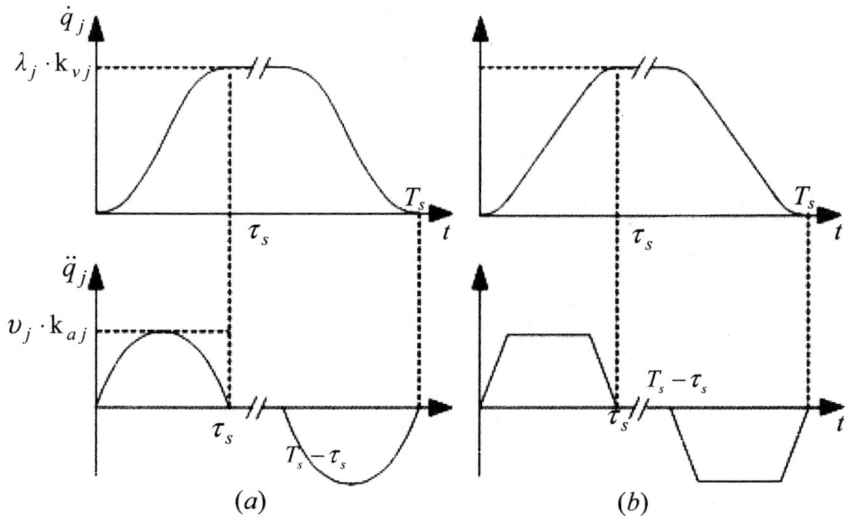

Figure 4.5. *Modification of the acceleration of the trapeze profile to ensure a continuous acceleration*

Let τ_s be the new duration of the acceleration phase and let λ_j k_{vj} be the maximum velocity in this case. The boundary conditions for joint j during the acceleration phase are defined as:

$$q_j(0)=q_j^{ini}\ ,\ \dot{q}_j(0)=0\ ,\ \dot{q}_j(\tau_s)=\lambda_j K_{vj}\, sign(D_j)\ ,\ \ddot{q}_j(0)=0\ ,\ \ddot{q}_j(\tau_s)=0 \qquad [4.23]$$

From these constraints, we derive the equations of position, velocity and acceleration of joint j for $1 \le j \le n$ and $0 \le t \le \tau_s$ as follows:

$$\begin{cases} q_j(t)=q_j^{ini}-\dfrac{1}{\tau_s^3}\cdot\lambda_j\cdot k_{vj}\cdot sign(D_j)\cdot(\dfrac{1}{2}\cdot t-\tau_s)\cdot t^3 & [4.24] \\ \dot{q}_j(t)=-\dfrac{1}{\tau_s^3}\cdot\lambda_j\cdot k_{vj}\cdot sign(D_j)\cdot(2\cdot t-3\cdot\tau_s)\cdot t^2 & [4.25] \\ \ddot{q}_j(t)=-\dfrac{6}{\tau_s^3}\cdot\lambda_j\cdot k_{vj}\cdot sign(D_j)\cdot(t-\tau_s)\cdot t & [4.26] \end{cases}$$

The joint position equation corresponding to the constant velocity phase, given a duration h_s, is as follows:

$$q_j(t)=q_j(\tau_s)+(t-\tau_s)\cdot\lambda_j\cdot k_{vj}\cdot sign(D_j) \ \text{ for } \tau_s\le t\le T_s-\tau_s \qquad [4.27]$$

By assuming that the acceleration and deceleration phases are symmetric ($T_s=2\tau_s+h_s$), the trajectory corresponding to the deceleration phase is defined in the interval $\tau_s+h_s\le t\le T_s$ as:

$$\begin{cases} q_j(t)=q_j^{fin}+\dfrac{1}{2}\left[\dfrac{1}{\tau_s^3}(t-\tau_s-T_s)(t+\tau_s-T_s)^3+(2t+\tau_s-2T_s)\right]\lambda_j k_{vj} sign(D_j) \\ \dot{q}_j(t)=\left[\dfrac{1}{\tau_s^3}(2t-\tau_s-2T_s)(t+\tau_s-T_s)^2+1\right]\lambda_j k_{vj} sign(D_j) \\ \ddot{q}_j(t)=\left[\dfrac{6}{\tau_s^3}(t-T_s)(t+\tau_s-T_s)\right]\lambda_j k_{vj} sign(D_j) \end{cases} \qquad [4.28]$$

The acceleration is maximal at $t = \tau_s/2$ and its magnitude is:

$$\left|\ddot{q}_{j\max}\right| = \frac{3}{2} \cdot \frac{\lambda_j \cdot \mathrm{k}_{vj}}{\tau_s} \qquad [4.29]$$

If we suppose $|\ddot{q}_{j\,max}|$ is equal to $\upsilon_j \mathrm{k}_{aj}$, and assume that all the joints have the same duration of acceleration, we obtain:

$$\tau_s = \frac{3}{2} \cdot \frac{\lambda_j \cdot \mathrm{k}_{vj}}{\upsilon_j \cdot \mathrm{k}_{aj}} \qquad [4.30]$$

According to equations [4.24] and [4.30], it should be noted that the distance traveled during the acceleration phase is equal to:

$$\left|q_j^{ini} - q_j(\tau_s)\right| = \frac{3}{4} \frac{(\lambda_j \cdot \mathrm{k}_{vj})^2}{\upsilon_j \cdot \mathrm{k}_{aj}} \qquad [4.31]$$

By calculating the area under the velocity curve, along the three parts of the motion, we verify that:

$$T_s = \tau_s + \frac{\left|D_j\right|}{\lambda_j \cdot \mathrm{k}_{vj}} \qquad [4.32]$$

This expression is similar to equation [4.12] giving the traveling time for the trapeze profile. This in turn suggests that λ_j and υ_j can be calculated with equations [4.22].

We also note that to saturate the velocity and the acceleration of the joint j trajectory, the distance to travel must be such that:

$$\left|D_j\right| > \frac{3}{2} \cdot \frac{\mathrm{k}_{vj}^2}{\mathrm{k}_{aj}}$$

If this condition is not verified, the maximum velocity k_{vj} of this joint will be taken equal to:

$$k_{vj} = \sqrt{\frac{2}{3} \cdot |D_j| \cdot k_{aj}} \qquad [4.33]$$

Now, if T_s and τ_s are chosen, the trajectory evolution of the jth joint is given as follows:

$$q_j(t) = \begin{cases} q_j^{ini} + D_j \dfrac{1}{2(T_s - \tau_s)} \left(\dfrac{2t^3}{\tau_s^2} - \dfrac{t^4}{\tau_s^3} \right) & 0 \le t \le \tau_s \\ q_j^{ini} + D_j \dfrac{(2t - \tau_s)}{2(T_s - \tau_s)} & \tau_s \le t \le T_s - \tau_s \\ q_j^{ini} + D_j \left(1 - \dfrac{1}{2(T_s - \tau_s)} \left(\dfrac{2(T_s - t)^3}{\tau_s^2} - \dfrac{(T_s - t)^4}{\tau_s^3} \right) \right) & T_s - \tau_s \le t \le T_s \end{cases} \qquad [4.34]$$

The above equations have the advantage that the maximum velocity and acceleration are not appearing explicitly.

4.3. Point-to-point trajectory in the task-space under kinematic constraints

Let ${}^0\boldsymbol{T}_E^{ini}$ and ${}^0\boldsymbol{T}_E^{fin}$ be the homogenous transformations describing the initial and final desired locations, respectively. For convenience, let us note:

$${}^0\boldsymbol{T}_E^{ini} = \begin{bmatrix} \boldsymbol{R}^{ini} & \boldsymbol{P}^{ini} \\ 0\ 0\ 0 & 1 \end{bmatrix} \text{ and } {}^0\boldsymbol{T}_E^{fin} = \begin{bmatrix} \boldsymbol{R}^{fin} & \boldsymbol{P}^{fni} \\ 0\ 0\ 0 & 1 \end{bmatrix}$$

The most common way to move from one location to the other is to split the motion into a linear translation between the origins of frames ${}^0\boldsymbol{T}_E^{ini}$ and ${}^0\boldsymbol{T}_E^{fin}$, and a rotation α around an axis of the end-effector ${}^E\boldsymbol{u}$ to align $\boldsymbol{R}^{ini}$ and $\boldsymbol{R}^{fin}$. The translation and rotation should be synchronized.

The distance the translation has to travel is given by:

$$D_t = \left\| \boldsymbol{P}^{fin} - \boldsymbol{P}^{ini} \right\| = \sqrt{(P_x^{fin} - P_x^{ini})^2 + (P_y^{fin} - P_y^{ini})^2 + (P_z^{fin} - P_z^{ini})^2} \qquad [4.35]$$

The terms $\boldsymbol{u}$ and α are calculated from the following relation:

$$\boldsymbol{R}^{ini}\, \boldsymbol{rot}(\boldsymbol{u}, \alpha) = \boldsymbol{R}^{fin} \quad [4.36]$$

where we recall that $\boldsymbol{rot}(\boldsymbol{u}, \alpha)$ is a (3×3) rotation matrix corresponding to a rotation of an angle α about a vector $\boldsymbol{u}$. Hence, we get:

$$rot(\boldsymbol{u},a) = \left[\boldsymbol{R}^{ini}\right]^T \cdot \boldsymbol{R}^{fin} = \begin{bmatrix} \left(\boldsymbol{s}^{ini}\right)^T \\ \left(\boldsymbol{n}^{ini}\right)^T \\ \left(\boldsymbol{a}^{ini}\right)^T \end{bmatrix} \cdot \left[\boldsymbol{s}^{fin} \quad \boldsymbol{n}^{fin} \quad \boldsymbol{a}^{fin}\right] = \begin{bmatrix} s_x & n_x & a_x \\ s_y & n_y & a_y \\ s_z & n_y & a_z \end{bmatrix} \quad [4.37]$$

Solving $\boldsymbol{u}$ and α we obtain [KHA 02]:

$$\begin{cases} C\alpha = \dfrac{1}{2} \cdot [s_x + n_y + a_z - 1] \\ S\alpha = \dfrac{1}{2} \cdot \sqrt{(n_z - a_y)^2 + (a_x - s_z)^2 + (s_y - n_x)^2} \\ \alpha = \operatorname{atan2}(S\alpha, C\alpha) \\ u = \dfrac{1}{2 \cdot S\alpha} \cdot \begin{bmatrix} n_z - a_y \\ a_x - s_z \\ s_y - n_x \end{bmatrix} \end{cases} \quad [4.38]$$

When $S\alpha$ is small, the vector $\boldsymbol{u}$ is calculated by identifying the terms of the main diagonal of $\boldsymbol{rot}(\boldsymbol{u},\theta)$ from [4.37] :

$$u_x = \pm\sqrt{\frac{s_x - C\theta}{1 - C\theta}}, \quad u_y = \pm\sqrt{\frac{n_y - C\theta}{1 - C\theta}}, \quad u_z = \pm\sqrt{\frac{a_z - C\theta}{1 - C\theta}}$$

The signs are determined from the expression of $\boldsymbol{u}$ given in [4.38]. We obtain:

$$\begin{cases} u_x = sign(n_z - a_y) \cdot \sqrt{\dfrac{s_x - C\theta}{1 - C\theta}} \\ u_y = sign(a_x - s_z) \cdot \sqrt{\dfrac{n_y - C\theta}{1 - C\theta}} \\ u_z = sign(s_y - n_x) \cdot \sqrt{\dfrac{a_z - C\theta}{1 - C\theta}} \end{cases}$$

Let k_{v1} and k_{a1} be the maximum velocity and acceleration for the translation motion, and let k_{v2} and k_{a2} be the maximum velocity and acceleration for the rotation motion. The methods described in section 4.2 can be used to generate a synchronized trajectory for the two variables D_t and α, resulting in the minimum time T. The trajectory of the end-effector frame is given by:

$$^{0}\boldsymbol{T}_E(t) = \begin{bmatrix} \boldsymbol{R}(t) & \boldsymbol{P}(t) \\ 0\ 0\ 0 & 1 \end{bmatrix}$$

with:

$$\boldsymbol{P}(t) = \boldsymbol{P}^{ini} + \frac{s(t)}{D_t} \cdot (\boldsymbol{P}^{fin} - \boldsymbol{P}^{ini}) = \boldsymbol{P}^{ini} + r(t) \cdot (\boldsymbol{P}^{fin} - \boldsymbol{P}^{ini}) \qquad [4.39]$$

$$\boldsymbol{R}(t) = \boldsymbol{R}^{ini} \cdot \mathbf{rot}(\boldsymbol{u}, r(t)\alpha) \qquad [4.40]$$

where $s(t) = D_t\, r(t)$ is the curvilinear distance traveled at time t and $r(t)$ is the interpolation function.

NOTE.– we can specify the rotation from $\boldsymbol{R}^{ini}$ to $\boldsymbol{R}^{fin}$ using the Euler angles ϕ, θ and ψ. Let (ϕ^{ini}, θ^{ini}, ψ^{ini}) and (ϕ^{fin}, θ^{fin}, ψ^{fin}) designate the angles corresponding to $\boldsymbol{R}^{ini}$ and $\boldsymbol{R}^{fin}$ respectively. Thus, equation [4.40] is replaced by:

$$\boldsymbol{R}(t) = \boldsymbol{R}^{ini}\, \boldsymbol{rot}(\boldsymbol{z}, \phi^{ini} + r(t)\ \phi)\ \boldsymbol{rot}(\boldsymbol{x}, \theta^{ini} + r(t)\ \theta)\ \boldsymbol{rot}(\boldsymbol{z}, \psi^{ini} + r(t)\ \psi) \qquad [4.41]$$

with $\phi = \phi^{fin} - \phi^{ini}$, $\theta = \theta^{fin} - \theta^{ini}$, $\psi = \psi^{fin} - \psi^{ini}$.

We can also choose to specify the rotation around an axis that is fixed with respect to frame R_0. In this case, $\boldsymbol{u}$ and α are calculated by solving:

$$rot(\boldsymbol{u}, \alpha)\ \boldsymbol{R}^{ini} = \boldsymbol{R}^{fin} \qquad [4.42]$$

The angular velocity ω of the end-effector, with respect to the frame where $\boldsymbol{u}$ is defined, is such that:

$$\omega = \boldsymbol{u} \cdot \dot{r}(t) \cdot \alpha = w\, \boldsymbol{u} \qquad [4.43]$$

4.4. Trajectory generation under kinodynamic constraints

This problem has been intensively studied during the last three decades and has been the focus of numerous papers. Concerning the special case of the optimal motion problem with geometric path constraints (OMPGPC), the interested reader is invited to consult the following references: [BOB 85], [SHI 85], [PFE 87], [TAN 88], [TAN 89], [JAQ 89], [SHI 91], [SHI 92], [JAM 96], [BET 99] and [CON 00]. Concerning the other case, which covers the optimal free motion planning problem (OFMPP), the following references deal with the simplified problem that consists of finding geometrically feasible paths: [LAT 91], [BAR 92], [OVE 92], [KAV 94], [YAM 94], [BAR 96], [KAV 96], [LAV 98], [LAT 99]. The techniques that handle the kinodynamic aspect of the optimal free motion planning problem are grouped in two families: *direct* and *indirect* methods [STR 93b], [HUL 97].

In general, indirect methods are applications of the optimal control theory. They use particularly Pontryagin's maximum principle (PMP) [PON 65], [BRY 75]. They state optimality conditions under the form of a boundary value problem [GEE 86], [CHE 90], [BES 92], [CHE 93], [LAZ 96], [DAN 98]. However, such techniques are applied usually to a limited number of cases and suffer from several drawbacks:

– they need to solve a non-linear multi-point shooting problem;

– they require an analytical form of the Hamiltonian gradient;

– they need a good initial guess as their region of convergence may be small;

– they use co-state variables that are, in general, quite difficult to estimate;

– they have difficulties in handling path constraints involving inequalities.

To overcome some of these drawbacks, direct methods have been proposed. They use a discretization of dynamic variables (state and/or control) to transform the difficult original infinite-dimension problem to a more tractable finite-parameter optimization problem. Then, non-linear programming [STR 93a], [STR 93b], [STE 95], [JAM 96], [BOB 01], [CHE 04] or sequential/parallel stochastic optimization techniques [RAN 96], [MAO 97], [CHE 02] are applied to calculate optimal values of design parameters. Although they have been used successfully to solve a large variety of problems, techniques based on non-linear programming have two essential shortcomings:

– they are easily attracted by local minima;

– differential equations of motion are integrated implicitly or explicitly at each iteration of the optimization process. Thus, the direct dynamic model is needed;

– continuity of second order is assumed, while realistic physical models may include discontinuous terms (for example, frictions).

In this section we present an efficient direct approach based on a discretization of joint position variables [CHE 02]. This method can be adapted to treat either the optimal motion problem with geometric path constraints or the optimal free motion planning problem with kinodynamic constraints. The resulting non-linear optimization problem is solved by a stochastic process. This versatile method offers several practical advantages:

– it makes use of the inverse dynamic model, therefore it does not integrate any set of differential equations and thus the reduction of computational burden can be quite significant;

– it can treat obstacle avoidance, provided a roadmap pre-processing phase is added to the basic scheme;

– it remains applicable in the case of discontinuous dynamic models such as those involving friction efforts;

– it can be adapted to handle various types of optimization criteria since the derivatives of the cost function are never evaluated;

– it is applicable for different systems such as wheeled mobile robots [HAD 05], closed chain robots [CHE 05] and bipedal robots [SAI 05].

A deterministic variant of this method has been proposed in [CHE 04] using the powerful sequential quadratic programming (SQP) technique of optimization. Solutions have been found for dynamically constrained problems involving obstacle avoidance and grasping moving objects. Here, however, we will focus on the general method of [CHE 02], which is characterized by a greater flexibility since additional constraints can be handled without inducing any significant changes in the procedure. In particular, we will give results obtained using trapezoidal velocity models which might be of interest in an industrial context.

4.4.1. *Problem statement*

Let us consider a serial manipulator with n dof required to move from an initial location $\boldsymbol{X}^{ini}$ to a final location $\boldsymbol{X}^{fin}$. Motion equation for the ith joint can be written as (see Chapter 1):

$$\sum_{j=1}^{n} A_{ij}\left(\boldsymbol{q}(t)\right)\cdot \ddot{q}_j(t) + c_i\left(\boldsymbol{q}(t),\dot{\boldsymbol{q}}(t)\right) + Q_i\left(\boldsymbol{q}(t)\right) = \Gamma_i(t) \qquad [4.44]$$

Recall that $\dot{\boldsymbol{q}}$ and $\ddot{\boldsymbol{q}}$ are, respectively, joint velocity and acceleration vectors. $\boldsymbol{A}(\boldsymbol{q})$ is the inertia matrix, $\boldsymbol{c}(\boldsymbol{q},\dot{\boldsymbol{q}})$ is the vector of centrifugal and Coriolis forces in which joint velocities appear under a quadratic form. $\boldsymbol{Q}(\boldsymbol{q})$ is the vector of gravity forces and $\boldsymbol{\Gamma}$ is the vector of actuator efforts.

The problem consists of determining the transfer time T, the joint trajectory vector $\boldsymbol{q}(t)$ and its derivatives with respect to time as well as the corresponding actuator efforts $\boldsymbol{\Gamma}(t)$, such that all constraints are respected and a given goal function F_{obj} is minimized.

4.4.1.1. *Constraints*

a) Boundary conditions

We suppose that locations $\boldsymbol{X}^{ini}$ and $\boldsymbol{X}^{fin}$ are characterized by zero joint velocities and that the corresponding joint positions $\boldsymbol{q}^{ini}$ and $\boldsymbol{q}^{fin}$ are given. Thus, we have:

$$\boldsymbol{q}(t=0) = \boldsymbol{q}^{ini}; \quad \boldsymbol{q}(t=T) = \boldsymbol{q}^{fin} \qquad [4.45a]$$

$$\dot{\boldsymbol{q}}(t=0) = 0; \quad \dot{\boldsymbol{q}}(t=T) = 0 \qquad [4.45b]$$

b) Geometric constraints

First, there are constraints imposed on joint positions:

$$|q_i(t)| \leq q_i^{max} \quad i = 1, \ldots, n \quad 0 \leq t \leq T \qquad [4.46a]$$

But if obstacles are present in the workspace, collisions must be avoided and the following additional constraint will hold during the transfer:

$$C(\boldsymbol{q}(t)) = \text{False} \quad 0 \leq t \leq T \qquad [4.46b]$$

Here, C denotes a Boolean function that indicates whether the robot at configuration $\boldsymbol{q}$ is in collision either with an obstacle or with itself.

Now, if the robot is required to move along a prescribed geometric path, then we suppose that:

– this path is already given in the operational space under a parametric form $\boldsymbol{X}(r)$ in terms of an independent parameter $r \in [0, 1]$;

– all geometric constraints as well as boundary conditions [4.45a] are already accounted for.

c) Kinematic constraints

Depending on the problem, any of the following constraints may have to be satisfied during the transfer ($0 \le t \le T$):

For $i = 1, \ldots, n$

$$-\textit{joint velocities}: \quad \left|\dot{q}_i(t)\right| \le \mathrm{k}_{vi} \qquad [4.47a]$$

$$-\textit{joint accelerations}: \quad \left|\ddot{q}_i(t)\right| \le \mathrm{k}_{ai} \qquad [4.47b]$$

$$-\textit{joint jerks}: \quad \left|\dddot{q}_i(t)\right| \le \mathrm{k}_{ji} \qquad [4.47c]$$

These constraints define bounds on the robot kinematics performances due to technological restrictions or to the nature of the task to be achieved. Of course, non-symmetrical bounds can also be treated.

d) Dynamic constraints

We consider here constraints imposed on the amplitude of joint torques, i.e.:

$$\left|\Gamma_i(t)\right| \le \Gamma_i^{max} \quad i = 1, \ldots, n \qquad [4.48]$$

4.4.1.2. *Objective function*

The objective function F_{obj} represents the cost of the task's achievement. It is generally an expression involving significant physical parameters related to the robot behavior and to the productivity of the robotic system. The following expression is a balance between transfer time and quadratic averages of actuator efforts (A) and power (B):

$$F_{obj} = \mu T + (1-\mu)(\eta\, A + (1-\eta)\, B) \qquad [4.49]$$

A and B have the dimension of time:

$$A = \int_0^T \sum_{i=1}^{n} \left(\frac{\Gamma_i}{\Gamma_i^{\max}} \right)^2 dt\, ; \quad B = \int_0^T \sum_{i=1}^{n} \left(\frac{\dot{q}_i \cdot \Gamma_i}{\mathrm{k}_{vi} \cdot \Gamma_i^{\max}} \right)^2 dt.$$

μ and η are dimensionless weight coefficients: $0 \leq \mu \leq 1$ and $0 \leq \eta \leq 1$. The limit case $\mu = 1$ corresponds to the minimum-time probiem (i.e. $F_{obj} = T$) while the case $\mu = 0$ corresponds to the minimum-effort problem (if $\eta = 1$) or to the minimum-power problem (if $\eta = 0$) . Additional coefficients w_i ($i = 1\ldots n$) may be incorporated in the above sums if it is desired to give more weight to some of the joints.

4.4.2. *Description of the method*

The method is described below in the general case of a free motion (OFMPP). The other problem with a prescribed path (OMPGPC) can be viewed as a particular case.

4.4.2.1. *Outline*

Let $\boldsymbol{q}(t)$ be any given trajectory with transfer time T. We may always time-scale this trajectory as follows:

$$\boldsymbol{q}(t) \equiv \tilde{\boldsymbol{q}}(\xi) \; o \; \xi(t) \qquad [4.50]$$

Here, $\xi(t) \equiv t/T$ is a linear time-scale function designed so that ξ will be in [0, 1]. The parametric form $\tilde{\boldsymbol{q}}(\xi)$ is what we call the trajectory profile which, in essence, gives the shape of the time history of the joint positions, from the beginning of the motion ($\xi = 0$) to its end ($\xi = 1$). We say that a trajectory profile is valid if it leads to a trajectory that is geometrically feasible (no co. sion) and that satisfies required boundary conditions. Hereafter, the prime symbol will be reserved exclusively to indicate derivatives with respect to ξ, while the dot symbol will refer as usual to derivatives with respect to t.

The problem requires solving for the optimal trajectory $\boldsymbol{q}^{best}(t)$ with unknown transfer time T^{best}, but with the above scaling this problem reduces to finding only the optimal profile $\tilde{\boldsymbol{q}}^{best}(\xi)$ because the transfer time may be viewed now as a dependent parameter. Indeed, suppose we choose arbitrarily a valid candidate profile $\tilde{\boldsymbol{q}}$. Then, the performance index [4.49] would become a function $F_{obj}(T \mid \tilde{\boldsymbol{q}})$ of the single variable T. Therefore, via a straightforward (albeit constrained) one-dimensional minimization, we may readily deduce (sometimes analytically) the critical transfer time T_q that would give the lowest possible cost $F_q \equiv F_{obj}(T_q \mid \tilde{\boldsymbol{q}})$ for the arbitrarily proposed $\tilde{\boldsymbol{q}}$. In other words, to any valid profile $\tilde{\boldsymbol{q}}$ will correspond a specific score F_q and an associated transfer time T_q, both of which are being dictated by kinodynamic constraints.

The proposed method uses a stochastic optimization scheme that evaluates the score of random candidate profiles for a global minimization of F_{obj}. This is in effect a nested optimization scheme. At the inner level, there is a slave clipping process that calculates the score of a single candidate by solving a one-dimensional constrained minimization problem using a standard non-linear programming approach. At the outer level there is a master hit-and-miss process carried out using, for example, a simulated annealing technique that selects valid candidates for a global minimization of F_{obj}. The output is the (approximate) overall lowest-cost profile $\tilde{\boldsymbol{q}}^{best}(\xi)$ which will be used next in [4.50], along with its associated transfer time T^{best}, to yield the (approximate) optimal solution of the problem.

Naturally, this quite difficult infinite dimensional constrained optimization is transformed to a more tractable finite dimensional parametric optimization via a discrete representation of trajectory profiles. Namely, each trial profile will be defined as a finite set of free nodes, continuity being insured by fitting a sufficiently smooth model that accounts for imposed boundary conditions. The major drawback is that the search space will be clearly restricted to the sub-space reachable via the selected fitting model. Hence, solutions by this technique are expected to be sub-optimal. On the other hand, for various forms of cost functions and types of kinodynamic constraints, the problem boils down to finding the optimal position of a few randomly disrupted nodes. As this can be treated quite efficiently by standard stochastic optimization routines, the proposed approach turns out to be remarkably flexible as well as computationally competitive. In general, it is able to provide reasonably good approximate solutions to a variety of practical dynamic problems.

4.4.2.2. *Construction of a random trajectory profile*

Choosing a trajectory profile amounts to specifying both a geometric path and a motion on this path. It is then convenient to express a trajectory profile as follows:

$$\tilde{\boldsymbol{q}}(\xi) \equiv \boldsymbol{p}(r) \circ r(\xi) \qquad [4.51]$$

The parametric form $\boldsymbol{p}(r)$, with $r \in [0, 1]$, is the robot path, while the monotonically increasing function $r(\xi)$ is the motion profile.

The former is a continuous time-independent sequence of joint positions that describes completely the geometry of the path in the n-dimensional joint space as r varies continuously from $r = 0$ (initial state) to $r = 1$ (final state). Evidently, a feasible path must satisfy all geometric constraints for any r in [0, 1] and must also satisfy boundary conditions [4.45a] imposed on joint positions, namely:

$$\boldsymbol{p}(r=0) = \boldsymbol{q}^{ini} \text{ and } \boldsymbol{p}(r=1) = \boldsymbol{q}^{fin} \qquad [4.52a]$$

Since a given feasible path can be executed differently in time, we need to specify the execution mode. This is done by choosing the motion profile $r(\xi)$ which gives the shape of the time history of parameter r. For obvious reasons, a valid motion profile must satisfy the following conditions of consistency:

$$r(\xi = 0) = 0, \quad r(\xi = 1) = 1 \qquad [4.52b]$$

$$r'(\xi) \geq 0 \qquad [4.52c]$$

In addition, it must satisfy boundary conditions [4.45b] imposed by joint velocities, namely:

$$r'(\xi = 0) = r'(\xi = 1) = 0 \qquad [4.52d]$$

Introducing now the discretization scheme, any trajectory profile will be represented using $n + 1$ distinct sets of nodes: one set for the motion profile $r(\xi)$ and n other sets for the n components of the path vector $p(r)$.

Figure 4.6 shows a set of N_m discretization nodes distributed in the (r, ξ) plan. The first and last nodes are held fixed according to [4.52b]. However, the free interior nodes can be randomly positioned to produce a trial motion profile $r(\xi)$ using any appropriate fitting model that accounts for conditions [4.52c] and [4.52d]. Similarly, with n other sets, each with N_p nodes, we can generate the n components of the path vector $\boldsymbol{p}(r)$ (see Figure 4.7).

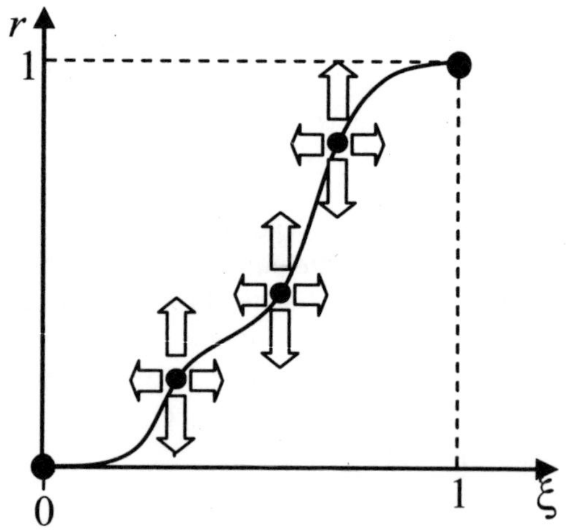

Figure 4.6. *Motion profile $r(\xi)$ built using N_m nodes*

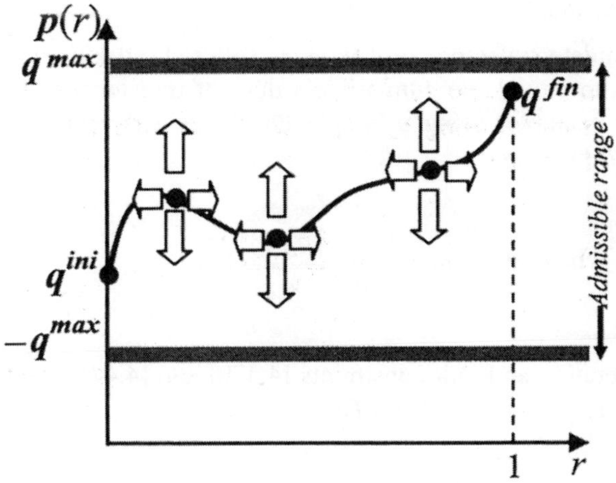

Figure 4.7. *A component of a path vector* $\mathbf{p}(r)$ *using* N_p *nodes*

One straightforward way to generate smooth curves from any given interpolation points is to use a cubic spline model (see section 4.4.3 for other models). A cubic spline model [EDW 82], [LIN 83] insures continuity up to second order. It also offers a good compromise between the cost of computations and the quality of the solutions. For the path $p(r)$, we may use a so-called natural cubic spline model [EDW 82], [LIN 83] which only requires $d^2\mathbf{p}/dr^2 = 0$ at the end points (but leaves $d\mathbf{p}/dr$). However, for $r(\xi)$, it would be more appropriate to use a clamped cubic spline model, which is particularly adapted to handle the first-order boundary conditions in [4.52d].

The creation of natural and clamped spline functions is straightforward and is done by using, for example, the Splines Matlab toolbox by calling the function CSAPE with appropriate end-conditions.

The problem is therefore reduced to finding the best positions of some free nodes. A stochastic optimization technique can scan randomly the available solution space of these nodes and interpolates between them to propose both a path and a motion profile. Then, the resulting trajectory profile $\tilde{\mathbf{q}}(\xi)$ is deduced and its corresponding score $F_{\mathbf{q}}$ and transfer time $T_{\mathbf{q}}$ are calculated. This cycle is repeated, generating new random candidate trajectory profiles, while keeping track of the overall best score and the associated transfer time.

4.4.2.3. *Handling kinodynamic constraints*

Once a candidate trajectory profile $\tilde{\boldsymbol{q}}$ is selected, all kinodynamic constraints simply translate to bounds on admissible values of transfer time for that candidate [CHE 02]. For example, using a simple chain-rule differentiation, joint velocity constraints [4.47a] become:

$$T \geq T_V \quad \text{where } T_V = \max_{i=1,\ldots,n}\left[\max_{\xi\in[0,1]}\frac{|\tilde{q}_i'(\xi)|}{k_{vi}}\right] \qquad [4.53a]$$

While acceleration and jerk constraints [4.47b] and [4.47c] yield two new lower bounds marked, respectively, T_A and T_J:

$$T \geq T_A \quad \text{where } T_A = \max_{i=1,\ldots,n}\left[\max_{\xi\in[0,1]}\frac{|\tilde{q}_i''(\xi)|}{k_{ai}}\right]^{1/2} \qquad [4.53b]$$

$$T \geq T_J \quad \text{where } T_J = \max_{i=1,\ldots,n}\left[\max_{\xi\in[0,1]}\frac{|\tilde{q}_i'''(\xi)|}{k_{Ji}}\right]^{1/3} \qquad [4.53c]$$

However, dynamic constraints [4.48] may translate to bilateral bounds T_L and T_R (see the example given below):

$$T \in [T_L, T_R] \qquad [4.54]$$

Intersecting [4.53] and [4.54] yields a final interval $[T_{Lower}, T_{Upper}]$ of admissible values of transfer time T that accounts for kinodynamic constraints. The optimal transfer time $T_{\boldsymbol{q}}$ of any selected trajectory profile $\tilde{\boldsymbol{q}}(\xi)$ is therefore deduced by minimizing $F_{obj}(T \mid \tilde{\boldsymbol{q}})$ within this final interval.

To illustrate this approach, let us introduce the following notation:

$$h_i(\xi) = \sum_{j=1}^{n} A_{ij}(\tilde{\boldsymbol{q}}(\xi)) \cdot \tilde{q}_j''(\xi) + c_i(\tilde{\boldsymbol{q}}(\xi), \tilde{\boldsymbol{q}}'(\xi))$$

$$\bar{\Gamma}_i(\xi) = \frac{\Gamma_i(\xi)}{\Gamma_i^{\max}}, \quad \bar{h}_i(\xi) = \frac{h_i(\xi)}{\Gamma_i^{\max}}, \quad \bar{Q}_i(\xi) = \frac{Q_i(\tilde{\boldsymbol{q}}(\xi))}{\Gamma_i^{\max}}$$

The motion equation [4.44] becomes:

$$\overline{\Gamma}_i(\xi) = \frac{1}{T^2} \cdot \overline{h}_i(\xi) + \overline{Q}_i(\xi) \qquad \xi \in [0,1]$$

Dynamic constraint [4.48] becomes:

$$-1 \leq \frac{1}{T^2} \cdot \overline{h}_i(\xi) + \overline{Q}_i(\xi) \leq 1$$

which can be written as:

$$-b(\xi) \leq \frac{1}{T^2} \cdot \overline{h}_i(\xi) \leq a(\xi)$$

where $a_i(\xi) = 1 - \overline{Q}_i(\xi)$ and $b_i(\xi) = 1 + \overline{Q}_i(\xi)$.

For every $\xi \in [0,1]$, the resulting bounds for T are summarized in the following table, depending on the sign of $\overline{h}_i(\xi)$, $a_i(\xi)$ and $b_i(\xi)$.

<table>
<tr><td colspan="3">$\overline{h}_i(\xi) \geq 0$</td><td colspan="3">$\overline{h}_i(\xi) < 0$</td></tr>
<tr><td>$a_i(\xi) < 0$</td><td colspan="2">$a_i(\xi) \geq 0$</td><td>$b_i(\xi) < 0$</td><td colspan="2">$b_i(\xi) \geq 0$</td></tr>
<tr><td rowspan="2">$\varnothing$</td><td>$b_i(\xi) < 0$</td><td>$b_i(\xi) \geq 0$</td><td rowspan="2">$\varnothing$</td><td>$a(\xi) < 0$</td><td>$a(\xi) \geq 0$</td></tr>
<tr><td>$T_{L_i} = \sqrt{\frac{\overline{h}_i(\xi)}{a_i}}$
$T_{R_i} = \sqrt{\frac{\overline{h}_i(\xi)}{-b_i}}$</td><td>$T_{L_i} = \sqrt{\frac{\overline{h}_i(\xi)}{a_i}}$
$T_{R_i} = +\infty$</td><td>$T_{L_i} = \sqrt{\frac{\overline{h}_i(\xi)}{a_i}}$
$T_{R_i} = \sqrt{\frac{\overline{h}_i(\xi)}{-b_i}}$</td><td>$T_{L_i} = \sqrt{\frac{\overline{h}_i(\xi)}{-b_i}}$
$T_{R_i} = +\infty$</td></tr>
</table>

Table 4.1. *Bounds on T due to dynamic constraints for a given* $\tilde{\boldsymbol{q}}(\xi)$

The cost function [4.49], restricted here for simplicity to the case where $\eta = 1$, becomes:

$$F_{obj} = T \cdot \left(S_0 + \frac{S_2}{T^2} + \frac{S_4}{T^4} \right) \quad [4.55]$$

where S_0, S_2 and S_4 are real coefficients given by:

$$S_0 = \mu + (1-\mu) \cdot \int_0^1 \sum_{i=1}^{n} \overline{Q}_i^2 d\xi \, ;$$

$$S_2 = 2 \cdot (1-\mu) \cdot \int_0^1 \sum_{i=1}^{n} \left(\overline{h}_i \cdot \overline{Q}_i \right) d\xi \, ;$$

$$S_4 = (1-\mu) \cdot \int_0^1 \sum_{i=1}^{n} \overline{h}_i^2 d\xi \, .$$

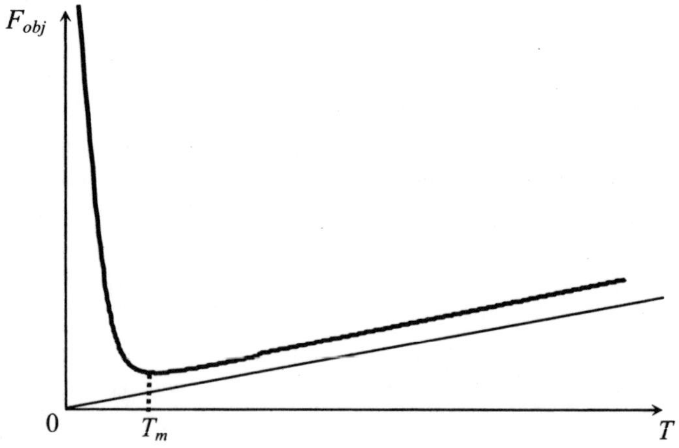

Figure 4.8. F_{obj} *vs* T

These coefficients depend implicitly on the selected candidate profile $\tilde{\boldsymbol{q}}(\xi)$ but do not depend on T. Also, S_0 and S_4 are always positive. Equation [4.55] represents a family of curves whose general shape is shown for any candidate in Figure 4.8. Such a curve exhibits a unique minimum at $T = T_m$ given by:

$$T_m = \left(\frac{S_2 + \sqrt{S_2^2 + 12 S_0 S_4}}{2 S_0} \right)^{1/2} = \left(\frac{6 \cdot S_4}{\sqrt{S_2^2 + 12 S_0 S_4} - S_2} \right)^{1/2} \qquad [4.56]$$

This result is obtained by setting to zero the derivative of [4.55] with respect to T. By doing so, we obtain the following equation: $S_0 T^4 - S_2 T^2 - 3 S_4 = 0$. The negative root is rejected, while the positive root is given above in [4.56]. Although the two forms given for T_m are mathematically equivalent we should use the first if S_2 is positive, and the second if S_2 is negative, in order to reduce rounding errors. Note that for some complicated cost functions, especially those that include discontinuous friction efforts, the expression of T_m might not be available explicitly, although its value would still be readily accessible numerically.

If kinodynamic constraints were not included, the critical value T_q that minimizes the cost function for the imposed profile $\tilde{q}$ would be simply $T_q = T_m$. But if kinodynamic constraints were included, then T_m might end up outside the admissible interval $[T_{Lower}, T_{Upper}]$. Therefore, following [CHE 02], three situations (a, b, c) must be distinguished (Figure 4.9) depending on which the critical value T_q is respectively, $T_q = T_{Upper}$, $T_q = T_m$ or $T_q = T_{Lower}$.

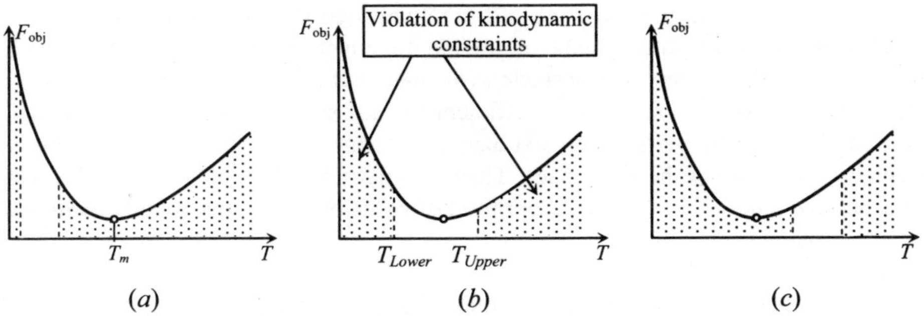

Figure 4.9. *Position of T_m relatively to $[T_{Lower}, T_{Upper}]$*

Note that for a minimum-time problem ($\mu = 1$), the curve simply reduces to a straight line, in which case we will have $T_q = T_{Lower}$ (Figure 4.10).

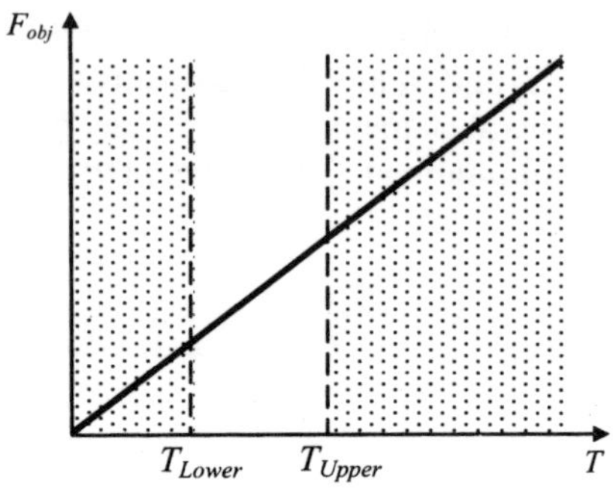

Figure 4.10. *Minimum-time cost function and the effect of kinodynamic constraints*

4.4.2.4. *Summary*

Figure 4.11 illustrates one cycle of the optimization process that is applied, in the case of an optimal free motion planning problem, to one candidate trajectory profile. In Step 1, a random path is generated and checked for all geometric constraints. Next, a feasible random motion profile is generated and a candidate trajectory profile is deduced. Then, kinodynamic constraints, which yield a search interval of admissible values of T, are included. Finally, the score and the optimal transfer time of the current candidate are calculated.

This cycle, which constitutes in fact a simple one-dimensional minimization of a cost function in a restricted interval, is repeated in a global stochastic optimization process. In practice, we can use a conventional simulated annealing technique, which is known for its efficiency in exploring large solutions spaces [KIR 83], [HAJ 85]. We can also use the modified hill-climbing algorithm detailed in [CHE 02], which includes an adaptive scheme of gradual node insertion that is particularly well adapted to problems involving obstacle avoidance. Let us note that, for this category of problems, it will be much more efficient to bias suitably the stochastic search. This can be done by performing first a preprocessing phase that builds a visibility graph or a roadmap of the workspace. Then, geometrically feasible path profiles will be generated randomly in the neighborhood of collision-free paths, readily available from the roadmap.

For the particular case of an optimal motion problem with geometric path constraints, the only modification in the procedure concerns the elimination of the first step since the path $\boldsymbol{p}(r)$ is no longer randomly generated. The path $\boldsymbol{p}(r)$ and its derivatives with respect to r are deduced via, respectively, the inverse geometric model and the inverse kinematic models, from the parametric form $\boldsymbol{X}(r)$ imposed in the operational space. Note that this operation only needs to be done once, since subsequent trajectory profiles $\tilde{\boldsymbol{q}}(\xi)$ will necessarily follow the same geometric path.

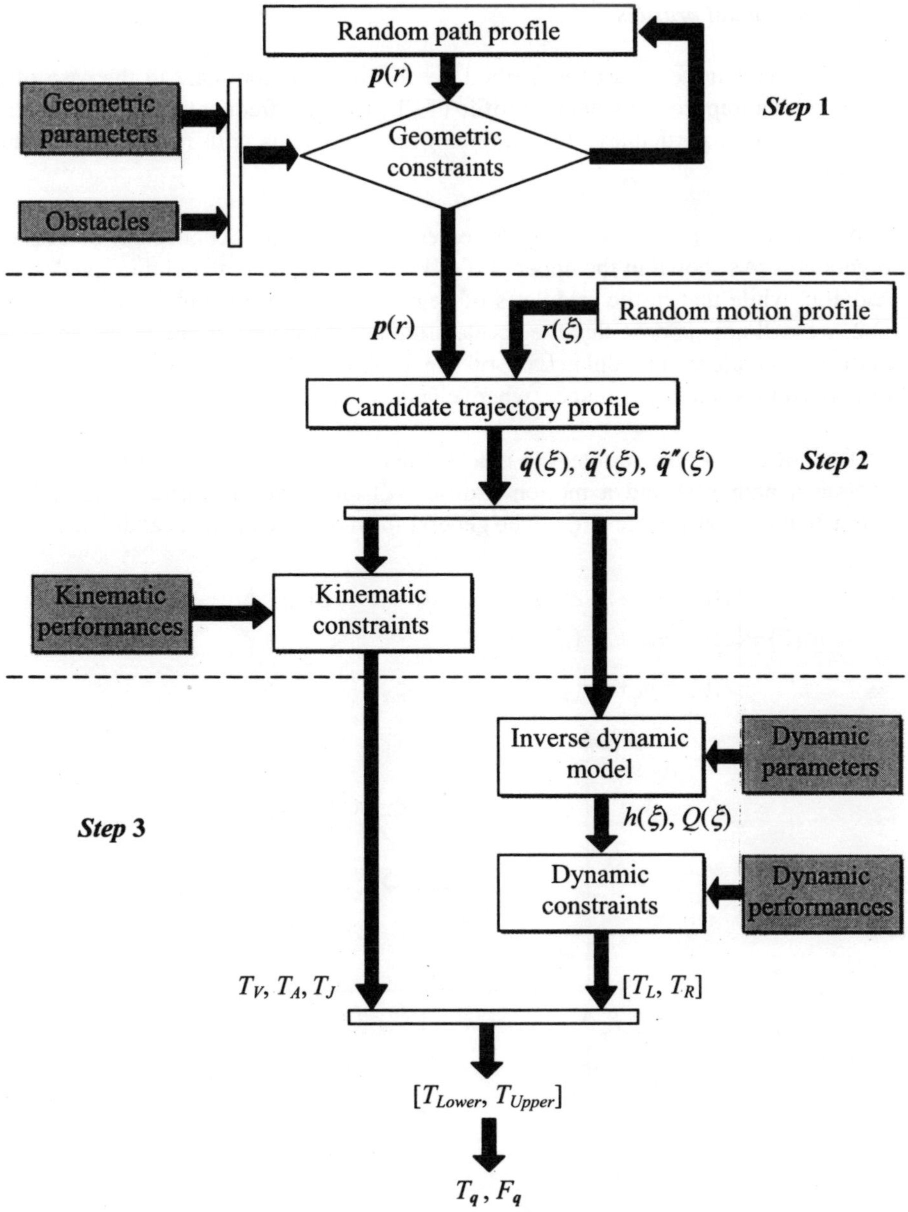

Figure 4.11. *Evaluation of the score F_q and the optimal transfer time T_q corresponding to one candidate trajectory profile, taking into account kinodynamic constraints (point-to-point motion)*

4.4.3. *Trapezoidal profiles*

In this section we adjust the method described in section 4.4.2 to the case of a motion with a trapezoidal velocity profile (TVP) which is frequently implemented in industrial robots, particularly for minimum-time problems with no obstacle in the workspace.

We recall that a TVP consists essentially of three connected pieces of polynomials. As shown in the upper part of Figure 4.12, the first and third pieces are quadratic, while that in the middle is of degree one. Here, in contrast with section 4.2.2, we will not suppose that the acceleration and deceleration phases are equal in duration. Therefore, the unknowns of the problem are the positions along the normalized time axes of ξ_a and ξ_b, where changes in acceleration take place.

Here, of course, the problem is much easier to solve. There will be no need to generate a path $\boldsymbol{p}(r)$ and a motion profile $r(\xi)$ since we can produce directly a candidate trajectory profile $\tilde{\boldsymbol{q}}(\xi)$. The general form of a trapezoidal candidate is:

$$\tilde{q}_j(\xi) = \begin{cases} P_1(\xi) = a_0 + a_1 \cdot \xi + a_2 \cdot \xi^2 & 0 \le \xi \le \xi_a \\ P_2(\xi) = a_3 + a_4 \cdot (\xi - \xi_a) & \xi_a \le \xi \le \xi_b \\ P_3(\xi) = a_5 + a_6 \cdot (\xi - \xi_b) + a_7 \cdot (\xi - \xi_b)^2 & \xi_b \le \xi \le 1 \end{cases}$$

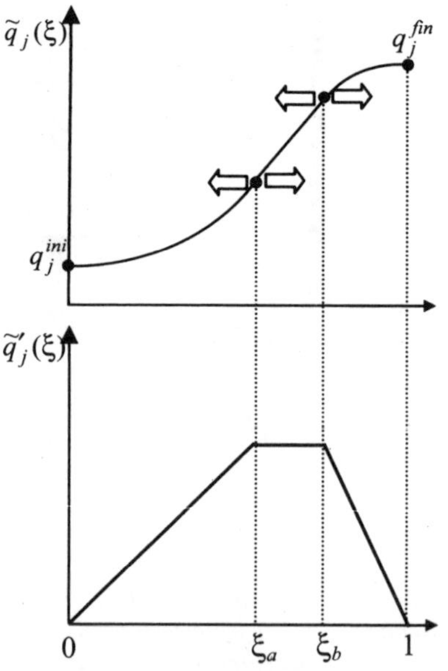

Figure 4.12. *Trapezoidal profile*

The constants a_0, a_1, …, a_7 are determined in terms of ξ_a and ξ_b from the following:

– *boundary conditions*:

$$\begin{cases} \tilde{q}_j(0) = q_j^{ini} \\ \tilde{q}'_j(0) = 0 \\ \tilde{q}_j(1) = q_j^{fin} \\ \tilde{q}'_j(1) = 0 \end{cases}$$

– *continuity conditions*:

$$\begin{cases} P_1(\xi_a) = P_2(\xi_a) \\ P'_1(\xi_a) = P'_2(\xi_a) \\ P_2(\xi_b) = P_3(\xi_b) \\ P'_2(\xi_b) = P'_3(\xi_b) \end{cases}$$

The solution of this system is:

$$\tilde{q}_j(\xi) = \begin{cases} q_j^{ini} + D_j \dfrac{\xi^2}{\xi_a \cdot (1 + \xi_b - \xi_a)} & 0 \le \xi \le \xi_a \\ q_j^{ini} + D_j \dfrac{(2 \cdot \xi - \xi_a)}{(1 + \xi_b - \xi_a)} & \xi_a \le \xi \le \xi_b \\ q_j^{ini} + D_j \left(1 - \dfrac{(1-\xi)^2}{(1-\xi_b) \cdot (1 + \xi_b - \xi_a)} \right) & \xi_b \le \xi \le 1 \end{cases} \quad [4.57]$$

Let us recall that: $D_j = q_j^{fin} - q_j^{ini}$.

By randomly selecting ξ_a and ξ_b in [0, 1] we produce a trial candidate using [4.57]. We then apply the same stochastic process described in the previous section to determine the optimal values of ξ_a and ξ_b, taking into account kinodynamic constraints.

In the case of an optimal motion problem with geometric path constraints the same technique remains applicable. However, since the path $\boldsymbol{p}(r)$ is being imposed, it is the motion profile $r(\xi)$ that is modeled using three pieces of polynomials. We obtain in this case trapezoidal profiles for curvilinear velocities along the imposed path. $r(\xi)$ is still expressed by equation [4.57] if we set $q_j^{ini} = 0$ and $q_j^{fin} = 1$.

We can modify [4.57] to avoid a discontinuity in accelerations. The resulting smoothed trapezoidal profile (Figure 4.13) is created by modeling both the acceleration and the deceleration phases by a quadratic form.

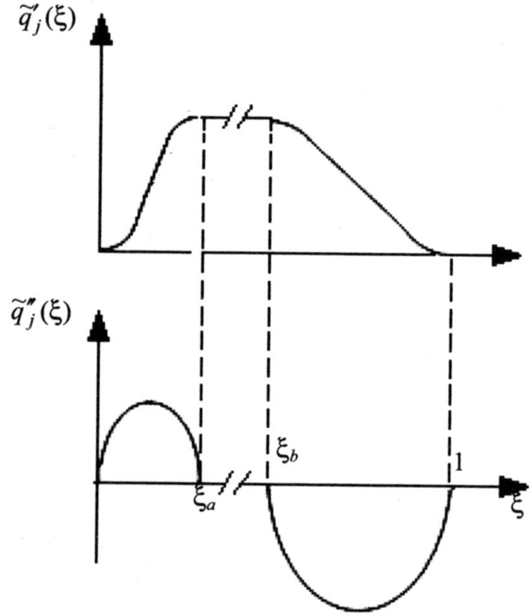

Figure 4.13. *Smoothed trapezoidal profiles*

The expression of the corresponding trajectory profile is given below, where the first and third pieces are fourth-order polynomials:

$$\tilde{q}_j(\xi) = \begin{cases} q_j^{ini} + D_j \dfrac{1}{(1+\xi_b-\xi_a)} \left(\dfrac{2\,\xi^3}{\xi_a^2} - \dfrac{\xi^4}{\xi_a^3} \right) & 0 \le \xi < \xi_a \\ q_j^{ini} + D_j \dfrac{(2\,\xi - \xi_a)}{(1+\xi_b-\xi_a)} & \xi_a \le \xi < \xi_b \\ q_j^{ini} + D_j \left(1 - \dfrac{1}{(1+\xi_b-\xi_a)} \left(\dfrac{2\,(1-\xi)^3}{(1-\xi_b)^2} - \dfrac{(1-\xi)^4}{(1-\xi_b)^3} \right) \right) & \xi_b \le \xi \le 1 \end{cases} \qquad [4.58]$$

4.5. Examples

We give numerical results obtained on a 2.5 GHz PC by a simulated annealing technique of optimization (see Appendix for more details). The annealing schedule uses a temperature reduction factor of 0.9. The maximum number of loops at constant temperature is 5. The maximum number of loops before step adjustment is set to 15. The number of annealing steps is fixed at 2 for trapezoidal profiles and 70 for spline profiles. Here, we focus mainly on the minimum-time problem that is of interest in industrial applications.

4.5.1. *Case of a two dof robot*

This case concerns the IBM SCARA robot. Its parameters are listed in Table 4.2 according to those reported in [LAZ 96].

Link i	$m_i[kg]$	$O_iG_i\,[m]$	$O_iO_{i+1}\,[m]$	$Izz_i\,[kg.m^2]$	$k_{vj}\,[rad/s]$	$\Gamma_i^{max}\,[N.m]$
1	5	0.35	0.7	15	1.2	3.5
2	10	0.50	0.625	4	2	3.5

Table 4.2. *Geometric and inertial parameters of the IBM SCARA*

4.5.1.1. *Optimal free motion planning problem*

The robot is required to move from the configuration (0, 1.2) to (2, 2.3) given in radians. The minimum-time motion problem has been treated using implementations of Pontryagin's maximum principle by [LAZ 96] under kinodynamic constraints. Results obtained using the method described in section 4.4 are listed in Table 4.3. The time evolution of joint velocities and torques are shown in Figures 4.14 and 4.15 using, respectively, trapezoidal and smoothed trapezoidal profiles.

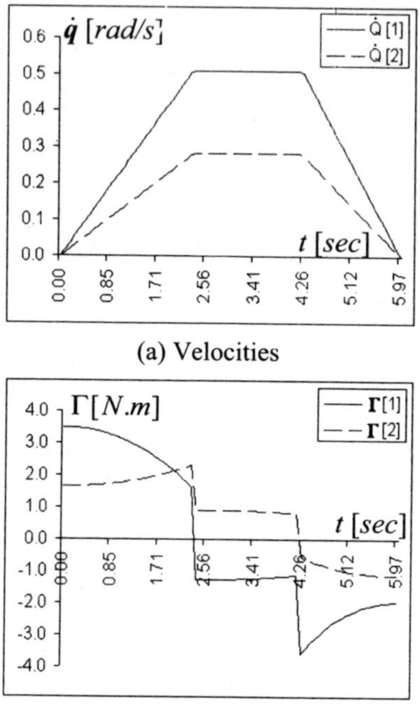

(a) Velocities

(b) Torques

Figure 4.14. *Motion obtained by a trapezoidal profile*

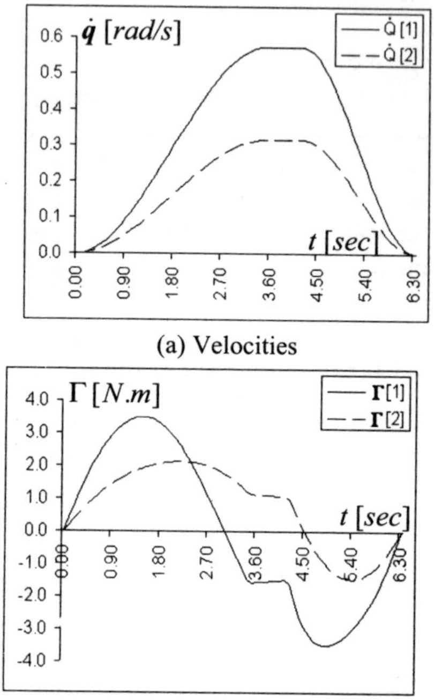

(a) Velocities

(b) Torques

Figure 4.15. *Motion obtained by a smoothed trapezoidal profile*

	Lazrek [LAZ 96]	Cubic spline (2 free nodes)	Trapezoidal	Smoothed trapezoidal
$T_q = F_q$ [*sec*]	4.23	4.29	5.97	6.30
Runtime [*sec*]	–	8.526	0.030	0.032

Table 4.3. *Simulation results for the minimum time transfer problem*

The same problem has been treated with a cost function that includes the quadratic average of joint torques. Setting $\eta = 1$ and $\mu = 0.5$ in expression [4.49], we have found that F_q = 3.517 sec and T_q = 5.275 sec in a runtime of 122 sec using a spline profile.

4.5.1.2. *Optimal motion problem with geometric path constraint*

This example concerns the minimum-time transfer problem with geometric path constraint. The path considered is a straight line from (0.8*m*, 0) to (0, 0.8*m*) in the horizontal *x-y* plan. Numerical results obtained using various types of profiles are summarized in Table 4.4. Figure 4.16 shows the optimized motion corresponding to the smoothed trapezoidal profile.

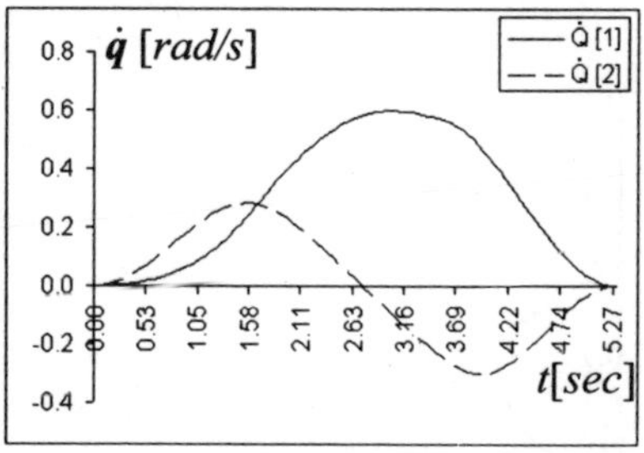

(a) Joint velocities

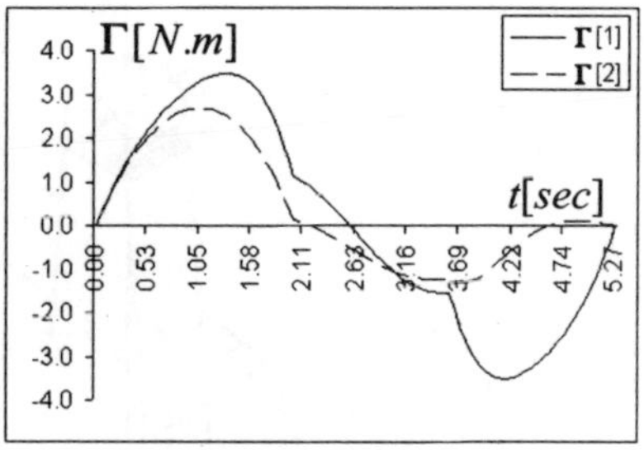

(c) Actuator torques

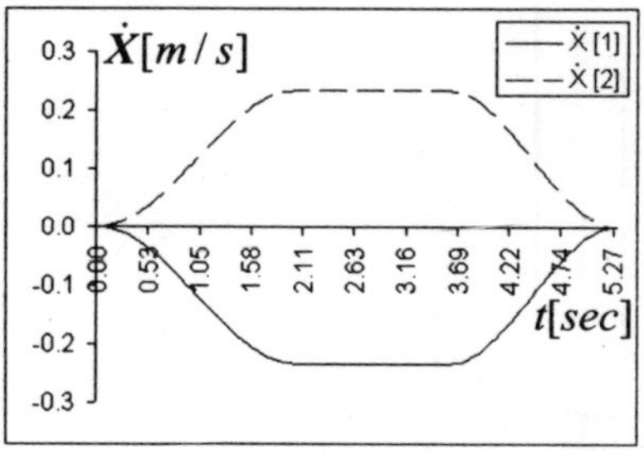

(b) Operational velocity

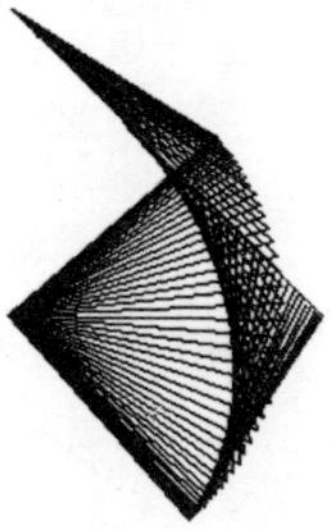

(d) Calculated linear path

Figure 4.16. *Motion obtained by a smoothed trapezoidal profile for a linear path*

	Cubic spline (4 free nodes)	Trapezoidal	Smoothed trapezoidal
$T_q = F_q$ [*sec*]	4.35	5.36	5.27
Runtime [*sec*]	21.425	0.469	0.475

Table 4.4. *Simulation results for the linear path*

4.5.2. *Case of a six dof robot*

This example concerns the standard six dof PUMA 560 robot (Figure 4.17). The robot geometric and inertial parameters are given in [DAN 98]. The limit performances used for this robot are listed in Table 4.5.

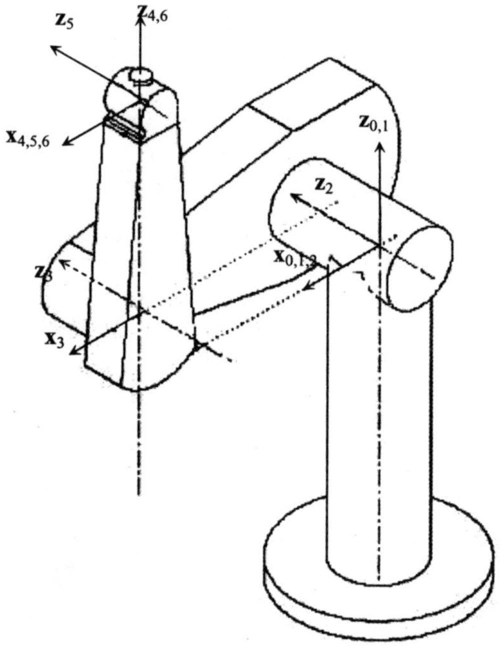

Figure 4.17. *Puma 560 robot shown here in its zero angle configuration*

Joint i	1	2	3	4	5	6
k_{vj} [*rad/s*]	6.5	6.5	6.5	4.5	4.5	4.5
Γ_i^{max} [*N.m*]	97.6	186.4	89.4	24.2	20.1	21.3

Table 4.5. *Limit performances used for PUMA 560*

4.5.2.1. *Optimal free motion planning problem*

The robot is asked to carry out a transfer between the configuration (–45°, –20°, –110°, –50°, 50°, 0) and (175°, 100°, –20°, 20°, –70°, 180°) with zero initial and final velocities.

The numerical results obtained are listed in Table 4.6 for various types of profiles and in Figure 4.18 for a smoothed trapezoidal profile. These solutions are obtained in a few seconds for trapezoidal profiles, while a cubic spline profile requires several minutes. In the first case, we have only two unknowns to solve, while in the second case the number of unknowns is $4 \times 6 = 24$.

	Cubic spline (4 free nodes)	Trapezoidal	Smoothed trapezoidal
$T_q = F_q$ [*sec*]	0.89	0.97	1.07
Runtime [*sec*]	571	1.655	1.680

Table 4.6. *Simulation results*

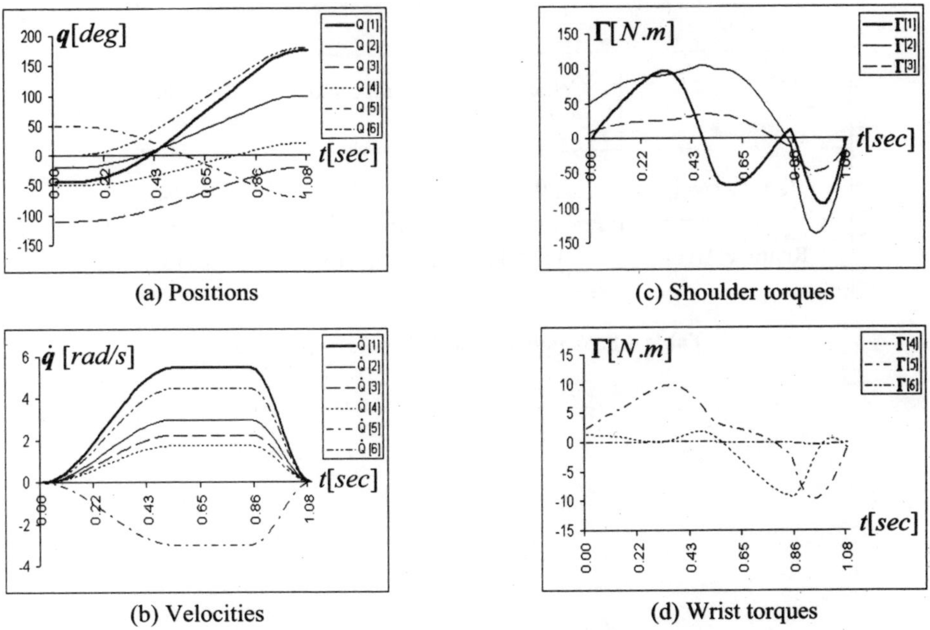

Figure 4.18. *Motion obtained by a smoothed trapezoidal profile*

4.5.2.2. *Optimal motion problem with geometric path constraints*

We now consider a problem with geometric path constraints. The fixed path is a straight line linking the initial and final configurations listed in Table 4.7. The angles φ, θ and ψ given in this table represent, respectively, the roll, pitch and yaw angles.

	Px [m]	Py [m]	Pz [m]	φ [deg]	θ [deg]	ψ [deg]
X^{ini}	0.8	0.2	0.1	0	70	0
X^{fin}	0.2	0.6	0.3	80	0	45

Table 4.7. *Initial and final configurations in the linear path*

The results for the calculated motion are listed in Table 4.8 for various types of profiles. Figure 4.19 illustrates the time evolution of velocities and torques for a smoothed trapezoidal profile.

	Cubic spline (4 free nodes)	Trapezoidal	Smoothed trapezoidal
$T_q = F_q$ [sec]	0.584	0.590	0.670
Runtime [sec]	113.844	3.310	3.313

Table 4.8. *Simulation results for the linear path*

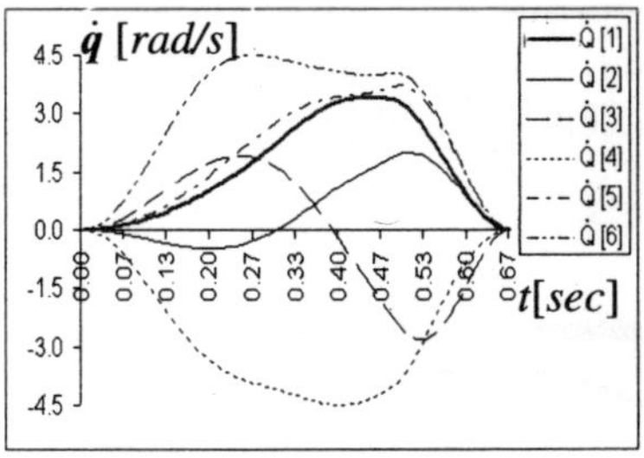

(a) Joint velocities

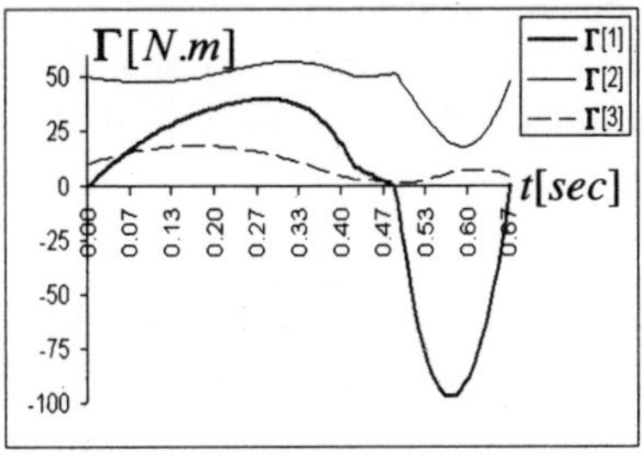

(c) Shoulder torques

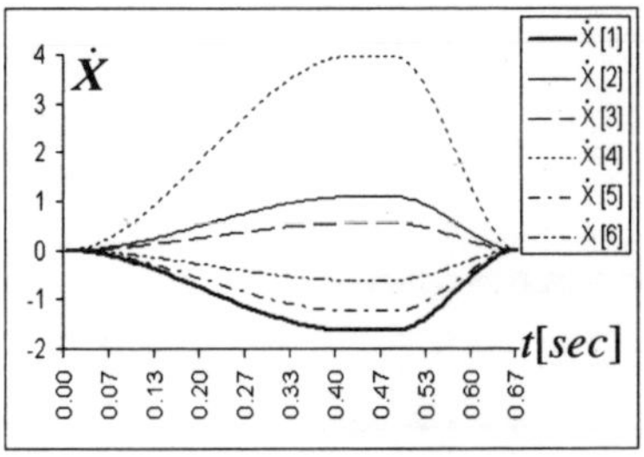

(b) Operational velocities

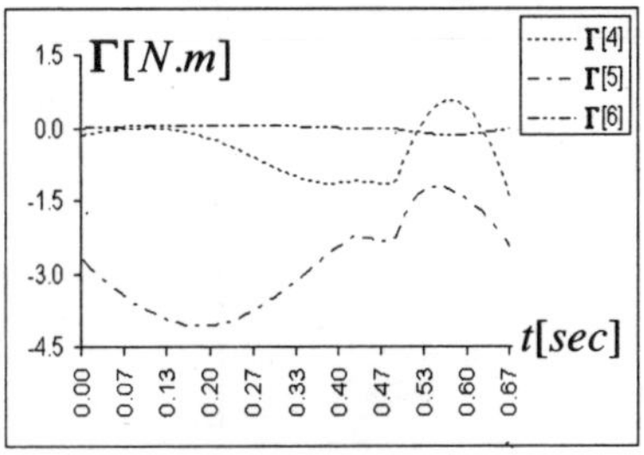

(d) Wrist torques

Figure 4.19. *Motion obtained by a smoothed trapezoidal profile for the linear path*

4.5.2.3. *Optimal free motion planning problem with intermediate points*

We now consider the case where the path is constrained to go through a given set of intermediate points (via points) defined in the joint space (if these via points are defined in the operational space, the corresponding configurations are determined by the inverse kinematic model). In practice, these via points are introduced to distort the robot trajectory in order to avoid collisions with the environment. The creation of the optimal trajectory using the method described in section 4.4.2 is still applicable. Here, the use of cubic spline functions is strongly recommended to ensure continuity up to the second order.

The objective is to construct a joint trajectory $\boldsymbol{q}(t)$ with transfer time T, interpolating all via points and satisfying limit conditions [4.45]. Here again, we can directly build a trial trajectory profile $\tilde{\boldsymbol{q}}\,(\xi)$ through the imposed via points. A straightforward technique consists simply of using the via points as free control nodes. As shown in Figure 4.20, these nodes are disrupted along the horizontal direction only. This is because the configuration of the via points is already imposed. Let us note that the process may be initialized using a uniform distribution of nodes along the ξ axis. The goal of the process is to determine their optimal positions. Alternatively, if the distance between via points is too large, we may include additional free control nodes. As shown in Figure 4.21, these additional nodes are free to move in all directions, since their role is essentially to give more freedom for the trajectory shape. In both cases, with or without additional nodes, the optimization process described in section 4.4 is directly applicable.

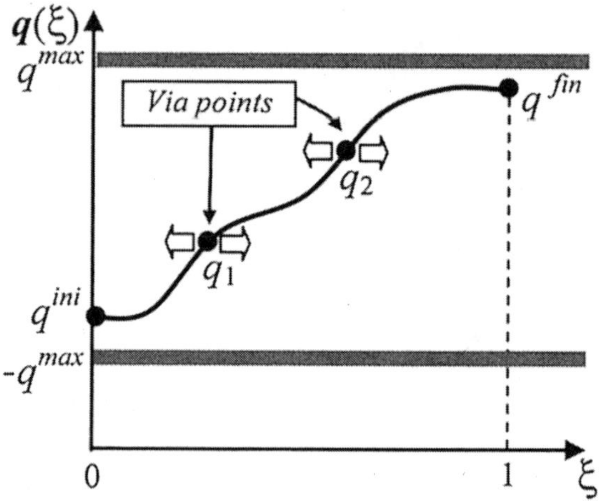

Figure 4.20. *A trajectory profile $q(\xi)$ built using via points*

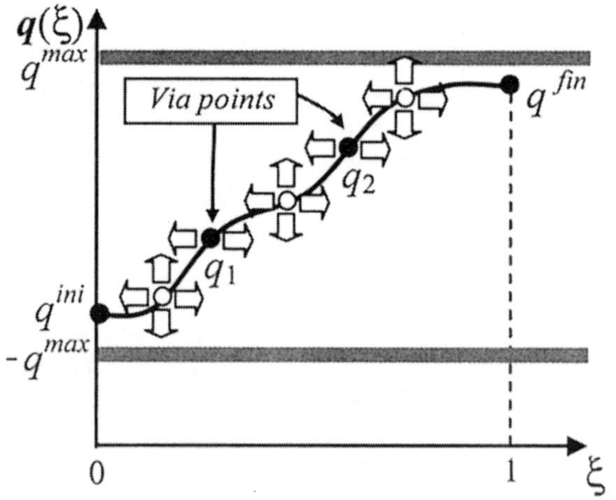

Figure 4.21. *A trajectory profile* $\mathbf{q}(\xi)$ *built using via points and additional nodes*

As an example, we put the minimum-time trajectory through the same transfer seen in section 4.5.2.1, but now with one via point at configuration $\boldsymbol{q}_0 = (0,0,0,0,0,0)$. This problem is solved using a trajectory profile $\tilde{\boldsymbol{q}}(\xi)$ constructed using the nodes $(\boldsymbol{q}^{ini}, \boldsymbol{q}_1, \boldsymbol{q}_0, \boldsymbol{q}_2, \boldsymbol{q}^{fin})$; $\boldsymbol{q}_1$ and $\boldsymbol{q}_2$ being free additional nodes. Results are illustrated in Figure 4.22. The calculated value of the cost function is: $F_{obj} = T = 1.18$ sec for a runtime of 457 sec.

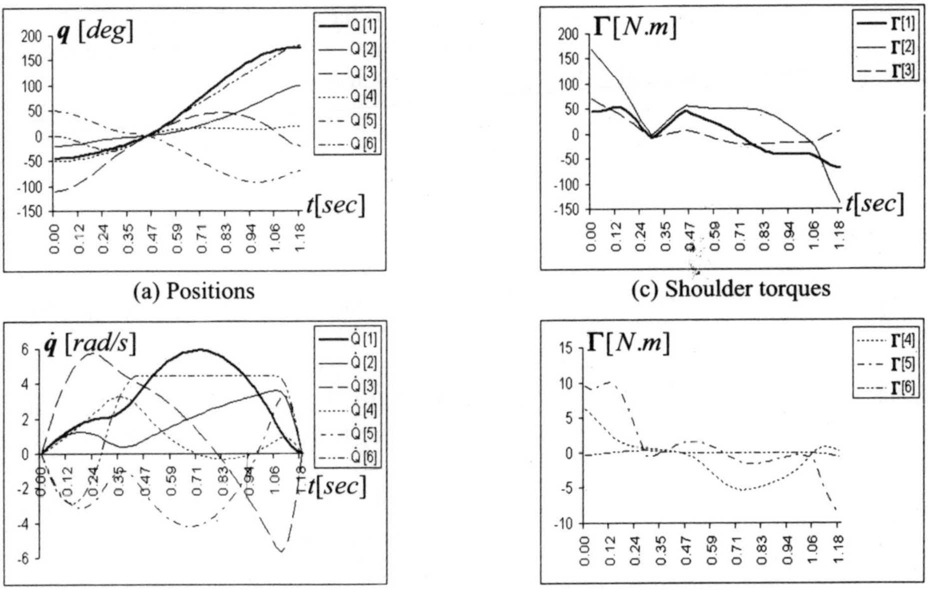

(a) Positions

(c) Shoulder torques

(b) Velocities

(d) Wrist torques

Figure 4.22. *Motion obtained by a cubic spline profile*

4.6. Conclusion

In this chapter, we presented several methods of trajectory generation under kinematic constraints that are commonly used in robotics. We started with point-to-point trajectories. Different models were studied, namely the polynomial model and the trapezoidal velocity profile, which is implemented in most of the industrial controllers. For each model, we calculated the minimum travel time from which it is possible to synchronize the joints, so that they reach the final position simultaneously. We then tackled the problem of point-to-point trajectory generation in the operational space and showed that the same trajectory generators can be applied for the straight line path motion.

Next, we introduced a general methodology for generating optimal trajectories under kinodynamic constraints. By time-scaling and discretizing the robot's trajectory, the original infinite-dimension problem was cast as a finite-dimension optimization problem and kinodynamic constraints were converted to bounds on the transfer time. Then, the resulting parametric constrained-profile problem was solved using a stochastic optimization technique. The simplicity of this method made it possible to generate optimal trajectories in both Cartesian and joint spaces, with or without intermediate points. In particular, we presented a trajectory generator, based on the classical trapezoidal velocity model, able to handle kinodynamic constraints. This generator was capable of producing, in a few seconds, approximate minimum-time trajectories for six dof robots, which makes it quite interesting for industrial applications.

4.7. Bibliography

[BAR 92] BARRAQUAND J., LANGLOIS B. and LATOMBE J.C., "Numerical potential field techniques for robot path planning", *IEEE Tr. on Sys., Man, and Cyb.*, 22(2):224-241, 1992.

[BAR 96] BARRAQUAND J., KAVRAKI L., LATOMBE J.C., LI T.Y., MOTWANI R. and RAGHAVAN P., "A random sampling scheme for path planning", *7th Int. Conf. on Rob. Research ISRR*, 1996.

[BES 92] BESSONNET G., "Optimisation dynamique des mouvements point à point de robots manipulateurs", PhD Thesis, Poitiers University, France, 1992.

[BET 99] BETTS J.T., "A direct approach to solving optimal control problems", *Computing in Science & Engineering*, pp. 72-75, 1999.

[BOB 85] BOBROW J.E., DUBOWSKY S. and GIBSON J.S., "Time-optimal control of robotic manipulators", *Int. J. Robot. Res.*, 4(3):3-17, 1985.

[BOB 01] BOBROW J.E., MARTIN B.J., SOHL G., WANG E.C., PARK F.C. and KIM J., "Optimal robot motions for physical criteria", *Jour. of Rob. Syst.* 18(12), 785-795, 2001.

[BRY 75] BRYSON A.E. and HO Y.C., *Applied Optimal Control*, Revised Printing Hemisphere, New York, 1975.

[CAS 84] CASTAIN R.H. and PAUL R.P., "An on-line dynamic trajectory generator", *The Int. J. of Robotics Research*, Vol. 3(1), pp. 68-72, 1984.

[CHE 90] CHEN Y. and DESROCHERS A., "A proof of the structure of the minimum time control of robotic manipulators using Hamiltonian formulation", *IEEE Trans. on Rob. and Aut.* 6(3), pp. 388-393, 1990.

[CHE 93] CHEN Y., HUANG J. and WEN J.T.Y., "A continuation method for time-optimal control synthesis for robotic point-to-point motion", *Proc. of the 32nd IEEE Conf. on Decision and Control*, 1993.

[CHE 02] CHETTIBI T. and LEHTIHET H.E., "A new approach for point to point optimal motion planning problems of robotic manipulators", in *Proc. of 6th Biennial Conf. on Engineering Syst. Design and Analysis*, APM10, 2002.

[CHE 04] CHETTIBI T., LEHTIHET H.E., HADDAD M. and HANCHI S., "Minimum cost trajectory planning for industrial robots", *European J. of Mechanics/A*, pp. 703-715, 2004.

[CHE 05] CHETTIBI T., HADDAD M., LABED A. and HANCHI S., "Generating optimal dynamic motions for closed-chain robotic systems", *European Journal of Mechanics A/Solids* 24, pp. 504-518, 2005.

[CON 00] CONSTANTINESCU D. and CORFT E.A., "Smooth and time-optimal trajectory planning for industrial manipulators along specified paths", *J. Rob. Sys.*, 17(5), pp. 233-249, 2000.

[DAN 98] DANES F., "Critères et contraintes pour la synthèse optimale des mouvements de robots manipulateurs, application à l'évitement d'obstacles", PhD Thesis, Poitiers University, France, 1998.

[EDW 82] EDWALL C.W., POTTINGER H.J., HO C.Y. "Trajectory generation and control of a robot arm using spline functions", *Proc. Robot-6*, Detroit, pp. 421-444, 1982.

[GEE 86] GEERING H.P., GUZZELLA L., HEPNER S.A.R. and ONDER C.H., "Time-optimal motions of robots in assembly tasks", *IEEE Trans. on Automatic Control*, Vol. AC-31, No. 6, 1986.

[HAD 05] HADDAD M., CHETTIBI T., LEHTIHET H.E., HANCHI S., "A new approach for minimum-time motion planning problem of wheeled mobile robots", *16th IFAC World Congress*, Prague, 2005.

[HAJ 85] HAJEK B., "A tutorial survey of theory and application of simulated annealing", *Proceedings of 24th Conference on Decision and Control*, 755-760, 1985.

[HEI 95] HEIMANN B. and KRÜGER M., "Optimal path-planning of robot-manipulators in the presence of obstacles", *9th World Congress on the Theory of Machines and Mechanisms*, Italy, 1995.

[HOL 83] HOLLERBACH J.M., "Dynamic scaling of manipulator trajectories", *MIT AI Lab Memo* 700, 1983.

[HUL 97] HULL D.G., "Conversion of optimal control problems into parameter optimization problems", *Jour. of Guidance, Cont. and Dyn.*, 20(1), 57-62, 1997.

[JAM 96] JAMHOUR E. and ANDRÉ P.J., "Planning smooth trajectories along parametric paths", *Mathematics and Computers in Simulation*, 41, pp. 615-626, 1996.

[JAQ 89] JAQUES J., SOLTINE E. and YANG H.S., "Improving the efficiency of time-optimal path following algorithms", *IEEE Trans. Robot. Aut.*, Vol. 5, No. 1, 1989.

[KAV 94] KAVRAKI L. and LATOMBE J.C., "Randomized preprocessing of configuration space for fast path planning", *Proc. of IEEE Int. Conf. on Rob. and Aut.*, pp. 2138-2139, San Diego, 1994.

[KAV 96] KAVRAKI L., SVESTA P., LATOMBE J.C. and OVERMARS M., "Probabilistic roadmaps for path planning in high dimensional configuration space", *IEEE Trans. on Rob. and Aut.*, 12:566-580, 1996.

[KHA 02] KHALIL W., *Commande des robots manipulateurs*, Hermes Science/Lavoisier Paris, 2002.

[KIR 83] KIRKPATRICK S., GELETT C. and VECCHI M., "Optimization by simulated annealing", *Science*, May 1983, 621-680, 1983.

[LAT 91] LATOMBE J.C., *Robot Motion Planning*, Kluwer Academic Publishing, 1991.

[LAT 99] LATOMBE J.C., "Motion planning: a journey of molecules, digital actors and other artefacts", *Int. Jour. of Rob. Research*, 18(11), pp. 1119-1128, 1999.

[LAV 98] LAVALLE S.M., "Rapidly exploring random trees: a new tool for path planning", R98-11, http://janowiec.cs.iastate.edu/papers/rrt.ps, Computer Science Dept., Iowa State University, 1998.

[LAV 99] LAVALLE S.M. and KUFFNER J.J., "Randomized kinodynamic planning", *Proc. IEEE Int. Conf. on Rob. and Aut.*, pp. 473-479, 1999.

[LAZ 96] LAZRAK M., "Nouvelle approche de commande optimale en temps final libre et construction d'algorithmes de commande de systèmes articulés", PhD Thesis, Poitiers University, France, 1996.

[LIN 83] LIN C.S., CHANG P.R., LUH J.Y.S., "Formulation and optimization of cubic polynomial joint trajectories for industrial robots", *IEEE Trans. on Automatic Control*, Vol. AC-28 (12), pp. 1066-1073, December 1983.

[MAO 97] MAO Z. and HSIA T.C., "Obstacle avoidance inverse kinematics solution of redundant robots by neural networks", *Robotica*, 15, 3-10, 1997.

[OVE 92] OVERMARS M.H., "A random approach to motion planning", Technical report RUU-CS-92-32, Utrecht University, 1992.

[PFE 87] PFEIFFER F. and RAINER J.A., "A concept for manipulator trajectory planning", *IEEE Jour. Robot. Aut.*, Ra-3(2), pp. 115-123, 1987.

[PON 65] PONTRYAGIN L., BOLTIANSKI V., GAMKRELIDZE R. and MICHTCHENKO E., *Théorie mathématique des processus optimaux*, Edition Mir, 1965.

[RAN 96] RANA A.S. and ZALAZALA A.M.S., "Near time optimal collision free motion planning of robotic manipulators using an evolutionary algorithm", *Robotica*, 14, pp. 621-632, 1996.

[RIC 93] RICHARD M.J., DUFOURN F. and TARASIEWICZ S., "Commande des robots manipulateurs par la programmation dynamique", *Mech. Mach. Theory*, 28(3), 301-316, 1993.

[SAI 05] SAIDOUNI T., "Synthèse numérique d'allures de marche optimales de robots bipeds anthropomorphes", PhD Thesis, Poitiers University, France, 2005.

[SHI 85] SHIN K.G. and MCKAY D.N., "Minimum-time control of robotic manipulators with geometric path constraints", *IEEE Trans. Aut. Cont.*, AC-30 (6), 1985.

[SHI 86] SHIN K.G. and MCKAY D.N., "A dynamic programming approach to trajectory planning of robotic manipulators", *IEEE Trans. Aut. Cont.*, AC-31(6), pp. 491-500, 1986.

[SHI 89] SHILLER Z. and DUBOWSKY S., "Robot path planning with obstacles, actuator, gripper and payload constraints", *Int. Jour. Rob. Res.*, 8(6), pp. 3-18, 1989.

[SHI 91] SHIN K.G. and MCKAY D.N., "Minimum-cost trajectory planning for industrial robots", *Cont. and Dyn. Sys.*, 39, pp. 345-375, 1991.

[SHI 92] SHILLER Z. and LU H.H., "Computation of path constrained time optimal motions with dynamic singularities", *ASME Jour. Dyn. Sys. Meas. Cont.*, 114, pp. 34-40, 1992.

[SHI 96] SHILLER Z., "Time-energy optimal control of articulated systems with geometric path constraints", *ASME Jour. Dyn. Sys. Meas. Cont.*, 118(1), pp. 139-143, 1996.

[STE 95] STEINBACH M.C., "Fast recursive SQP methods for large scale optimal control problem", PhD Thesis, Heidelberg University, 1995.

[STR 93a] STRYK O.V., "Numerical solution of optimal control problems by direct collocation", *Optimal control theory and numerical methods, Int. Series of Numerical Math.*, Vol. 111, pp. 129-143, 1993.

[STR 93b] STRYK O.V. and BULIRSCH R., "Direct and indirect methods for trajectory optimization", *Annals of Operations Research*, Vol. 37, pp. 357-373, 1993.

[TAN 88] TAN H.H. and POTTS R.B., "Minimum time trajectory planner for the discrete dynamic robot model with dynamic constraints", *IEEE J. Rob. Aut.*, 4(2), pp. 174-185, 1988.

[TAN 89] TAN H.H. and POTTS R.B., "A discrete trajectory planner for robotic arms with six degrees of freedom", *IEEE Trans. Robot. Aut.*, 5(5), pp. 681-690, 1989.

[YAM 94] YAMAMOTO M., ISSHIKI Y. and MOHRI A., "Collision free minimum time trajectory planning for manipulators using global search and gradient method", *Proc. IEEE/RSJ/GI*, IROS, 2184-2191, 1994.

Appendix:
Stochastic Optimization Techniques

An optimization problem consists of finding, within a space Ω of possible solutions S, the optimal solution S^{opt} that minimizes (or maximizes) a given cost function $F(S)$. From now on, we consider the minimization problem.

The classical deterministic optimization techniques that are devoted to solve this problem start with an initial guess S_0, then this guess is iteratively improved until a given convergence criterion is satisfied. The output is an approximation S^* of the optimal solution S^{opt}. These techniques are based on the construction of a privileged search direction in Ω that is determined using information on variations of F. Although they can be very effective, they usually converge to a local minimum in the vicinity of the starting guess S_0. Hence, when the cost function F has numerous local minima, these techniques can easily miss the targeted global minimum. For such problems, stochastic optimization techniques are often preferred.

Stochastic optimization techniques are based on a random search that decreases the risk for the process to become stuck in a local minimum. They require only information on values of F, which makes their implementation easy. In what follows, we will first introduce some basic concepts and then we will focus on the Hill Climbing and the Simulated Annealing methods.

Basic concepts

The basic scheme of a stochastic technique consists of generating random trial solutions S in the search space Ω. Then, a comparison between these solutions is made in order to retain the best one. This is implemented in the following pseudo-code:

```
Generate randomly an initial solution S0 in Ω
S* = S0
REPEAT
Generate a random solution S in Ω;
If F(S) < F(S*) then S* = S
Until a convergence criterion is satisfied
```

Obtaining the optimal solution of the problem requires, in theory, an infinite runtime. For practical reasons, the process is stopped after a finite number of iterations, considered as sufficient to obtain a good approximation of the optimal solution. Other stochastic optimization techniques derive from this basic scheme but they try to guide the process so that it is not completely a blind random search. This reduces the number of calculations and accelerates convergence.

The basic idea consists of concentrating the search in the vicinity $V(S^*)$ of the current best solution S^* obtained at each iteration. As the solution S^* is improved, the position in Ω of the sub-space $V(S^*)$ is modified such that the search remains centered on the new S^*. This makes it possible to increase the probability of solution improvement and thus accelerates the convergence. This idea is exploited both in the Hill Climbing and the Simulated Annealing methods.

a) The Hill Climbing method

The algorithm of a Hill Climbing method is as follows:

Generate randomly an initial solution S_0 in Ω

$S^* = S_0$

$V\_Center = S^*$

REPEAT

  Generate a random solution S in $V(S^*)$;

  If $F(S) < F(S^*)$ then

    $S^* = S$

    $V\_Center = S^*$

  End if

Until a convergence criterion is satisfied

With this method, the size δ of $V(S^*)$ remains fixed. This technique accelerates significantly the search process. The major inconvenient concerns the choice of δ. If δ is too small, the method can easily get stuck in a local minimum. While if δ is too large, the runtime will increases unnecessarily.

An improved version of this method is the Hill Climbing method with variable size. The principle consists of starting the process using a large sized δ. Then, as the process converges and the solution is improved, the value of δ is continuously reduced. In other words, the method starts as a blind search technique and ends up as a targeted search technique. The difficulty, however, concerns the reduction schedule for δ. The following method avoids this difficulty.

b) The simulated annealing method

This method is based on an analogy with thermodynamics, specifically with the way that liquids freeze and crystallize, or metals cool and anneal. At high temperatures, the molecules of a liquid move freely with respect to one another. If the liquid is cooled slowly, atoms are often able to line themselves up to form a pure and ordered crystal which corresponds to a state of minimum energy for the system. The amazing fact is that, for slowly cooled systems, nature is able to find this minimum energy state. Indeed, if a liquid metal is cooled quickly it does not reach this state but rather ends up in a polycrystalline or amorphous state having somewhat higher energy. So the essence of the process is slow cooling, allowing ample time for the redistribution of atoms as they lose mobility and ensuring that a low energy state will be achieved.

In the simulated annealing (SA) method, each trial solution S is compared to a state of some physical system, and the function F(S) to be minimized is interpreted as the internal energy of the system in that state. Therefore, the goal is to bring the system from an arbitrary initial state S_0 to a state S^{opt} having the minimum energy.

At each step, the SA heuristically considers a candidate S in the vicinity $V(S^*)$ of the current state S^*. Then, it probabilistically decides between two possibilities: moving the system to state S or remaining in state S^*. The decision is made according to the Metropolis algorithm (see below) designed so that the system will ultimately move to states of lower energy. Typically this step is repeated until the system reaches a state which is good enough or until a given runtime budget has been exhausted.

The probability of making the transition from the current state S^* to the new state S is a function $P(\Delta F, \Theta)$, where $\Delta F = F(S) - F(S^*)$ is the energy difference between the two states and Θ is a time-varying parameter called the temperature of the system. This probability is generally given by the Boltzman formula:

$$P = \exp\left(-\frac{\Delta F}{\Theta}\right)$$

One essential feature of the SA method is that the transition probability P is defined to be non-zero even if ΔF is positive, meaning that the system may move to the new state even when it is worse (has a higher energy) than the current one. It is this feature that prevents the method from becoming stuck in a local minimum. Also, when the temperature tends to zero and ΔF is positive, the probability $P(\Delta F, \Theta)$ tends to zero. Therefore, for sufficiently small values of Θ, the system will increasingly favor moves that go downhill (to lower energy values) and avoid those that go uphill.

Another feature of the SA method is that the temperature is gradually reduced as the optimization proceeds. Initially, Θ is set to Θ_0, a value calculated so that the system is expected to wander initially towards a broad region of the search space containing good solutions, ignoring small features of the energy function. Then, as the temperature is decreased, the system will drift towards low-energy regions that become narrower and narrower.

The following pseudo-code implements the simulated annealing heuristically as described above.

Generate randomly an initial solution S_0 in Ω;

$S^* = S_0$

Calculate the initial temperature Θ_0 (see below)

$\Theta = \Theta_0$;

Repeat

k = 0;

For $k = 1$ to N_{it}

Generate a random solution S in $V(S^*)$

$\Delta F = F(S) - F(S^*)$

If **Accept**(ΔF, Θ) then $S^* = S$

Next k

Reduce Θ (see below)

Until convergence

Boolean Function **Accept**(ΔF, Θ) *"Metropolis"*

If Random (0,1) < $\exp(-\Delta F / \Theta)$ then

Accept = True

Else

Accept = False

End if

A standard way to choose Θ_0 consists of carrying out a number of random samplings of values of F in order to calculate the median M of F increases. Then, the initial temperature Θ_0 is set so that the initial probability of acceptance is equal to 0.5:

$$P = \exp\left(-\frac{M}{\Theta_0}\right) = 0.5 \Rightarrow \Theta_0 = -\frac{M}{\mathrm{Ln}(0.5)} \approx 1.44M$$

The SA algorithm envisages carrying out a number N_{it} of iterations at constant temperature before reduction. The annealing schedule has a limited series of temperatures, decreasing by level (quasi-static procedure). Generally, a geometric progression is adopted:

$$\Theta_{i+1} = \alpha \cdot \Theta_i \qquad 0 < \alpha < 1$$

We can show that, for any given finite problem, the probability that the simulated annealing algorithm terminates with the global optimal solution approaching 1 as the annealing schedule is extended. This theoretical result is, however, not particularly helpful, since the annealing time necessary to ensure a significant probability of success will usually exceed the time required for a complete search of the solution space.

Further reading

A.1 KIRKPATRICK S., GELATT C.D. and VECCHI M.P., "Optimization by simulated annealing" *Science*, Vol. 220, No. 4598, pp. 671-680, 1983.

A.2 CERNY V., "A thermodynamical approach to the travelling salesman problem: an efficient simulation algorithm", *Journal of Optimization Theory and Applications*, 45:41-51, 1985.

A.3 PRESS W. *et al.*, *Numerical Recipes in C: The Art of Scientific Computing*, 2nd Edition, Cambridge University Press, 1992.

A.4 LAM J. and DELOSME J.-M., "Simulated annealing: a fast heuristic for some generic layout problems", *Proc. International Conf. on Computer Aided Design ICCAD-88*, Santa Clara, CA, pp. 510-513, November 1988.

A.5 LAM J., DELOSME J.-M. and SECHEN C., "A new simulated annealing schedule for row-based placement", *Proc. International Workshop on Placement and Routing*, Research Triangle Park, NC, May 1988.

Chapter 5

Position and Force Control of a Robot in a Free or Constrained Space

5.1. Introduction

The object of this chapter is to provide the reader with basic knowledge on control techniques of a serial rigid robot manipulator in a free or constrained space. The various solutions use synthesis methods that are established depending on the implementation criteria, the complexity of the task to be carried out (surface following, object transportation), the desired static or dynamic behavior (velocity and accuracy) and the insensitivity to parameter variations.

This chapter is divided into two main parts. In the first part, we will look at the case of the robot evolving in a free space, i.e. without any contact with its environment. We take this chance to specify the hypotheses selected throughout this chapter and to detail the various elements of the complete dynamic model of the robot. Within this framework, we will introduce the control laws that are most commonly used in robotics (decentralized, dynamic, adaptive and robust). In the second part, we will present some control methods that enable the robot to carry out tasks in a constrained space, i.e. involving a contact with the environment.

Obviously, we will not be able to give all the details and present all the subtleties of the robot manipulators control within a few dozen of pages; for a finer study of such solutions, the reader may, for example, refer to [KEL 05, KHA 99, SCI 96, SCI 01, SPO 89, ZOD 96]. However, we will discuss here certain modeling or

---

Chapter written by Pierre DAUCHEZ and Philippe FRAISSE.

calculation aspects that are very rarely addressed in other works. It is for example the question of showing why we can limit ourselves to the modeling of the mechanical part of the robot in order to elaborate its control; the question of clearly explaining the principle of non-linear decoupling in a constrained space; the question of showing the inconsistency between a parallel hybrid control structure and Mason's description [MAS 81]; or also the question of treating the problem of control robustness versus environmental rigidity.

5.2. Free space control

5.2.1. *Hypotheses applying to the whole chapter*

Although robots are now generally controlled by digital computers, therefore with a sampling period, we consider that this period is sufficiently short to assimilate the hybrid system (discrete and continuous) thereby constituted to a continuous system and we will present and use continuous-time equations.

On the other hand, for the clarity of the discussion, we will only consider the mechanical part of the robot. In other words, the robot, considered as a system in automatics, will have as input variable the vector $\boldsymbol{\tau}$ of the torques applied to the joints, and as output variable, the vector $\mathbf{q}$ of the joint positions. Moreover, it is the approach that is adopted in most of the books. But in fact, this is a restricted view of the system, which is only acceptable because of the presence of electronic control circuits of the robot actuators, as we will explain in the following paragraph.

Finally, we will suppose that the number of degrees of mobility of the robot is equal to the number of constraints imposed by the task (number of variables of the operational space that is necessary to specify the task). In other words this means that the Jacobian matrix $\mathbf{J}$ that links the joint velocities and operational velocities is square and that we therefore see its inverse appear in certain formulae, like in section 5.2.4. However, this hypothesis is not restrictive as far as the concepts discussed in this chapter go. If the robot was redundant towards the task, i.e. if the number of its degrees of mobility were higher than the number of necessary operational variables, the inverse of $\mathbf{J}$ (and possibly other matrices) would have to be replaced by a pseudo-inverse or a generalized inverse [BOU 71]. We could also then take advantage of redundancy in order to optimize certain criteria or to impose supplementary constraints while performing the task, provided, of course, that this number of constraints, added to the number of operational variables necessary to perform the basic task, does not exceed the number of degrees of mobility of the robot [DAU 85, SCI 01].

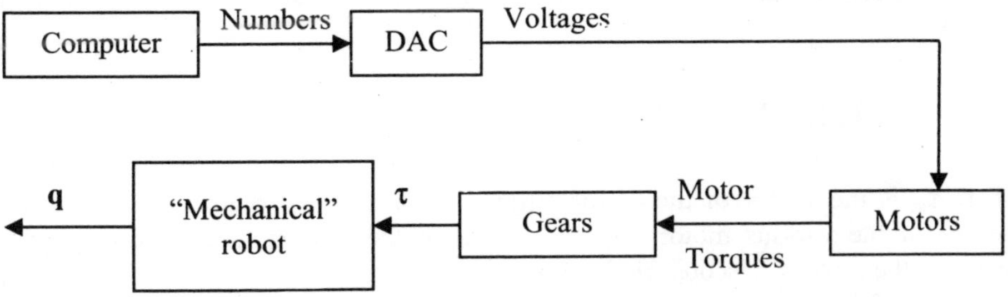

Figure 5.1. *Simplified control chain of a robot manipulator*

5.2.2. *Complete dynamic modeling of a robot manipulator*

In order to illustrate our comments, we will consider here a PUMA 560 manipulator as presented in [DEL 92]. A complete dynamic modeling was rendered necessary in this work by the constraint to implement a simulator which should be as close to reality as possible. But this also makes it possible to understand what is "hidden" when the dynamic equation relating $\boldsymbol{\tau}$ and $\mathbf{q}$ is used.

If the purpose of the control is to calculate the torques $\boldsymbol{\tau}$ that make it possible to obtain the positions $\mathbf{q}$, it has to be well understood that it is not practically possible to directly apply these torques to the joints of the robot. In fact, we will have to calculate, on the computer controlling the robot, a set of numbers which will be transformed into analog values by digital-to-analog converters, these values being the voltage inputs of the motors operating the robot. These motors will produce torques that will then provide the actuating torques $\boldsymbol{\tau}$ through gear ratios except for the direct-drive motors (Figure 5.1).

The equations that govern the chain represented in Figure 5.1 are *a priori* very complex, but we are going to show here that it is not necessary to use all their details to determine a control diagram of the robot. Let us first go back to the relation between $\boldsymbol{\tau}$ and $\mathbf{q}$, given by the following dynamic equation [KHA 99]:

$$\boldsymbol{\tau} = \mathbf{M}\ddot{\mathbf{q}} + \mathbf{C}\dot{\mathbf{q}} + \mathbf{G} + \boldsymbol{\tau}_f = \mathbf{M}\ddot{\mathbf{q}} + \mathbf{H} \qquad [5.1]$$

For a robot with n joints, $\mathbf{M}$ is the $(n \times n)$ inertia matrix of the robot, $\mathbf{C}$ the $(n \times 1)$ vector of Coriolis forces/torques and of centrifugal forces, $\mathbf{G}$ the $(n \times 1)$ vector of the torques/forces of gravity, and $\boldsymbol{\tau}_f$ the $(n \times 1)$ vector of the friction torques/forces. Let us now focus on the mechanical modeling of the robot actuators that are direct-current motors in the case of a PUMA 560. If we ignore the gyroscopic effects due to the rotation of the motors, we obtain the dynamic equation [DEL 92]:

$$\boldsymbol{\tau}_m = \mathbf{J}_m \ddot{\mathbf{q}}_m + \mathbf{f}_m \dot{\mathbf{q}}_m + \mathbf{C}_m \qquad [5.2]$$

where $\boldsymbol{\tau}_m$ is the vector of the motor torques, $\mathbf{J}_m$ a diagonal matrix containing the inertias of the various motors, $\ddot{\mathbf{q}}_m$ the vector of the motors' acceleration, $\mathbf{f}_m$ the vector of the motors' viscous frictions, $\dot{\mathbf{q}}_m$ the vector of the motors' velocity and $\mathbf{C}_m$ the vector of the resisting torques at the motor level.

If we group in a matrix **N** the transmission velocity reducing ratios (the extra diagonal terms represent the possible couplings between joints, which is the case for the three motors of the wrist joint in a PUMA 560), we can write:

$$\dot{\mathbf{q}}_m = \mathbf{N}\,\dot{\mathbf{q}} \qquad [5.3]$$

$$\mathbf{C}_m = \mathbf{N}^{-T}\,\boldsymbol{\tau} \qquad [5.4]$$

By substituting relations [5.3] and [5.4] in equation [5.2], and by replacing $\boldsymbol{\tau}$ by its expression given by equation [5.1], we obtain:

$$\boldsymbol{\tau}_m = \mathbf{M}'\ddot{\mathbf{q}} + \mathbf{H}' \qquad [5.5]$$

where:

$$\mathbf{M}' = \mathbf{J}_m\mathbf{N} + \mathbf{N}^{-T}\mathbf{M}$$

$$\mathbf{H}' = \mathbf{f}_m\mathbf{N}\,\dot{\mathbf{q}} + \mathbf{N}^{-T}\,\mathbf{H}$$

Let us now consider the actuator electric equations. They can be written under the following vector form [DEL 92]:

$$\mathbf{U}_m = \mathbf{r}_m\,\mathbf{I}_m + \mathbf{L}_m \frac{d\mathbf{I}_m}{dt} + \mathbf{E}_m \qquad [5.6]$$

where $\mathbf{U}_m$ is the voltage vector at the motor inputs, $\mathbf{r}_m$ the vector of the motors' armature resistance, $\mathbf{I}_m$ the vector of the motors' armature current, $\mathbf{L}_m$ the vector of the inductances in the motors, $\mathbf{E}_m$ the vector of the motors' counter-electromotive forces.

Given that for direct-current motors we additionally have:

$$\mathbf{E_m} = \mathbf{K_e}\,\dot{\mathbf{q}}_\mathbf{m} \qquad [5.7]$$

$$\boldsymbol{\tau}_\mathbf{m} = \mathbf{K_c}\,\mathbf{I_m} \qquad [5.8]$$

where $\mathbf{K_e}$ is the vector of the motors' electrical constants and $\mathbf{K_c}$ is the vector of the motors' torque constants.

If we seek the relation between $\mathbf{U_m}$ and $\mathbf{q}$, we notice that since equation [5.6] displays the derivative of $\mathbf{I_m}$, which is a vector proportional to $\boldsymbol{\tau}_\mathbf{m}$ according to relation [5.8], we will obtain a term in $\dddot{\mathbf{q}}$, because equation [5.5] displays a term in $\ddot{\mathbf{q}}$. Therefore, we do not have, *a priori,* the same degree of derivation of $\mathbf{q}$ in this equation and in equation [5.1] which is generally used to establish the control diagrams. However, in reality, Figure 5.1 is incomplete. There are in fact electronic control circuits, called power amplifiers, between the digital-to-analog converters and the motors, which, in the case of the PUMA 560 for example, ensure that the $\mathbf{I_m}$ currents, for small variations, are proportional to the $\mathbf{U_{ev}}$ input voltage of the power amplifiers:

$$\mathbf{U_{ev}} = \mathbf{K_v}\mathbf{I_m}$$

The control chain is therefore more accurately represented in Figure 5.2. So, according to equations [5.5] and [5.8], and based on the fact that the DACs bring about, in a linear working range, a simple coefficient of proportionality between their inputs and their outputs, the relation between the output numbers of the computer and $\mathbf{q}$ has the same form as equation [5.1], which we will therefore use for simplicity reasons and physical meaning of the phenomena.

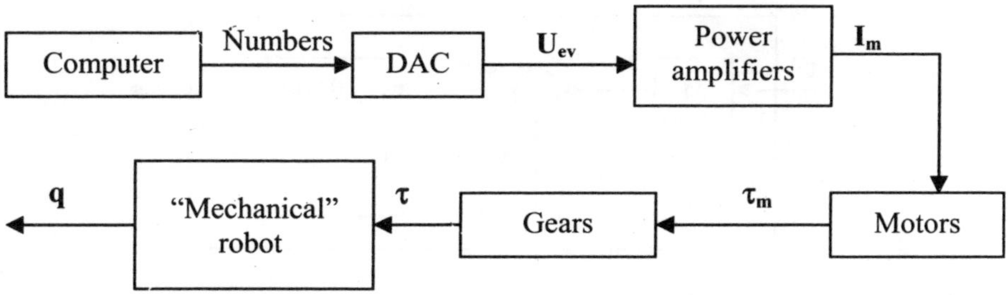

Figure 5.2. *Control chain of a robot manipulator with its power amplifiers*

5.2.3. *Ideal dynamic control in the joint space*

Considering what we have just discussed, we can return here to the image of the robot reduced to its mechanics and present an ideal control solution that relies on the perfect knowledge of the dynamic parameters of equation [5.1]. We will also discuss the influence of an incorrect knowledge of these parameters. The method summarized here can also be called non-linear decoupling or "computed torque method". The first term is used because, due to non-linear terms, we are going to decouple the complex dynamic equation [5.1], and the second one because we are going to implement a control law that calculates the vector $\boldsymbol{\tau}$ of the torques to be applied to the robot.

A general description of the non-linear decoupling can be found in books on automatics such as [KHA 99, SCI 96, SLO 91, SPO 89], but we prefer presenting here a simple explanation of its principle: knowing the non-linear dynamic coupled equation [5.1], we write $\boldsymbol{\tau} = \boldsymbol{\alpha\tau}' + \boldsymbol{\beta}$ and we choose $\boldsymbol{\alpha} = \mathbf{M}$ and $\boldsymbol{\beta} = \mathbf{H}$. We thus obtain:

$$\boldsymbol{\tau}' = \ddot{\mathbf{q}} \qquad [5.9]$$

This represents n linear equations corresponding to n "double integrators", if n is the number of actuators of the robot. In other words, this non-linear decoupling transformed the non-linear coupled system in Figure 5.3 into the linear decoupled system in Figure 5.4.

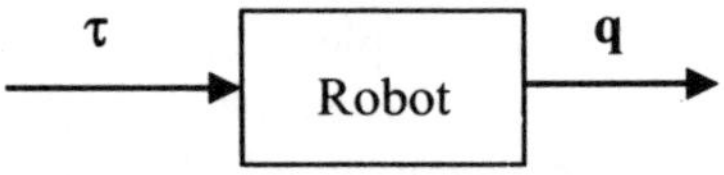

Figure 5.3. *Schematic representation of equation [5.1]*

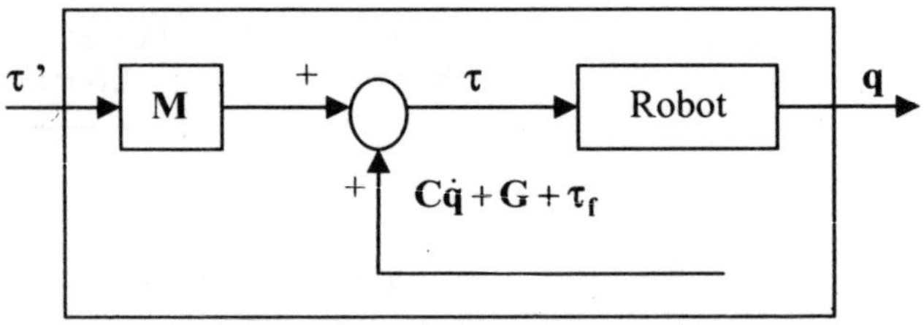

Figure 5.4. *Schematic representation of equation [5.9]*

The new physical system now consisting of n double integrators can be controlled by n appropriate and independent control laws, and in this case, n times of the same type. A solution for such a control is given by equation [5.10]:

$$\boldsymbol{\tau}' = \ddot{\mathbf{q}}_d + \mathbf{K}_v\dot{\mathbf{e}} + \mathbf{K}_p\mathbf{e} \qquad [5.10]$$

where $\mathbf{q_d}$ is the vector of the desired joint positions, $\mathbf{e} = \mathbf{q_d} - \mathbf{q}$, $\mathbf{K_v}$ the diagonal matrix of the velocity gains (or derivative), and $\mathbf{K_p}$ the diagonal matrix of the position gains (or proportional).

We thus have, on the one hand, a modified physical system represented by Figure 5.4 and, on the other hand, the calculation of $\boldsymbol{\tau}'$ represented by Figure 5.5.

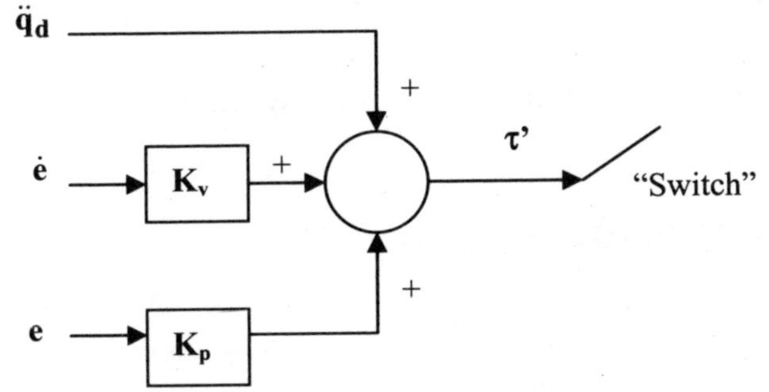

Figure 5.5. *Schematic representation of equation [5.10]*

When we use the output $\mathbf{q}$ of Figure 5.4 to calculate the vector $\mathbf{e} = \mathbf{q_d} - \mathbf{q}$ at the input of Figure 5.5 (servoing) and we "turn off the switch" at the output of Figure 5.5 on the input of Figure 5.4, the looped system is then governed by equation [5.11]:

$$\ddot{\mathbf{q}} = \ddot{\mathbf{q}}_d + \mathbf{K}_v\dot{\mathbf{e}} + \mathbf{K}_p\mathbf{e} \qquad [5.11]$$

that is, $\ddot{\mathbf{e}} + \mathbf{K}_v\dot{\mathbf{e}} + \mathbf{K}_p\mathbf{e} = \mathbf{0}$

Hence, we obtained a system which, rather than being modeled by a differential equation relating its input and output, is represented by a differential equation on the error of the variable to be controlled. We can thus, by choosing the gains $\mathbf{K_p}$ and $\mathbf{K_v}$, make the error $\mathbf{e}$ evolve towards $\mathbf{0}$ in different ways. For example, the choice

$\mathbf{K_v} = 2\sqrt{\mathbf{K_p}}$ will give us a critical damping (for more details, please refer to any book dealing with the control of linear systems).

What happens if we do not have a perfect knowledge of the dynamic parameters **M**, **C**, **G** and τ_f, but an approximate knowledge such as $\hat{\mathbf{M}}, \hat{\mathbf{C}}, \hat{\mathbf{G}}$ and $\hat{\tau}_f$? Then, turning off the switch at the output of Figure 5.5 on the input of Figure 5.4, which now uses approximate values, will give:

$$\hat{\mathbf{M}}\left(\ddot{\mathbf{q}}_d + \mathbf{K}_v\dot{\mathbf{e}} + \mathbf{K}_p\mathbf{e}\right) + (\hat{\mathbf{C}}\dot{\mathbf{q}} + \hat{\mathbf{G}} + \hat{\tau}_f) = \tau = \mathbf{M}\ddot{\mathbf{q}} + (\mathbf{C}\dot{\mathbf{q}} + \mathbf{G} + \tau_f) \quad [5.12]$$

Hence, the equation of the system in closed loop is:

$$\ddot{\mathbf{e}} + \mathbf{K}_v\dot{\mathbf{e}} + \mathbf{K}_p\mathbf{e} = \hat{\mathbf{M}}^{-1}\left\{(\mathbf{M} - \hat{\mathbf{M}})\ddot{\mathbf{q}} + (\mathbf{C} - \hat{\mathbf{C}})\dot{\mathbf{q}} + (\mathbf{G} - \hat{\mathbf{G}}) + (\tau_f - \hat{\tau}_f)\right\} \quad [5.13]$$

which will lead to a non-ideal behavior of the system and in particular to a steady-state error $\mathbf{e_p}$ such that:

$$\mathbf{K}_p\mathbf{e} = \hat{\mathbf{M}}^{-1}\left\{(\mathbf{G} - \hat{\mathbf{G}}) + (\tau_f - \hat{\tau}_f)\right\} \quad [5.14]$$

This error can be reduced or even suppressed by adding an integral term to the control law represented by equation [5.10]. As far as the error in transient regime is concerned, it can be compensated by the use of robust or adaptive controls (see sections 5.2.6, 5.2.7 and 5.2.8).

5.2.4. *Ideal dynamic control in the operational working space*

In certain cases, such as that for which the desired trajectory is described in the operational space (which we will also call Cartesian space in this chapter) or that presented in section 5.3 where certain Cartesian directions must be force controlled, it will be necessary to develop the control in the operational space. The principle of the non-linear decoupling then remains identical to the one presented in section 5.2.3, but by using the operational dynamic model of the robot [KHA 99], i.e.:

$$\boldsymbol{F} = \mathbf{M}_x\ddot{\mathbf{X}} + \mathbf{C}_x\dot{\mathbf{X}} + \mathbf{G}_x + \mathbf{F}_x \quad [5.15]$$

where:

– $\boldsymbol{F}$ is the vector of the 6 forces and torques (in general) acting on the frame to be controlled (note that this frame does not necessarily belong to the robot, see for example [DAU 90]);

– $\mathbf{X}$ is a variable representing the Cartesian position and orientation of the frame to be controlled;

– $\mathbf{M_x}$ is the Cartesian inertia matrix of the robot;

– $\mathbf{C_x}\dot{\mathbf{X}}$ is the vector representing the centrifugal and Coriolis effects in the Cartesian space;

– $\mathbf{G_x}$ is the Cartesian vector of gravity terms;

– $\mathbf{F_x}$ is the Cartesian vector of the frictions.

A simple calculation using equations [5.1] and [5.15], plus the direct kinematic model $\dot{\mathbf{X}} = \mathbf{J}\,\dot{\mathbf{q}}$ (relating operational velocities $\dot{\mathbf{X}}$ and joint velocities $\dot{\mathbf{q}}$ through the Jacobian matrix $\mathbf{J}$) and the static relation $\boldsymbol{\tau} = \mathbf{J}^T \boldsymbol{F}$ makes it possible to show that:

$$\mathbf{M_x} = (\mathbf{J}^\mathrm{T})^{-1}\,\mathbf{M}\,\mathbf{J}^{-1}, \quad \mathbf{C_x}\dot{\mathbf{X}} = (\mathbf{J}^\mathrm{T})^{-1}(\mathbf{C}\,\dot{\mathbf{q}} - \mathbf{M}\,\mathbf{J}^{-1}\,\dot{\mathbf{J}}\,\dot{\mathbf{q}})$$

$$\mathbf{G_x} = (\mathbf{J}^\mathrm{T})^{-1}\,\mathbf{G}, \quad \mathbf{F_x} = (\mathbf{J}^\mathrm{T})^{-1}\,\boldsymbol{\tau}_\mathbf{f}$$

Therefore, we are capable of calculating the operational dynamic parameters based on the dynamic joint parameters and we can apply the principle of non-linear decoupling to equation [5.15]. In this instance:

– we write $\boldsymbol{F} = \boldsymbol{\alpha}_\mathbf{x}\boldsymbol{F}' + \boldsymbol{\beta}_\mathbf{x}$;

– we choose $\boldsymbol{\alpha}_\mathbf{x} = \mathbf{M_x}$ and $\boldsymbol{\beta}_\mathrm{x} = \mathbf{C_x}\dot{\mathbf{X}} + \mathbf{G_x} + \mathbf{F_x}$;

– we then obtain $\boldsymbol{F}' = \ddot{\mathbf{X}}$

– which we control by $\boldsymbol{F}' = \ddot{\mathbf{X}}_\mathbf{d} + \mathbf{K_{vx}}\dot{\mathbf{E}} + \mathbf{K_{px}}\mathbf{E}$.

Obviously, when $\boldsymbol{F}'$ is calculated in this way and $\boldsymbol{F}$ is obtained by $\boldsymbol{F} = \mathbf{M_x}\boldsymbol{F}' + \mathbf{C_x}\dot{\mathbf{X}} + \mathbf{G_x} + \mathbf{F_x}$, we still have to transform $\boldsymbol{F}$ in control torque $\boldsymbol{\tau} = \mathbf{J}^\mathrm{T}\boldsymbol{F}$, since we cannot directly apply a force on the frame to be controlled. Figure 5.6 summarizes this principle.

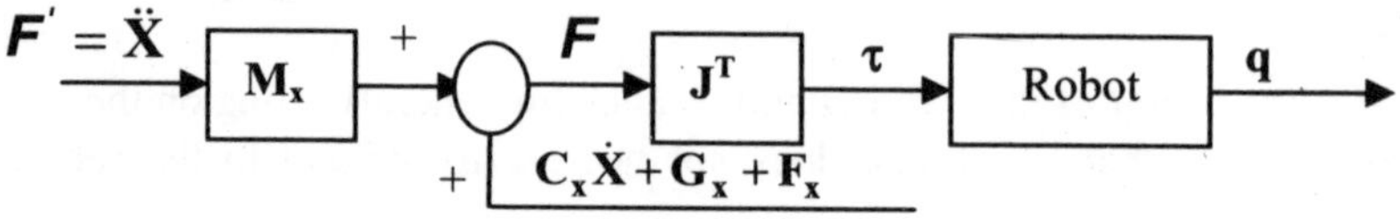

Figure 5.6. *Dynamic control in the operational space*

5.2.5. *Decentralized control*

The design of industrial robots is carried out according to the constraints linked to their future usage (maximum velocity, maximum acceleration, transported mass, workspace, etc.). For the design of serial type robots with n degrees of mobility, constructors often use motors with gears in order to decrease the influence of the inertia of the segment controlled compared to the one of the motor. The choice of a velocity reducing ratio such that the inertia of the link at the motor level equals that of the motor makes it possible to ensure a maximum acceleration at the start. This choice then enables us to do a first approximation at the level of control structure. In fact, we consider that we control n linear and decoupled subsystems of the second order (in the case of electrical actuators):

$$\tau_{mi} = j_i \ddot{q}_i + f_{vi} \dot{q}_i + \tau_{exti} \quad \text{with} \quad i = 1..n \qquad [5.16]$$

τ_{mi} is the control torque of the motor, j_i the average inertia from the point of view of the motor, f_{vi} the average friction, and τ_{exti} the disturbance torque consisting of non-linear coupling and gravity effects.

This is a decentralized control of n independent linear subsystems. The control of the joint position variable q_i is therefore ensured by PID type linear controllers:

$$\tau_{mi} = K_{Pi} e_i + K_{Vi} \dot{e}_i + K_{Ii} \int_t e_i \, d\tau \qquad [5.17]$$

$e_i = q_{di} - q_i$ being the tracking error and K_{Pi}, K_{Vi}, K_{Ii} being respectively the proportional, derivative and integral gains. These gains are non-zero positive constants. They can be determined with the classical continuous linear automatics tools. The representation under the form of a transfer function by using Laplace transform makes it possible to determine the coefficients under the stability constraint and the desired bandwidth. We can also, through a state representation of each subsystem, carry out a location of the poles which will ensure the desired dynamics [LAR 91].

This type of control structure is commonly used in the industrial field because it is simple to realize and use. In fact, apart from the theoretical determinations of the coefficients, we can apply a simple experimental method of PID adjustment. This method was proposed by Ziegler and Nichols. The first step consists of adjusting the proportional term only as a response to a position step in order to obtain an over-shooting of the order of 20 to 30%. Afterwards, we move to the adjustment of the integral term according to the same procedure. Finally, we associate the two terms previously identified and we adjust the derivative term in order to obtain a response with an over-shooting of around 10%. It is a first approximation adjustment that will not always be satisfactory in the case of a trajectory tracking.

This type of decentralized control will not be usable in the case of a complex task where the manipulator will have to carry loads of different masses at different velocities with the requirement of a small tracking error.

5.2.6. *Sliding mode control*

Numerous studies have been conducted on sliding mode control algorithms, by [HAR 85, HAS 86, SLO 91, UTK 87] intended for the control of robots. If we consider the dynamic model of a serial rigid manipulator arm that has n degrees of mobility (see equation [5.1]), we then define the following sliding surfaces s_i $(i = 1,\ldots, n)$ [SLO 91]:

$$s_i = \left(\frac{d}{dt} + \lambda_i\right) e_i \text{ with } i = 1,\ldots, n \qquad [5.18]$$

$e_i = q_{di} - q_i$ representing the tracking error. The selected sliding surfaces represent the frequency response of a low-pass linear system of the first order having a cross-over frequency equal to $f_c = \lambda_i/2\pi$. The objective is to reach the sliding mode for each controlled variable, i.e. $s_i = 0$. The entire system would then have the behavior of n linear decoupled first order systems. This mode is achieved by ensuring the sliding condition which is also a necessary and sufficient stability condition [UTK 87].

The control vector τ_i must be such that:

$$\lim_{s_i \to 0} s_i \dot{s}_i < 0 \qquad [5.19]$$

This condition implies a discontinuous element in the control vector. In fact, the derivative of the sliding surface must have an opposite sign to the one of the surface.

In the case of a manifold of variable switching in the time t, $\mathbf{s}(\mathbf{q},t) = \mathbf{0}$, the condition must be replaced by:

$$\lim_{s_i \to 0} s_i \dot{s}_i < -\eta \left| s_i \right| \qquad [5.20]$$

with η being an increasing monotonic function or a positive constant. In the case of a manipulator that is a coupled non-linear system of dimension n, the synthesis of the control vector is divided into two parts. The first one consists of obtaining an equivalent control vector that ensures the condition of ideal sliding mode, i.e. $\dot{s}_i = 0$, and the second one consists of adding a control term that corresponds to the non-modeled dynamics.

When modifying the presentation of equation [5.1], we obtain:

$$\ddot{\mathbf{q}} = \mathbf{M}^{-1}(\boldsymbol{\tau} - \mathbf{H}) \qquad [5.21]$$

By carrying this expression into the following one, from [5.18]:

$$\dot{\mathbf{s}} = (\ddot{\mathbf{q}}_\mathbf{d} - \ddot{\mathbf{q}}) + \boldsymbol{\Lambda}\,\dot{\mathbf{e}} = \mathbf{0} \quad \text{with} \quad \boldsymbol{\Lambda} = \text{diag}(\lambda_1, \ldots, \lambda_n) \qquad [5.22]$$

We obtain:

$$\boldsymbol{\tau}_{eq} = \mathbf{M}(\ddot{\mathbf{q}}_d + \boldsymbol{\Lambda}\dot{\mathbf{e}}) + \mathbf{H} \qquad [5.23]$$

This equivalent control corresponds to a control vector that ensures the non-linear decoupling of the system. In this specific case, it makes it possible to avoid the use of a discontinuous control element, completely ensuring the sliding condition, and thus risks creating strong oscillations around the switching surface. The switching element will then contribute to the lack of knowledge of non-modeled dynamics.

The control vector of estimated total control $\hat{\boldsymbol{\tau}}_{eq}$ becomes:

$$\hat{\boldsymbol{\tau}}_{eq} = \boldsymbol{\tau}_{eq} + \Delta\boldsymbol{\tau}_{eq} = \hat{\mathbf{M}}(\ddot{\mathbf{q}}_\mathbf{d} + \boldsymbol{\Lambda}\dot{\mathbf{e}}) + \hat{\mathbf{H}} \qquad [5.24]$$

$\Delta\boldsymbol{\tau}_{eq}$ corresponds to the non-modeled dynamics of the system. The final objective is to obtain a control vector $\boldsymbol{\tau}$ that ensures the sliding mode regardless of the nature of the desired trajectory. Therefore, it is necessary that the control vector $\boldsymbol{\tau}$ is such that:

$$\boldsymbol{\tau} = \boldsymbol{\tau}_{eq} + \Delta\boldsymbol{\tau} \qquad [5.25]$$

$\Delta\boldsymbol{\tau}$ is used to oppose to the action of the non-modeled dynamics. By transferring the vector $\boldsymbol{\tau}$ into equations [5.24], [5.23] and [5.22] we obtain:

$$\dot{\mathbf{s}} = -\mathbf{M}^{-1}(\Delta\boldsymbol{\tau}_{eq} + \Delta\boldsymbol{\tau}) \qquad [5.26]$$

We represent the non-modeled dynamics by: $\mathbf{F}(\mathbf{q},\dot{\mathbf{q}},\ddot{\mathbf{q}}) = -\mathbf{M}^{-1}\Delta\boldsymbol{\tau}_{eq}$. The control vector can be decomposed in: $\Delta\boldsymbol{\tau} = \mathbf{M}\,\Delta\mathbf{v}$ (we consider negligible the modeling errors on the inertia matrix).

We thus obtain the following result:

$$\dot{\mathbf{s}} = \mathbf{F}(\mathbf{q},\dot{\mathbf{q}},\ddot{\mathbf{q}}) - \Delta\mathbf{v} \qquad [5.27]$$

The function $\mathbf{F}(\mathbf{q},\dot{\mathbf{q}},\ddot{\mathbf{q}})$ can be upper bounded by known functions of the vectors $\mathbf{e}, \dot{\mathbf{e}}$:

$$|\mathrm{F}_i| \le \alpha_i |e_i| + \beta_i |\dot{e}_i| + \gamma_i \text{, with } \alpha_i, \beta_i, \gamma_i > 0 \qquad [5.28]$$

The components of $\Delta\mathbf{v}$ must be chosen with a discontinuous term in order to meet the sliding conditions. Therefore, it is interesting to propose:

$$\Delta \mathrm{v}_i = (\phi_i |e_i| + \sigma_i |\dot{e}_i| + \kappa_i)\,sign(s_i) \qquad [5.29]$$

By reporting equation [5.29] into equation [5.27], we obtain the following sliding conditions:

$$\dot{s}_i s_i = s_i \left(\mathrm{F}_i - (\phi_i |e_i| + \sigma_i |\dot{e}_i| + \kappa_i)\,sign(s_i)\right) \quad < -\eta\, |s_i|$$

$$\dot{s}_i s_i \le |s_i| \left((\alpha_i - \phi_i)|e_i| + (\beta_i - \sigma_i)|\dot{e}_i| + (\gamma_i - \kappa_i)\right) \; < -\eta\, |s_i|$$

We will obtain the coefficients α_i, β_i, γ_i through an *a priori* estimation of the modeling error and then through a more accurate adjustment on the controller during the experiments. The total control vector τ becomes:

$$\tau = \hat{\mathbf{M}}\left(\ddot{\mathbf{q}}_d + \Lambda\dot{\mathbf{e}}\right) + \hat{\mathbf{H}} + \hat{\mathbf{M}}\left(\phi|\mathbf{q}| + \sigma|\dot{\mathbf{q}}| + \gamma\right)\mathbf{sign(s)}$$

The final control vector consists of a first part that ensures the dynamic decoupling and a second part constituted of a discontinuous switching element that compensates for the non-modeled elements. The discontinuity generates undesired oscillations around the balance point (chattering phenomenon). Slotine proposed to linearize the sign function in order to ensure a steady state around the balance point [SLO 91].

It is nevertheless interesting in certain cases to use this discontinuity as the only element of control [FRA 94, HAR 85] in order to apply it directly in the control of switching transistors constituting the power amplifier (Pulse Width Modulator (PWM)).

5.2.7. *Robust control based on high order sliding mode*

The robustness obtained through the use of sliding mode control leads to the appearance of the chattering phenomenon. Hence, numerous studies focused on this aspect and an original proposal appeared [LEV 93, BAR 98, PER 02]. In fact, the discontinuity that ensures the dynamics of the system in the vicinity of the surface is shifted towards a derivative of this surface. The effect of this change is that the main advantages of the sliding mode control are kept, while the negative chattering effects are considerably reduced.

As part of these developments, a new concept appears: the sliding order. In fact, this order represents the derivative order of the surface where the discontinuity occurs due to the control vector. If we call r the sliding order, we can express the continuity of the sliding surface and its derivatives by:

$$s = \dot{s} = \ddot{s} = \ldots = s^{(r-1)} = 0$$

If we consider a controlled non-linear system such that:

$$\dot{\mathbf{x}} = f(\mathbf{x},\ t, u),\quad \mathbf{s} = \mathbf{s}(t,\ \mathbf{x}) \in \mathrm{R},\ u = \mathbf{U}(t,\ \mathbf{x}) \in \mathrm{R}$$

The successive differentiation of the surface $s(t, \boldsymbol{x})$ with respect to the control vector u makes it possible to obtain the following relations:

1) $r = 1, \ \dfrac{\partial \dot{s}}{\partial u} \neq 0$

2) $r \geq 2, \ \dfrac{\partial s^{(i)}}{\partial u} = 0 \ \ (i = 0,1,2,\ldots,r-1), \ \ \dfrac{\partial s^{(r)}}{\partial u} \neq 0$

The first case represents the classical sliding mode (see section 5.2.6). We can nevertheless process a first order system with a second order controller in order to eliminate the chattering phenomenon.

Twisting algorithm

The twisting algorithm is the first algorithm of second order to be used in the implementation of this approach [LEV 93]. Its main characteristics are represented in the space of phases of surface **s**. The phase trajectories perform decreasing amplitude rotations (geometric progression) that converge, in a given time, at the origin. In the case of a second order system, we can define the control vector u such that:

$$\dot{u}(t) = \begin{cases} -u & \text{if} \quad |u| > 1 \\ -V_m sign(s) & \text{if} \ s\dot{s} \leq 0, \ |u| \leq 1 \\ -V_M sign(s) & \text{if} \ s\dot{s} > 0, \ |u| \leq 1 \end{cases}$$

The choice of the surface s is identical to that in section 5.2.6. However, the convergence conditions in a given time are obtained through constraints on the constant values V_m and V_M. More detailed information is accessible in [PER 02].

5.2.8. *Adaptive control*

It is sometimes necessary to modify the parameters of a control law when certain parameters inherent to the robot or its environment have evolved. In fact, the wear or the modification of the mechanism adjustments lead to modifications of the robot dynamic behavior. It is therefore interesting in certain cases to use an adaptive control law. Much research work has been published on this topic. Among the most interesting ones we can cite the works of [CON 87, CRA 87, DUB 84, LAN 88].

The objective is to update online the estimated dynamic model $(\hat{\mathbf{M}}, \hat{\mathbf{H}})$ in order to minimize the error with the real model.

If we refer to the paragraph on dynamic control in the joint space (see equation [5.13]) we obtain the following result:

$$\ddot{\mathbf{e}} + \mathbf{K}_v \dot{\mathbf{e}} + \mathbf{K}_p \mathbf{e} = \mathbf{M}^{-1}(\tilde{\mathbf{M}}\ddot{\mathbf{q}} + \tilde{\mathbf{H}}) \qquad [5.30]$$

With $\tilde{\mathbf{M}} = \mathbf{M} - \hat{\mathbf{M}}$ and $\tilde{\mathbf{H}} = \mathbf{H} - \hat{\mathbf{H}}$ that represent the dynamic modeling errors. We choose to express the vector $\tilde{\mathbf{P}} = \mathbf{P} - \hat{\mathbf{P}}$ that represents the estimation errors of the dynamic parameters (inertial and frictional) of dimension r. There is a matrix $\mathbf{\Phi}(\mathbf{q}, \dot{\mathbf{q}}, \ddot{\mathbf{q}})$ of dimension n×r such that equation [5.31] is satisfied:

$$(\tilde{\mathbf{M}}\ddot{\mathbf{q}} + \tilde{\mathbf{H}}) = \mathbf{\Phi}\, \tilde{\mathbf{P}} \qquad [5.31]$$

The use of Lyapunov's method ensures the creation of an asymptotically stable adaptation law such that:

$$\dot{\tilde{\mathbf{P}}} = -\mathbf{Q}\mathbf{\Phi}^{\mathbf{T}} \hat{\mathbf{M}}^{-1} \mathbf{e}_1$$

With $\mathbf{Q}$ as a non-negative matrix of dimension r×r such that $\mathbf{Q} = \mathrm{diag}(q_1, ..., q_r)$ which is called adaptation gain. The adaptation is not carried out on the position vector but on the vector $e_1 = \dot{e} + \psi e$ which is a filtering of the position error.

If we derive $\tilde{\mathbf{P}}$, we obtain: $\dot{\tilde{\mathbf{P}}} = -\dot{\hat{\mathbf{P}}}$. In fact, only the estimate of parameters evolves in time. Hence, finally we have the adaptation law which follows:

$$\dot{\hat{\mathbf{P}}} = \mathbf{Q}\mathbf{\Phi}^{\mathbf{T}}(\mathbf{q}, \dot{\mathbf{q}}, \ddot{\mathbf{q}}) \hat{\mathbf{M}}^{-1} \mathbf{e}_1$$

The advantage of this method proposed by [CRA 87] is using a simple dynamic model and performing an asymptotically stable adaptation not on the physical parameters inherent to the system but on a set of parameters representing the model. One of the problems that we can highlight is that it is necessary to know the joint accelerations in order to carry out the estimation. Other more elaborate methods based on passivity are exposed in [LAN 88, SLO 91].

5.3. Control in a constrained space

5.3.1. *Interaction of the manipulator with the environment*

The realization of a robotic task often requires an interaction between the manipulator and the environment. Typical examples are surface following, fettling or insertion of mechanical parts. During the interaction, the environment somehow imposes some constraints on the trajectories that the robot tool can follow. Therefore, the use of position controls like those described in section 5.2 is not possible unless the trajectory of the tool is planned extremely accurately and the control ensures a perfect monitoring of this trajectory. To reach these two objectives, it is essential to have an accurate model of a manipulator (geometric, kinematic and dynamic) but also of the environment (geometry and mechanical characteristics). It is clear that this second constraint can be met very rarely. It is therefore necessary to implement controls that are not "pure" controls in position. In this section, we will present two main types of such controls. The first type does not use a force closed loop control and we will be interested here in the most popular solution which is the impedance control. The second type gathers all the controls that use, in a closed loop control structure, desired and measured values of the contact forces besides the position variables of the tool. We will name this approach the position/force hybrid control and will present two of its possible implementations.

5.3.2. *Impedance control*

The impedance control is a control strategy that ensures that the robot, submitted to an external force, has the behavior of an impedance constituted of a set: mass ($\mathbf{M_d}$), spring ($\mathbf{K_d}$), damper ($\mathbf{D_d}$). The linear behavior desired in the operational space can be written as an equation as follows:

$$\mathbf{F} = \mathbf{M_d}\,\ddot{\mathbf{E}} + \mathbf{D_d}\dot{\mathbf{E}} + \mathbf{K_d}\mathbf{E} \qquad [5.32]$$

$\mathbf{E} = \mathbf{X_d} - \mathbf{X}$ is the tracking error vector. There are two very different types of approach to obtain this behavior. The first one is an implicit force control that does not require external force measurement and does not take into account the dynamic model of the robot (except for the gravity compensation). In this case, the control vector is chosen in the following way:

$$\boldsymbol{\tau} = \mathbf{J}^{\mathrm{T}}\left(\mathbf{D_d}\,\dot{\mathbf{E}} + \mathbf{K_d}\mathbf{E}\right) + \mathbf{G(q)} \qquad [5.33]$$

This control, which is simple to implement, can be related to a proportional and derivative control in the operational space. This control law is particularly interesting when we want to obtain impedance behavior during small motions (insertion). In fact, since the robot dynamic parameters are negligible, we obtain the desired compliance (flexibility) in the directions defined by the operator, without having to use force sensors.

The second method presents two formulations of the control vector. Let us remember the dynamic model of the robot in the joint space when it is submitted to an external force:

$$\boldsymbol{\tau} = \mathbf{M}(\mathbf{q})\ddot{\mathbf{q}} + \mathbf{H}(\mathbf{q}, \dot{\mathbf{q}}) + \mathbf{J}^{\mathrm{T}}\mathbf{F} \qquad [5.34]$$

By expressing the vector $\ddot{\mathbf{X}} = \mathbf{J}\ddot{\mathbf{q}} + \dot{\mathbf{J}}\dot{\mathbf{q}}$ with joint variables in equation [5.34] and then replacing the vector $\ddot{\mathbf{q}}$ by this expression in equation [5.32], we obtain:

$$\boldsymbol{\tau} = \hat{\mathbf{M}}\mathbf{J}^{-1}\left(\ddot{\mathbf{X}}_{\mathbf{d}} + \mathbf{M}_{\mathbf{d}}^{-1}\left(\mathbf{D}_{\mathbf{d}}\,\dot{\mathbf{E}} + \mathbf{K}_{\mathbf{d}}\,\mathbf{E} - \mathbf{F}\right) - \dot{\mathbf{J}}\dot{\mathbf{q}}\right) + \hat{\mathbf{H}} + \mathbf{J}^{\mathrm{T}}\mathbf{F} \qquad [5.35]$$

A second formulation consists of expressing equation [5.35] in the operational space. We then obtain, by using the relations in section 5.2.4:

$$\boldsymbol{\tau} = \left(\mathbf{J}^{\mathrm{T}}\hat{\mathbf{M}}_{\mathbf{x}}\ddot{\mathbf{X}}_{\mathbf{d}} + \mathbf{J}^{\mathrm{T}}\hat{\mathbf{M}}_{\mathbf{x}}\mathbf{M}_{\mathbf{d}}^{-1}\left(\mathbf{D}_{\mathbf{d}}\,\dot{\mathbf{E}} + \mathbf{K}_{\mathbf{d}}\,\mathbf{E}\right) + \mathbf{J}^{\mathrm{T}}\mathbf{F}\left(\mathbf{I} - \hat{\mathbf{M}}_{\mathbf{x}}\mathbf{M}^{-1}\right) + \mathbf{J}^{\mathrm{T}}\hat{\mathbf{H}}_{\mathbf{x}}\right)$$

These two formulations require measuring the force as well as a complete calculation of the estimated dynamic model. This type of impedance control can be used as part of a task that requires impedance operation in the entire workspace and during a displacement of the arm such as the one for the transportation of fragile objects. This class of control law includes the active stiffness control [SAL 80] and the admittance control [WHI 77] which are particular cases of the impedance control law. In the first case, the desired impedance model is reduced to a stiffness matrix $\mathbf{K_d}$ and in the second case to a damping matrix $\mathbf{D_d}$.

5.3.3. *Force control of a mass attached to a spring*

In section 5.2.4, we saw how a non-linear decoupling in the operational space made it possible to transform a robot manipulator (coupled non-linear system) into 6 decoupled linear systems which are easy to control. But how does this method apply when the robot interacts with the environment and when we want to explicitly control the contact forces? In order to prepare the answer to this question, we will study in this section the simple case of a manipulator with one degree of freedom. For this, let us consider the mechanical system drawn in Figure 5.7.

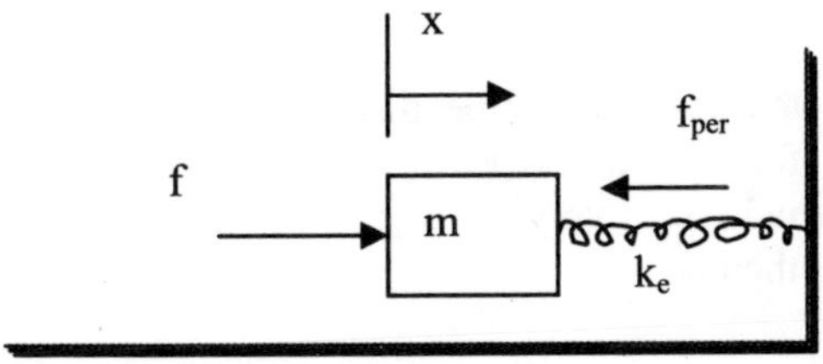

Figure 5.7. *Mass attached to a spring*

We have mass m traveling on a horizontal plane whose interaction with the environment is modeled by a spring whose stiffness is k_e. The input force f makes it possible to move the mass. f_{per} represents a set of forces disturbing the motion (for a robot, this corresponds, for example, to frictions but also to other effects, as we will see in the following section). The objective is to control the interaction force $f_e = k_e x$. The equation of motion of the open loop system is:

$$f = m\ddot{x} + k_e x + f_{per} \tag{5.36}$$

This equation can be rewritten to display the variable to be controlled f_e:

$$f = m k_e^{-1} \ddot{f}_e + f_e + f_{per} \tag{5.37}$$

Let us talk about a control of this system through perfect non-linear decoupling. We then write $f = \alpha f' + \beta$, then we choose $\alpha = mk_e^{-1}$ and $\beta = f_e + f_{per}$. This leads to $f' = \ddot{f}_e$, controlled by:

$$f' = \ddot{f}_{ed} + k_{vf} \dot{e}_f + k_{pf} e_f, \text{ with } e_f = f_{ed} - f_e$$

However, if β is not well known, which is almost always the case since it includes a disturbance f_{per}, the system will be described by equation [5.37] and the control will impose equation [5.38]:

$$f = mk_e^{-1} (\ddot{f}_{ed} + k_{vf} \dot{e}_f + k_{pf} e_f) + \hat{f}_e + \hat{f}_{per} \tag{5.38}$$

where $\hat{f}_e$ and $\hat{f}_{per}$ are the estimated values of f_e and f_{per}, respectively. Let us equal equations [5.37] and [5.38] and then analyze different cases:

1. $\hat{f}_e = f_e$ and $\hat{f}_{per} = 0$

This case means that we ignore the disturbances and that we can accurately measure the contact forces (this measure will be carried out on a robot with a force/torque sensor which, like every measuring device, has a certain accuracy). The steady-state error will then be:

$$e_{fp} = \frac{f_{per}}{m\,k_e^{-1}\,k_{pf}}$$

This error can be very significant if k_e is very high, which is the case when a rigid robot interacts with a rigid environment.

2. $\hat{f}_e = f_{ed}$ and $\hat{f}_{per} = 0$

The estimation of f_e does not use a measure of the contact forces but their desired value, which is therefore perfectly known. However, this case is interesting since the steady-state error:

$$e_{fp} = \frac{f_{per}}{1 + m\,k_e^{-1}\,k_{pf}}$$ is smaller than in the first case.

3. $\hat{f}_e = f_{ed}$ and $\hat{f}_{per} \neq 0$

$$e_{fp} = \frac{f_{per} - \hat{f}_{per}}{1 + m\,k_e^{-1}\,k_{pf}}$$ is a smaller error than in the second case.

4. $\hat{f}_e = f_{ed}$ and $\hat{f}_{per} = f_{per}$ then $e_{fp} = 0$

This brief analysis shows firstly that in the ideal case, where the disturbances are perfectly estimated, using $\hat{f}_e = f_{ed}$ is equivalent to using $\hat{f}_e = f_e$ (classical non-linear decoupling) in terms of steady-state error (which is then zero). But more importantly, this analysis also shows that if the disturbances are not well estimated, using $\hat{f}_e = f_{ed}$ rather than $\hat{f}_e = f_e$ reduces the steady-state error. Besides, there is no noise on f_{ed}, which will not be the case in practice on the measure of f_e.

To conclude, the control structure with anticipative term presented in Figure 5.8 is preferable to the "classical" one presented in Figure 5.9. In both cases, the "control law" unit corresponds to:

$$f'_1 = k_e^{-1} (\ddot{f}_{ed} + k_{vf}\, \dot{e}_f + k_{pf}\, e_f)$$

Hence, we note that the calculation of f'_1 requires knowing the stiffness k_e. If this not possible in practice, we can adjust the gains k_{vf} and k_{pf} in such a way that the system is stable in case of high stiffness. It will then be less efficient in terms of response time if the stiffness is lower, but it will remain stable. Other, much more sophisticated solutions will be mentioned in section 5.3.5.

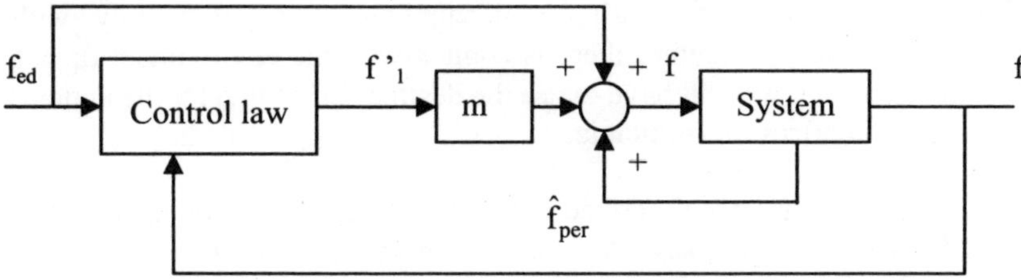

Figure 5.8. *Non-linear decoupling with an anticipative force term*

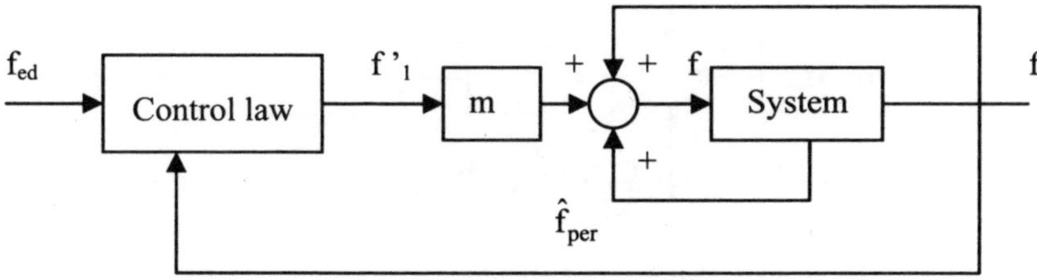

Figure 5.9. *Classical non-linear decoupling*

5.3.4. *Non-linear decoupling in a constrained space*

During the phase of contact with the environment, the operational dynamic model of the robot is written:

$$\boldsymbol{F} = \mathbf{M}_x\ddot{\mathbf{X}} + \mathbf{C}_x\dot{\mathbf{X}} + \mathbf{G}_x + \mathbf{F}_x + \mathbf{F}_{ex} \qquad [5.39]$$

where $\mathbf{F}_{ex}$ represents the interaction forces between the robot and the environment and the other terms are those of equation [5.15]. If we compare equations [5.37] and [5.39], the analogy is that $f_e = \mathbf{F}_{ex}$ and $f_{per} = \mathbf{C}_x\dot{\mathbf{X}} + \mathbf{G}_x + \mathbf{F}_x$.

During a complex manipulation task, theoretically there are some phases where the robot is not in contact with the environment. We will then develop a position control, for example the dynamic control described in section 5.2.4. There are also phases where the robot interacts with the environment and we will have to develop a force control in certain operational directions. However, we saw in section 5.3.3 that we could replace the feedback of f_e by an anticipative term f_{ed}. By analogy, we will be able to carry out the non-linear decoupling of a robot only with the terms $\mathbf{C}_x\dot{\mathbf{X}}$, $\mathbf{G}_x$ and $\mathbf{F}_x$, whether there is contact with the environment or not, and when there is contact, we will have to use the desired forces $\mathbf{F}_{exd}$ for the anticipation. Figure 5.10 summarizes this principle.

Note that unlike the case described in Figure 5.6, here $\boldsymbol{F'}$ is not equal to $\ddot{\mathbf{X}}$, since $\mathbf{F}_{exd}$ and $\mathbf{F}_{ex}$ are not rigorously identical (for example there always is a response time, as small as it is, so that $\mathbf{F}_{ex}$ duplicates $\mathbf{F}_{exd}$).

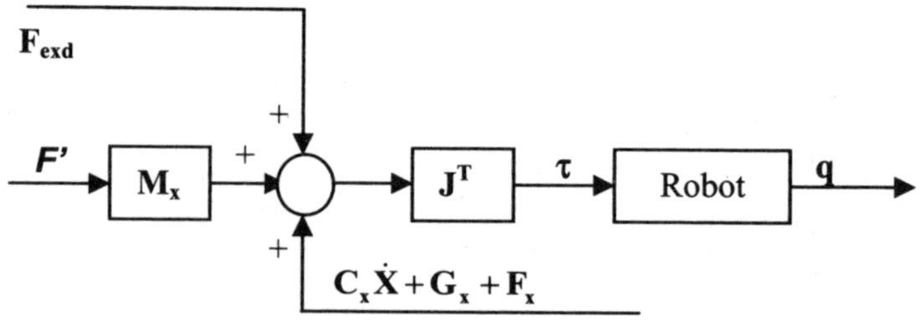

Figure 5.10. *Non-linear decoupling in a constrained place*

We now have the basic information to propose complete control structures in which certain operational directions are controlled in position and other ones in force/torque. This corresponds to the issue of the position/force hybrid control for which two solutions are presented in the following section.

5.3.5. *Position/force hybrid control*

Once again our purpose here is not to provide an exhaustive list of all hybrid control structures and give their details, this can be found in more specialized books such as [SIC 00]. On the contrary, we thought it would be interesting to discuss a few points that are not often presented in other works and this, as we have already mentioned, for two structures very commonly used.

5.3.5.1. *Parallel structure*

5.3.5.1.1. General principle

First of all, we have to clarify that what we mean by parallel structure is the equivalent of what is sometimes encountered in other books under the more general term of hybrid structure, or even hybrid control. However, actually the concept of hybrid control covers, according to us, the hybrid structure on one hand, i.e. the way of combining the position and force servo loops, and on the other hand the position and force control laws that are not the subject of a detailed study in this section.

So why use the term "parallel"? Simply because the position and force feedback loops could be drawn in parallel in a representation having the form of block-diagrams. This means that in this approach, certain operational directions will be controlled in position and the other ones in force, these directions mutually excluding each other. To be more accurate, while performing a task in a constrained space, there are some constraints that are referred to as "natural", defined by Mason 25 years ago [MAS 81]. These constraints correspond to the fact that the robot cannot move in certain directions because of the contact with the environment (natural position constraints) and cannot apply forces in other directions because nothing opposes to its displacement (natural force constraints). The directions which we are talking about are the ones of a frame called "of constraints" that the operator must intelligently choose in order to facilitate later the implementation of the control (frame linked to the end-effector, to the tool, to the environment, etc., depending on the task to be carried out). More information on the frame of constraints and on the definition of natural constraints can be found in [CRA 86]. Hence, the problem that arises is to define a control structure which enables:

– to control the robot in position in the directions where there are force natural constraints;

– to control the robot in force in the directions where there are natural position constraints;

– to implement whichever association of these control modes along the orthogonal directions of any frame of constraints. In fact, if that was not the case, a specific control structure would have to be developed for each specific case of constrained task.

To meet these requirements, "selection" matrices are used. We change their values according to the task to be carried out. Figure 5.11 shows a small part of a parallel hybrid control structure, like the one proposed by Raibert and Craig [RAI 81].

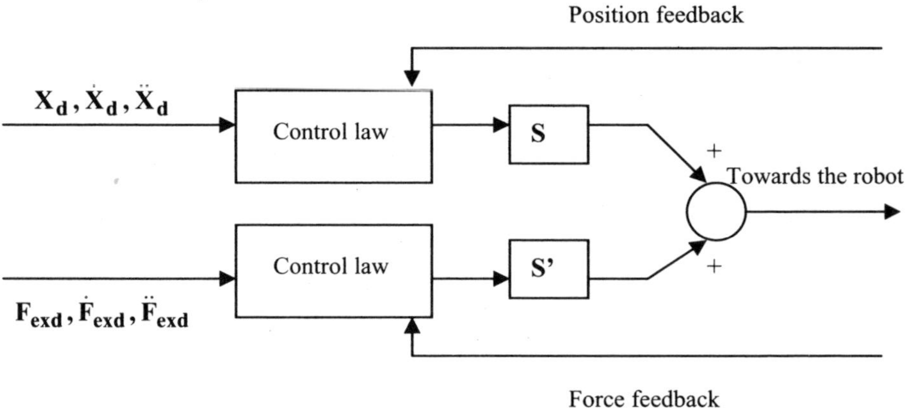

Figure 5.11. *Part of a parallel hybrid control structure*

The selection matrices **S** and **S'** must then make it possible to simply choose the variables to be controlled in position and those to be controlled in force. In order to have all possible choices, the position and force loops both act on the 6 components of the operational position and force variables. Thus, the simplest and most efficient solution consists of defining **S** as a diagonal matrix of 1 and 0. If the ith component of the position variable must be effectively controlled, we will specify a 1 on the element (i,i) of **S**. And since the ith component of the force variable must not be controlled, we will need a 0 on the ith line of **S'**. By generalizing this, we naturally come to the conclusion that **S' = I – S, I** being the identity matrix of dimension 6. Actually, the correct use of these selection matrices is a bit subtler, as it is explained in [PER 91]. Certain stability problems can thus arise from an incorrect use of the selection matrices [FIS 92]. If we deal with a complete block-diagram of the parallel hybrid control, we can use, for example, the non-linear decoupling in a constrained space seen in section 5.3.4 and the control laws seen in sections 5.2.4 and 5.3.3. We then obtain the dynamic parallel hybrid control of Figure 5.12.

In this figure, we note that the selection matrix **(I – S)** must be applied to the anticipative term of force. We must also specify that in this type of control, everything must be calculated in the constraint frame defined by the operator, i.e. the robot dynamic model, the Jacobian matrix **J**, the direct kinematic model DKM, etc.

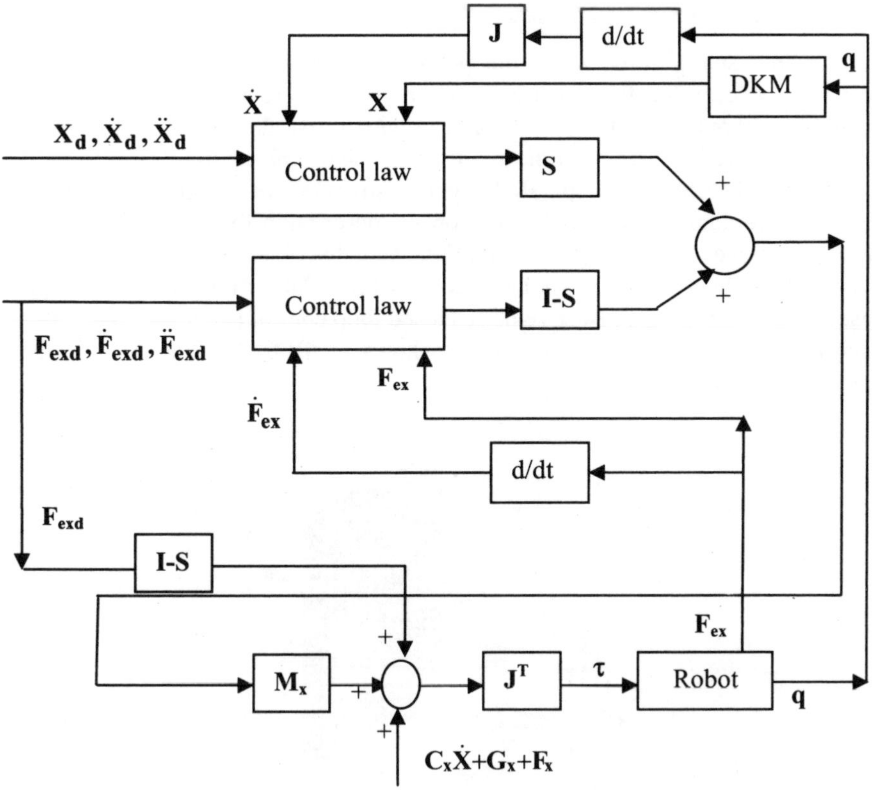

Figure 5.12. *Diagram of the dynamic parallel hybrid control*

Finally, we have to underline the fact that the diagram in Figure 5.12 is only an implementation solution for the parallel hybrid control. In fact, because of the calculation time or because the model is not known, we can choose not to decouple the robot or to decouple it only partially. We can also choose control laws in position or in force which are different, particularly adaptive or robust, but the general structure introduced in Figure 5.11 will remain the same.

This parallel structure has been widely studied since Mason's work and very often implemented on manipulators with 1 or even 2 arms [DAU 90]. However, it has certain drawbacks, some of which are well known [KHA 87], and some which are less known, like the two we will present, taken from [BER 95].

5.3.5.1.2. Inconsistency with respect to Mason's description

In a free space, it is firstly the position of the manipulator that has to be controlled, except for the specific case of tasks where it has to be positioned manually, without any a priori set point, in which case the closed loop control set point will not refer to its position in space. In a constrained space, forces must be controlled. Therefore, the parallel hybrid structure of [RAI 81] displays a position/force duality. However, it lies on the description formulated in [MAS 81] which introduces a velocity/force duality. Actually, a position set point can be defined from the velocity set point. This is what results from most of the hybrid closed loop controls, because a velocity closed loop control is unthinkable without a control of the position. However, Berthe shows in [BER 95] that the velocity set point is more pertinent.

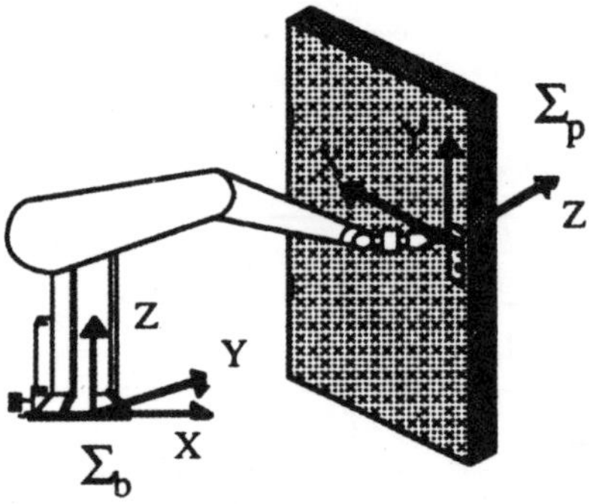

(a) general view

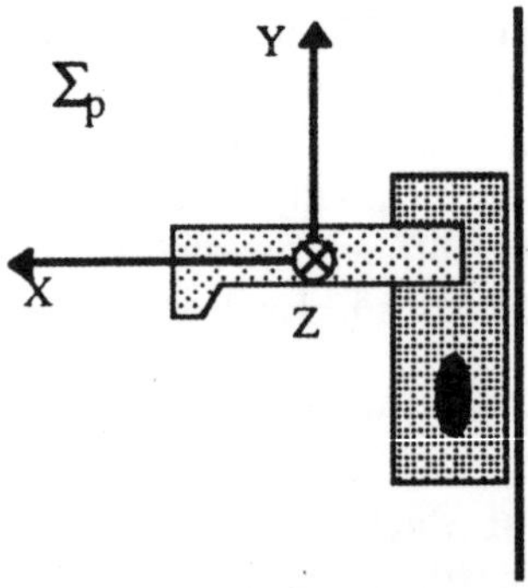

(b) view of the gripper

Figure 5.13. *Door opening: definition of reference frames Σ_b and Σ_p*

For that, we consider the example of door opening carried out by a manipulator. A frame Σ_b attached to the base of the robot is defined as the frame. The operational frame Σ_p is attached to the gripper. The constraints are defined in this frame. The orientation of this frame changes during the execution of the task. Therefore, the use of a velocity set point in this frame enables to carry out the task in good conditions even though the position of the robot is disturbed. This would be unthinkable in the case of a position set point directly defined in the frame Σ_b (we would have arcs of a circle). Figure 5.13 defines the frames Σ_b and Σ_p and clarifies the explanations that have just been given.

The action of opening the door is divided into two successive tasks. The first one consists of turning the handle and the second one of making the door move around its hinges. As the robot grips the handle firmly, only the direction **Y** of the frame Σ_p is velocity controllable during the first task, while it is the **Y** and **Z** directions during the second task. The other directions of the constraint frame Σ_p are force controllable. Let us consider for example the second task. A non-zero velocity set point along the **Z** axis makes it possible to open the door. Since this set point is expressed in the operational frame, the direction of the displacement depends on its current position. If this frame varies with impunity because of disturbances, the motion can still be carried out without generating constraints because the trajectory is automatically adapted. A closed loop position control where the motion generation is directly calculated in the basic frame, as arcs of a circle, cannot be adapted in such a way. This result arises from the task description in terms of velocity.

Nevertheless, this velocity set point is used to build a closed loop control of the position. The first possible approach consists of integrating the velocity set point (Figure 5.14). The displacement is therefore dealt with from the last position set point ${}^{b}\mathbf{X}_d(t - \Delta t)$. We assign the exponent b (respectively p) to a vector expressed in the base frame (respectively the gripper of the robot). Δt is the sampling period. ${}^{b}\mathbf{R}_p$ is the transformation matrix from Σ_b to Σ_p, and ${}^{p}\mathbf{R}_b$ is the inverse matrix. The selection of directions by matrix **S** is done in the constraint frame Σ_p. ${}^{b}\varepsilon_X(t)$ is the Cartesian position error at the time t, expressed in the frame Σ_b, and which is used in the control law in position. It has to be noted that when the closed loop control is entirely carried out in position, the creation of motion cannot be built in that way, because this calculation defines a succession of set points that move away from the real trajectory: the system is physically constrained to turn around its axis while the position's increment ${}^{p}\Delta\mathbf{X}_d$ is tangent to this movement of rotation.

But actually, since we calculate a displacement set point ${}^{b}\varepsilon_X(t)$, it is more logical to count it only from the last position reached, rather than from the last desired position. In Figure 5.14, ${}^{b}\mathbf{X}_d(t - \Delta t)$ must therefore be replaced by ${}^{b}\mathbf{X}(t - \Delta t)$ and as a result, the operations of the two summing elements can be simplified, as well as the two first frame changes. It is noticeable that there is no longer a position closed loop control; the system is in open loop on the velocity (Figure 5.15). This solution is attractive for its simplicity. In addition, in the full control situation, this solution will explicitly generate the velocity set point in parallel to the force set point. However, in practice, it is very important to consider the position in order to compensate for disturbances and to guarantee the position of the robot once the task is completed.

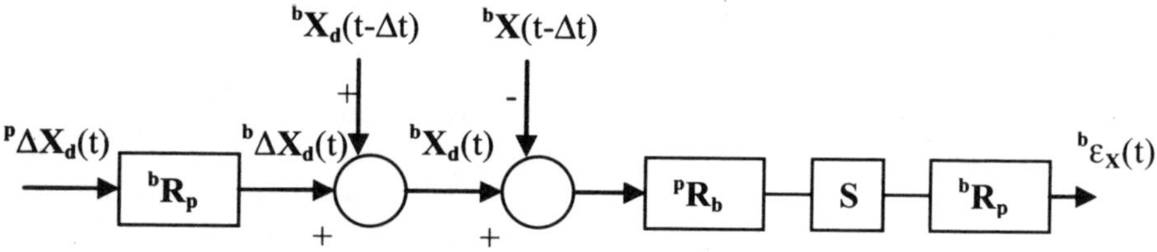

Figure 5.14. *Calculation of the set point position and closed loop control*

A lower level position loop must therefore be added, which will not contradict the description of the task used, therefore located after having combined the two contributions in velocity and in force, for example at the joint level where the closed loop control calculations can be carried out very quickly. Rigorously, a parallel hybrid structure must include a velocity/force control completed by a closed loop control of the robot joint position.

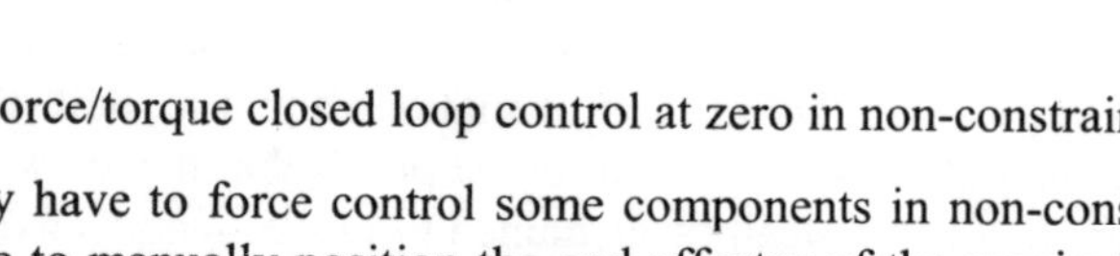

Figure 5.15. *Case where the displacement is calculated from the last position reached*

5.3.5.1.3. Force/torque closed loop control at zero in non-constrained directions

We may have to force control some components in non-constrained directions, for example to manually position the end-effector of the manipulator by measuring the forces applied on it by the operator. In this case, the robot balance is obtained for force set points that are zero or that compensate for the effects of gravity. With a "pure" parallel hybrid control, the directions corresponding to this closed loop control are not controlled in position and the experiment highlights the following disastrous phenomenon: if a disturbance occurs (or if is purposely applied) the force sensor, the robot moves in the directions controlled for a zero force, while the force sensor has not sensed the disturbance. The problem that appears is not one of a response quality in an operating point and to be maintained despite the variations of operating conditions, but one of a total absence of compensating effect by the closed loop control. It is the case of a robust behavior with respect to the possible operating modes. The solution to this problem is to use other sensors than the wrist force sensor. An excellent solution is once again to use the robot position sensors in an additional joint closed loop control.

When dealing with these disadvantages, it is legitimate to try to find another hybrid control structure which would keep, if possible, the advantage of the parallel control (i.e. the possibility of controlling the force value in certain directions) while introducing what seems to be a proof of safety (i.e. a permanent position closed loop control) regardless of the constraints of the task during the contact. A solution that seems to have proved its value in numerous and varied applications [DEG 94], is the external hybrid structure [PER 91, SCH 88], which we will summarize in the following section.

5.3.5.2. *External structure*

The name external structure stems from the fact that it consists of a force external loop on a position internal loop. The key idea of this solution is that an error on the force can be seen as a displacement to be conducted. To do so, from the force error, we calculate a position increment that modifies an initial position set point. In other words, the operational position of the robot is controlled to follow a set point that is automatically corrected by force closed loop control so that there is contact with the environment in the desired conditions. Figure 5.16 illustrates this principle.

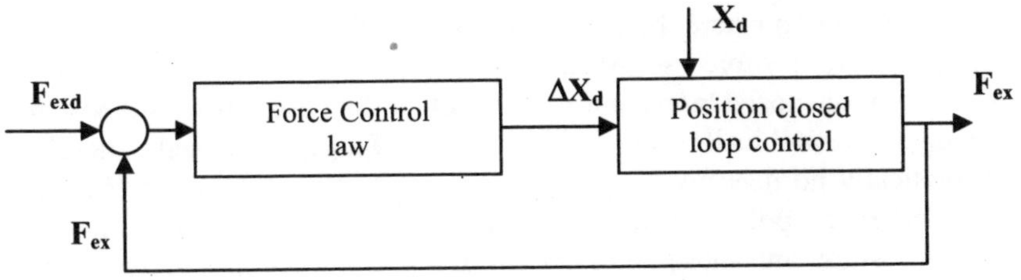

Figure 5.16. *Principle of the external hybrid structure*

The displacement set point $\mathbf{\Delta X_d}$ adds to the position set point $\mathbf{X_d}$ to constitute the input vector of the position closed loop control. We did not display this explicitly in Figure 5.16 because several possibilities can be envisaged. First of all, the position closed loop control can be Cartesian, i.e. the position/orientation $\mathbf{X}$ is compared to a total set point $\mathbf{X_d}^* = \mathbf{X_d} + \Delta\mathbf{X_d}$. Control laws such as the dynamic control seen in section 5.2.4 can then be applied to the lower part. The closed loop control can also be in the joint space and apply to the joint position $\mathbf{q}$ which is compared to the desired position $\mathbf{q_d}^*$ coming from the set points $\mathbf{X_d}$ and $\Delta\mathbf{X_d}$. So there are two ways to calculate $\mathbf{q_d}^*$. The first one consists of using the inverse kinematic model on $\mathbf{X_d}$ to obtain $\mathbf{q_d}$ and the pseudo-inverse of the robot Jacobian matrix on $\Delta\mathbf{X_d}$ to obtain $\Delta\mathbf{q_d}$. We then calculate $\mathbf{q_d}^* = \mathbf{q_d} + \Delta\mathbf{q_d}$. The second solution, which is simpler, consists of using the inverse kinematic model on $\mathbf{X_d}^*$. The joint closed loop control that follows can of course be more or less sophisticated. It is interesting to note that if it is about the closed loop control implemented in an industrial robot controller, and therefore generally impossible to modify, the external structure makes it possible by all means to develop a hybrid control for an industrial robot, which is impossible with a parallel structure [DEG 93].

Like in an impedance control structure based on the position, the purpose of a force measure is to correct a position set point. However, the essential difference is that here it is possible to impose a desired force. It must nevertheless be specified that it is the force set point that will be respected if the force loop during is hierarchically higher than the position loop. Berthe shows that this is true only if the force control law contains an integral term [BER 95]. If this is not the case, the force closed loop control is altered by the position set point $\mathbf{X_d}$. According to the principle of external structure, all directions of the constraint frame are affected by a position closed loop control. If it is the same for the force, then all directions are controlled in the same manner and when the robot is not constrained in all directions, the hybrid aspect of the task does not clearly appear. An analysis of this hybrid control based on Mason's formalism [MAS 81] would lead to place a selection matrix in the force loop, so that the force closed loop control concerns only the directions naturally constrained in position. Actually, this matrix is not always applied [PER 91]. In fact, in a direction where there is no detected force (direction of natural force constraint) and where the force set point must therefore be zero, $\Delta\mathbf{X_d} = 0$ and we have a simple position closed loop control. In the other directions, if we suppose that the force loop is hierarchically higher, the set point $\mathbf{F_{exd}}$ is respected. Therefore, there is theoretically no need to use a selection matrix, it is sufficient to correctly choose the force set points. But this supposes that the directions naturally constrained in position and those naturally constrained in force are really different (and orthogonal). However, in a surface following for example, the tangential forces due to frictions disturb the displacement of the tool. Without a selection matrix, they will have an impact on the closed loop control and will thus disturb the trajectory tracking in a direction that is controllable in position. Since frictions impede motion, the zero force closed loop control in the tangential directions contributes even to slow down the displacement. Because of this, the selection matrix is only strictly necessary so that the closed loop controls are correctly separated, i.e. the measure of the forces does not alter the directions that we want to control in position. Figure

5.17 illustrates a partial view of an external hybrid control structure, with Cartesian closed loop control of the position. **I** represents the identity matrix of dimension 6, CLF means "control law in force" and CLP "control law in position".

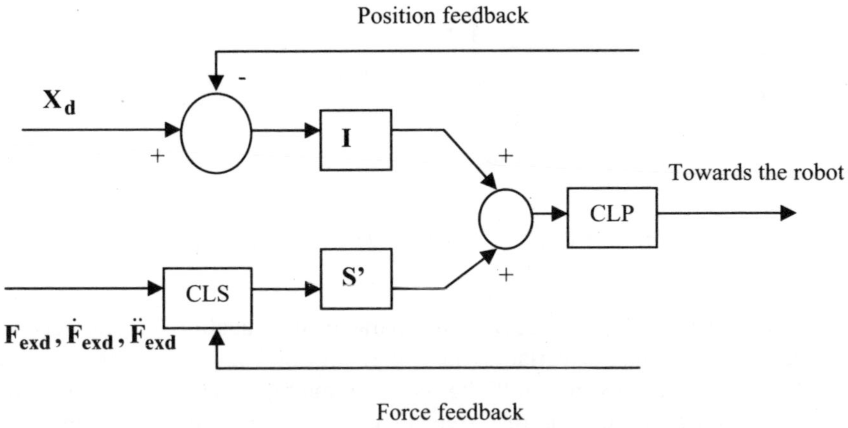

Figure 5.17. *Part of an external hybrid control structure*

This figure is purposely drawn in a similar way as Figure 5.11 in order to show that finally the external structure can be seen as a parallel structure in which the control law in position is displaced after summing the contributions of the two loops. Let us end by specifying that this solution is consistent with Mason's description and does not present the flaw described in section 5.3.5.1.3.

5.3.6. *Specificity of the force/torque control*

The force/torque control of a robot manipulator presents some characteristics that are different from the control in position. The dynamics of the contact phenomena between the robot and the environment are much higher than the one associated to the control in position. We can therefore encounter real stability problems. The system can also abruptly be in open loop because the contact with the environment depends on the position of the manipulator.

Hence, unlike the position variable, the force variable does not necessarily exist for whichever configuration of the robot. To this is added, as we mentioned in section 5.3.1, a real lack of knowledge of the environment's characteristics. In fact, if we consider the simplified dynamic model of a robot (see Figure 5.18) in contact with an environment, according to the force direction, we have a differential equation which depends on the parameters of the environment.

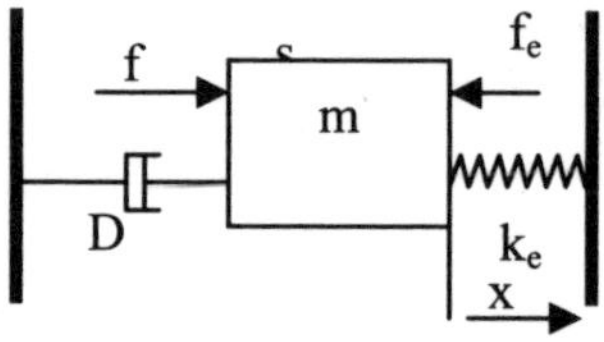

with

$$f - f_e = m\ddot{x} + D\dot{x} \quad \text{with } f_e = k_e\, x$$

$$F_e(p) = \frac{F(p)}{\frac{m}{k_e}p^2 + \frac{D}{k_e}p + 1}$$

Figure 5.18. *Simplified mode: mass, spring, damper*

This result shows that the roots of the characteristic equation directly depend on k_e if we consider that the variables m and D are known. The result of a model simulation in Figure 5.18 (m = 0.01 kg, D = 1 N/m/s) is given in Figure 5.19 and it shows the significant variation of a step force response f = 10 N, in open loop for three different values of k_e.

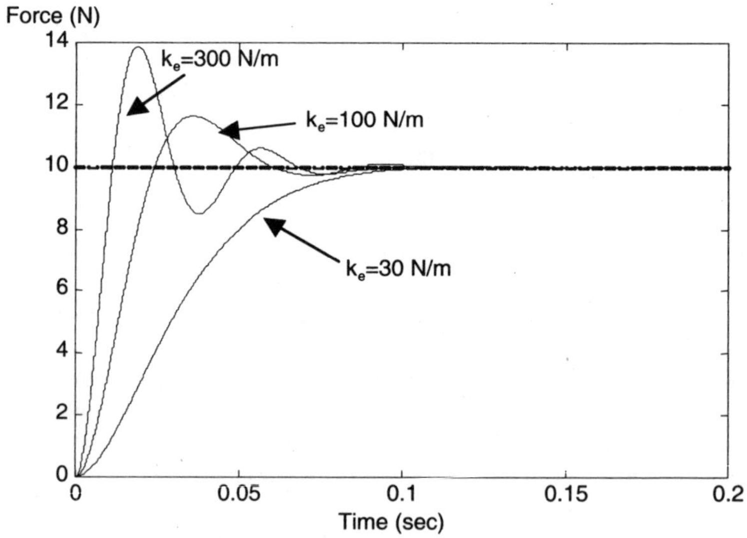

Figure 5.19. *Response to a force step*

The poles placement does not bring any solution because the new resulting poles will also depend on k_e. Also, there are solutions that have been explored, such as the online estimation of the stiffness [YOS 93], the use of adaptive control algorithms [HOS 98, WHI 97], the acquisition of the task [KAW 85] or even the use of a control vector using a virtual environment in order to minimize the influence of the variations of k_e [FRA 94].

Another solution suggests the use of an active observer (AOB) based on a Kalman filter that makes it possible to estimate the stiffness of the environment and thus to adapt the displacement of the robot manipulator controlled in position [COR 03]. This method makes it possible to reach a certain insensitivity of the control law in relation to stiffness variations and to require no structure modification for the passage between the free space and the constrained space.

The direct consequence of this observation does not concern the stability, which is ensured in this specific case, but the response to a desired force. In fact, the dynamics of the contact phenomena are very high during the collision of the effector with the environment and directly depend on its stiffness as well as on the velocity of approach [TAR 96].

A second simulation result is presented based on the model in Figure 5.20 ($m = 0.01$ kg, $D = 1$ N/m/s, $k_e = 100$ N/m, $\Delta x = 0.3$ m).

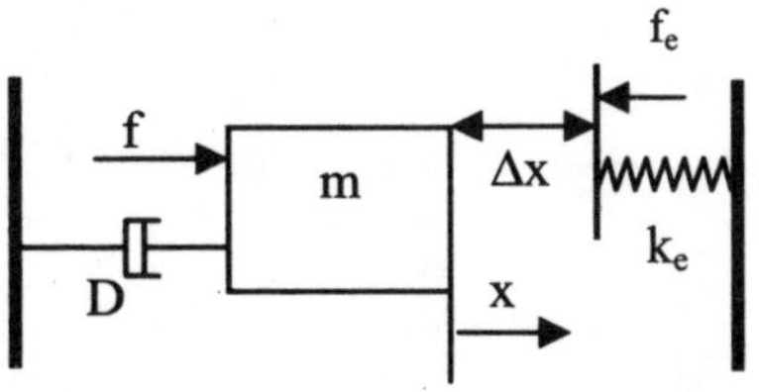

Figure 5.20. *Mass, spring, damper model in contact*

A 10N force is applied to the system. The mass m is moved and shifts towards the environment that is 30 cm away. During the contact, we notice a rebound that generates an oscillation which is illustrated in Figure 5.21. The consequence of this discontinuous model is, for example, the impossibility of ensuring, in closed loop, a closed loop control around zero with an integral term.

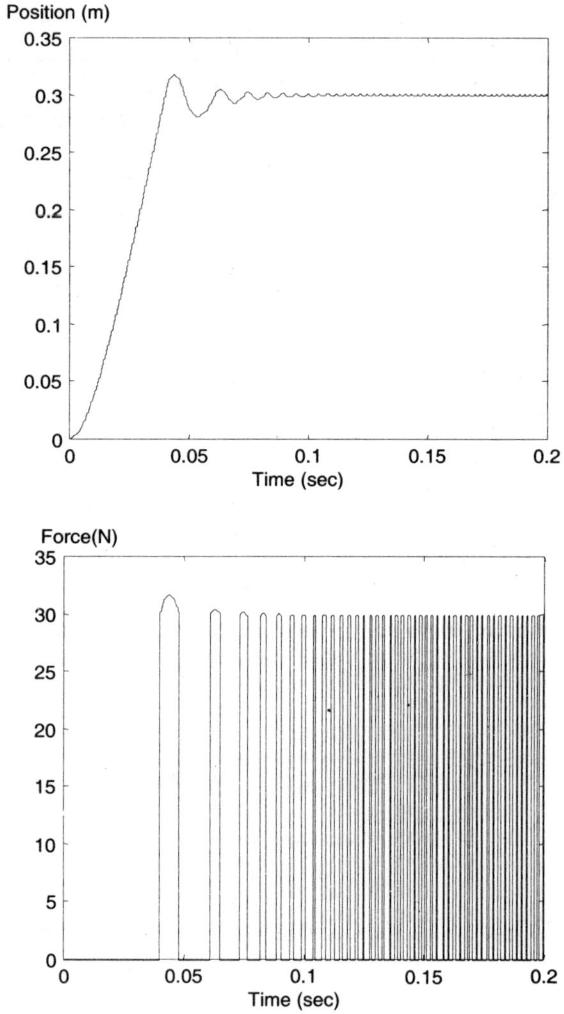

Figure 5.21. *Response to a force step with contact*

A possible solution to avoid this problem is to add, on the end-effector of the robot, a sensor capable of evaluating the distance with the environment (laser range finder, ultrasonic probe, cameras) in order to generate a force even when the effector is in a free space, in order to ensure the continuity of the force loop [NGU 92]. Other studies attempted to reduce the force transient during the contact by using a discontinuous theoretical model capable of ensuring transient management in closed loop [SAR 98].

5.4. Conclusion

The position and force control of robots is a very vast topic that we only approached it here very lightly. In fact, it is a subject that is the object of many studies. We notice a huge gap between the control in position of the industrial robots, the majority of which use PID decentralized solutions, and the advanced controls (dynamic, adaptive, variable structures, etc.) which are studied and tested in the research centers. This gap can be explained by the fact that the industrial robots are designed in such a way that linearity and decoupling hypotheses are verified. The advanced controls are often used in tasks outside the industrial context where the robot, for example, performs high velocity displacements with a load that is close to or even heavier than its theoretical maximum load, then moves with no-load at low velocity and finally carries out an insertion without a force sensor. These control laws try to generate the best performances with respect to more and more complex and varied tasks. The reader may refer to the books of [SAM 91, SCI 01, SLO 91, SPO 89, ZOD 96], that discuss this entire subject, with various approaches.

However, force control started being used in the industrial field a few years ago, even if it is true that the force sensor is a sensitive material that has a reduced lifespan in this medium. The most commonly used force control structure is that of the external control associated to an integral or proportional term, depending on whether we want to apply a specific force on an environment or to control it around a zero force in the direction of the displacement in order to contribute, for example, to the transportation of heavy and bulky objects. The force control is also the subject of numerous studies because its discontinuous structure related to the punctual contact with the environment can generate destructive instabilities. The current difficulty in the field is to control the force transient during contact or the temporary loss of the contact (displacement of the environment) [YOS 00].

5.5. Bibliography

[BAR 98] BARTOLINI G., FERRRARA A, USAI E., "Chattering avoidance by second-order sliding mode control", *IEEE Trans. Automat. Control*, Vol. 43, No. 2, pp. 241-246, 1998.

[BER 95] BERTHE A., "Etude de la robustesse de schémas de commande position/force pour robots à deux bras", PhD Thesis, Montpellier II University, 29 June 1995.

[BOU 71] BOUILLON T., ODELL P., *Generalized Inverse Matrices*, Wiley-Interscience, 1971.

[CON 87] CONTI P., "Contribution à la commande dynamique adaptive des robots manipulateurs", PhD Thesis, Paul Sabatier University, Toulouse, 3 December 1987.

[COR 03] CORTESAO R., PARK J., KHATIB O., "Real-time adaptive control for haptic manipulation with active observers", *Proc. of the Int. Conf. on Intelligent Robots and Systems (IROS'03)*, USA, 2003.

[CRA 86] CRAIG J., *Introduction to Robotics: Mechanics and Control*, Addison-Wesley, 1986.

[CRA 87] CRAIG J.J., HSU P., SASTRY S.S., "Adaptive control of mechanical manipulator", *The International Journal of Robotics Research*, Vol. 6, No. 2, 1987.

[DAU 85] DAUCHEZ P., COIFFET P., FOURNIER A., "Cooperation of two robots in assembly tasks", Chapter 11, *Computing Techniques for Robots*, Kogan Page, 1985.

[DAU 90] DAUCHEZ P., "Descriptions de tâches en vue de la commande hybride symétrique d'un robot manipulateur à deux bras", Thesis, Montpellier II University, 8 February 1990.

[DEG 93] DEGOULANGE E., "Commande en effort d'un robot manipulateur à deux bras: application au contrôle de la déformation d'une chaîne cinématique fermée", PhD Thesis, Montpellier II University, December 1993.

[DEG 94] DÉGOULANGE E., DAUCHEZ P., "External force control of an industrial PUMA 560 robot", *Journal of Robotic Systems*, vol. 11, No. 6, pp. 523-540, 1994.

[DEL 92] DELEBARRE X., "Commandes position et force de deux bras manipulateurs pour l'exploration planétaire", PhD Thesis, Montpellier II University, June 1992.

[DUB 84] DUBOWSKI S., KORNBLUH R.. "On the development of high performance adaptive control algorithms for robotic manipulators", *2nd International Symposium of Robotics Research*, Japan, August 1984.

[FIS 92] FISCHER W.D., MUJTABA M.S. "Sufficient stability condition for hybrid position/force control". *Proc. of the IEEE International Conference on Robotics and Automation*, Nice, France, May 1992, pp. 1336-1341.

[FRA 94] FRAISSE P., "Contribution à la commande robuste position/force des robots manipulateurs à architecture complexe. Application à un robot à deux bras", PhD Thesis, Montpellier II University, February 1994.

[HAR 85] HARASHIMA F., HASHIMOTO H., KONDO S., "Mosfet Converted-fed position servo system with sliding mode control", *IEEE Transactions on Industrial Electronics*, Vol. IE-32, No. 3, August 1985.

[HAS 86] HASHIMOTO H., MARUYAMA K., ARAI Y. and HARASHIMA F., "Practical realization of VSS controller for robotic arm", *Proc. of IECON'86*, pp.34-40, 1986.

[HOS 98] HOSODA K., IGARASHI K. and ASADA M., "Adaptive Hybrid Control for Visual and Force Servoing in an Unknown Environment", *IEEE Robotics and Automation Magazine*, pp.39-43, December 1998.

[KAW 85] KAWAMURA S., MIYAZAKI F., and ARIMOTO S., "Hybrid Position/Force Control of Robot Manipulators Based on Learning Method", *Proc. ICAR*, pp. 235-242, 1985.

[KEL 05] KELLY R., SANTIBÁÑEZ V. and LORÍA A., *Control of Robot Manipulators in Joint Space*, Springer, 2005.

[KHA 87] KHATIB O., "A unified approach for motion and force control of robot manipulators: the operational space formulation", *IEEE Journal of Robotics and Automation*, Vol. RA-3, No. 1, pp.43-53, February 1987.

[KHA 99] KHALIL W., DOMBRE E., *Modélisation, identification et commande des robots*, Hermès, 1999.

[LAN 88] LANDAU I.D., *Identification et commande de systèmes*, Hermès, 1998.

[LAR 91] DE LARMINAT P., "A survey on robust control", *Proc. of IMACS-IFAC Symposium*, Lille, France, pp.243-253, 1991.

[LEV 93] LEVANT A. "Sliding order and sliding accuracy in sliding mode control", *International Journal of Control*, Vol. 58, No.6, pp. 1247-1263, 1993.

[MAS 81] MASON M., "Compliance and force control for computer controlled manipulators", *IEEE Transactions on Systems, Man and Cybernetics*, Vol. SMC-11, No. 6, pp. 418-432, 1981.

[NGU 92] NGUYEN C.C. *et al.*, "Virtual force control of a six-degree-of-freedom parallel manipulator", *Proceedings of International Symposium on Robotics and Manufacturing*, Santa Fé, USA, pp. 357-363, 1992.

[PER 91] PERDEREAU V., "Contribution à la commande hybride force-position; application à la coopération de deux robots", PhD Thesis, Paris VI University, February 1991.

[PER 02] PERRUQUETTI W., BARBOT J.P., *Sliding Mode Control in Engineering*, Marcel Dekker Inc., 2002.

[RAI 81] RAIBERT M., CRAIG J., "Hybrid position/force control of manipulators", Transactions ASME, *Journal of Dynamic Systems, Measurement and Control*, Vol. 103, No. 2, pp. 126-133, 1981.

[SAL 80] SALISBURY J.K., "Active stiffness control of a manipulator in Cartesian coordinates", *Proc. IEEE Conf. Decision and Control*, pp. 95-100, 1980.

[SAM 91] SAMSON C., LE BORGNE M., ESPIAU B., *Robot Control*, Oxford University Press, Oxford, 1991.

[SAR 98] SARKAR N., YUN X., ELLIS R., "Live-Constraint Based Control for Transitions". *IEEE Trans. on Robotics and Automation,* Vol. 14, No. 5, pp. 743-754, 1998.

[SCH 88] DE SCHUTTER J., VAN BRUSSEL H., "Compliant robot motion II. a control approach based on external loop", *The International Journal of Robotics Research*, Vol. 4, No. 4, pp. 18-33, 1988.

[SCI 96] SCIAVICCO L., SICILIANO B., *Modeling and Control of Robot Manipulators*, McGraw-Hill, 1996.

[SCI 01] SCIAVICCO L., SICILIANO B., *Modeling and Control of Robot Manipulators*, Springer, 2001.

[SIC 00] SICILIANO B., VILLANI L., *Robot Force Control*, Kluwer, 2000.

[SLO 91] SLOTINE J.J., LI W., *Applied Non-Linear Control*, Prentice Hall, 1991.

[SPO 89] SPONG M., VIDYASAGAR M., *Robot Dynamics and Control*, John Wiley & Sons, 1989.

[TAR 96] TARN T.J., WU Y., XI N., ISHIDORI A., "Force regulation and contact transition control". *IEEE Control Systems Magazine*, Vol. 16, No. 1, pp. 32-40, 1996.

[UTK 87] UTKIN V.I.I., "Discontinuous control systems: state of art in theory and applications", *Proceedings of IFAC World Congress*, Munich, pp. 135-147, 1987.

[WHI 77] WHITNEY D.E., "Force feedback control of manipulator fine motions", *Trans. ASME, Journal of Dynamic Systems, Measurement and Control*, Vol. 99, No. 2, pp. 91-97, 1977.

[WHI 97] WHITCOMB L.L., ARIMOTO S., NANIWA T., OZAKI F., "Adaptive model-based hybrid control of geometrically constrained robot arms". *IEEE Trans. on Robotics and Automation*, Vol. 13, No. 1, pp. 105-116, 1997.

[YOS 93] YOSHIKAWA T., SUDOU A., "Dynamic hybrid position/force control of robot manipulators: estimation of unknown constraint". *IEEE Trans. on Robotics and Automation*, Vol. 9, No. 2, pp. 220-226, 1993.

[YOS 00] YOSHIKAWA T., "Force control of robot manipulators", *Proceedings of the 2000 IEEE International Conference on Robotics and Automation*, San Francisco, pp. 220-226, 2000.

[ZOD 96] THE ZODIAC, *Theory of Robot Control*, Springer-Verlag, 1996.

Chapter 6

Visual Servoing

6.1. Introduction

Robotic systems are more and more often equipped with exteroceptive sensors which, by definition, provide information on the environment in which they operate. These sensors are of course essential when a task has to be performed in an environment that is not completely rigid or not perfectly well known. They also make it possible to consider errors or inaccuracies that may occur in the robot's geometric (and therefore kinematic) models. Aside from force sensors, the purpose and applications of which were discussed in the previous chapter, there are many other sensors available that provide localization of the system in its environment, or give it a generally local perception of its surroundings. To give a few examples, road marking, passive beacon or radio-based systems, as well as GPS, all make it possible to localize a mobile robot, by determining either its absolute position or its movement. When it comes to perception, proximity sensors provide measurements on the distances to the closest objects. They are therefore particularly well suited for obstacle avoidance tasks. As for computer vision and telemetry sensors, they have a rather wide range of applications since they can be used for localization, navigation, and exploration.

For a long time 3-D reconstruction was considered an unavoidable, independent module, a prerequisite to any motion planning module for a robot in a not perfectly well known environment. In computer vision, this state of things, which used to be justified by the prohibitive computation time required by image processing algorithms, led to a number of successful studies, notably in the field of 3-D vision [FAU 93, MA 03]. The algorithmic and technological progress achieved over the past 15 years

---

Chapter written by François CHAUMETTE.

has made it possible to more closely link the aspects of perception with those of action, by directly integrating the measurements provided by a vision sensor into closed-loop control laws. This approach, known as visual servoing, shares some aspects with the studies on sensor-based control and is the focus of this chapter.

Visual servoing techniques consist of using image measurements provided by one or several cameras, in order to control the motions of a robotic system. This allows for the achievement of a wide variety of tasks designed to position a system with respect to its environment, or to track mobile objects, by controlling from one to all of the system's n degrees of freedom of the robot. Whatever the sensor's configuration, which can range from a camera mounted on the robot's end-effector to several cameras located in the environment and observing the robot's end-effector, the objective is to select as best as possible a set of k visual features, in order to control the m desired degrees of freedom, and to develop a control law so as to make these features $\mathbf{s}(t)$ reach a desired value $\mathbf{s}^*$ that defines when a task is suitably achieved. It is also possible to follow a desired trajectory $\mathbf{s}^*(t)$. The idea of control therefore amounts to regulating the error vector $\mathbf{s}(t) - \mathbf{s}^*(t)$ (i.e. making $\mathbf{s}(t) - \mathbf{s}^*(t)$ reach zero and maintaining it there).

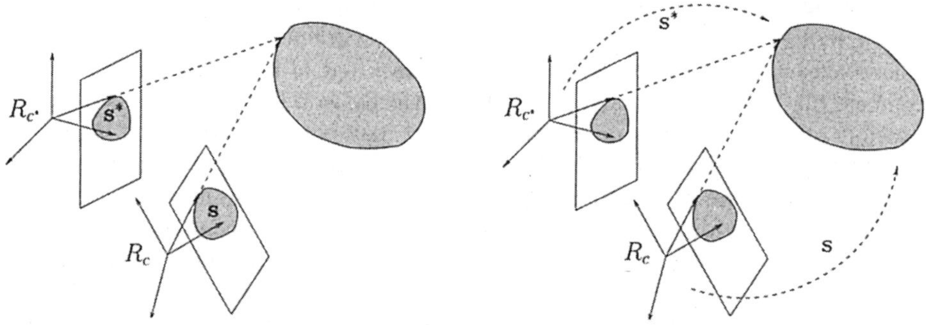

Figure 6.1. *2-D and 3-D visual servoing: in 2-D visual servoing the camera is moved from R_c to R_{c^*}, based on features* $\mathbf{s}$ *extracted directly from the image (left). With 3-D visual servoing,* $\mathbf{s}$ *is comprised of 3-D features estimated after a localization process (right)*

With a vision sensor, which provides 2-D measurements, the nature of the potential visual features is extremely rich, since it is possible to design visual servoing using both 2-D features, such as the coordinates of characteristic points in the image for example, and 3-D features, provided by a localization module operating on the extracted 2-D measurements (see Figure 6.1). This wide range of possibilities is the reason behind the major difficulty in visual servoing, that is to build and select as best as possible the visual features needed for a suitable behavior of the system, based on all the available measurements. A number of qualities are important: local or even global stability, robust behavior when facing measurement or modeling errors, absence of singularities and local minima, suitable trajectories for the robot, but also for the measurements in the image, and finally a maximum decoupling between the visual features and the controlled degrees of freedom. In short, visual servoing is basically a non-linear control problem. The goal is to modify the basic problem so that it becomes as linear as possible.

To study the behavior of the resulting system, a modeling phase is necessary to describe the relation between the visual features $\mathbf{s}(t)$ that were chosen and the control variables. This essential phase of model design will now be described. However, in this chapter we will not be dealing with aspects of image processing, crucial to extracting useful 2-D measurements from a digital image and tracking them at each iteration of the control law. For readers interested in knowing more, we suggest turning to works specializing in this field [VIN 00, KRA 05].

6.2. Modeling visual features

6.2.1. *The interaction matrix*

In order to be taken into account in a visual servoing scheme, a set $\mathbf{s}$ of k visual features needs to be defined by an application differentiable from the special Euclidean group SE_3 into $\mathbb{R}^k$:

$$\mathbf{s} = \mathbf{s}(\mathbf{p}(t)) \qquad [6.1]$$

where $\mathbf{p}(t)$, an element of the space of reference frames and rigid bodies SE_3, describes the pose at the instant t between the camera and its environment. Hence only the movements of the camera, or of the objects it perceives, can modify the value of a visual feature.

The differential of $\mathbf{s}$ allows us to know how the variations in the visual features are related to the relative movements between the camera and the scene, since by differentiating [6.1], we get:

$$\dot{\mathbf{s}} = \frac{\partial \mathbf{s}}{\partial \mathbf{p}}\, \dot{\mathbf{p}} = \mathbf{L_s}\, \mathbf{v} \qquad [6.2]$$

where:

– $\mathbf{L_s}$ is a $k \times 6$ matrix, referred to as the *interaction matrix* related to $\mathbf{s}$;

– $\mathbf{v}$ is the relative instantaneous velocity (also called kinematic screw vector) between the camera and the scene, expressed in the camera's frame R_c in its origin C. More accurately, if $\mathbf{v_c}$ and $\mathbf{v_o}$ are, respectively, the kinematic screws of the camera and of the scene it perceives, both expressed in R_c and in C, then let:

$$\mathbf{v} = \mathbf{v_c} - \mathbf{v_o} \qquad [6.3]$$

From now on, except if noted otherwise, we will write that a screw expressed in a frame of reference has its value given in the origin of this frame. Also, we will denote by $\boldsymbol{\upsilon}$ the translational velocity at the origin of the coordinate system, and by $\boldsymbol{\omega}$ the angular velocity, such that $\mathbf{v} = (\boldsymbol{\upsilon}, \boldsymbol{\omega})$. If ${}^o\mathbf{R}_c$ describes the rotation matrix from the frame R_o bound to the object to R_c, we have by definition [SAM 91]:

$$[\boldsymbol{\omega}]_\times = {}^o\dot{\mathbf{R}}_c{}^o\mathbf{R}_c^\top = -{}^c\dot{\mathbf{R}}_o{}^c\mathbf{R}_o^\top = {}^c\mathbf{R}_o{}^o\dot{\mathbf{R}}_c \qquad [6.4]$$

where $[\boldsymbol{\omega}]_\times$ is the skew symmetric matrix defined from $\boldsymbol{\omega}$.

COMMENT.– In more formal terms [SAM 91], the transpose of the interaction matrix can be defined as the matrix representation of the subspace generated by a family of k screws expressed in R_c. This is due to the fact that each component of s can be decomposed as the product of two screws, one called the interaction screw, and the other being of course the kinematic screw. We will see the practical advantage of this definition in section 6.3.3.1.

6.2.2. *Eye-in-hand configuration*

If we consider a camera mounted on the end-effector of a robot arm observing a static object, the relation between $\dot{\mathbf{s}}$ and the speed of the robot's joint variables $\dot{\mathbf{q}}$ can easily be obtained:

$$\dot{\mathbf{s}} = \mathbf{J_s}\dot{\mathbf{q}} = \mathbf{L_s}{}^c\mathbf{V}_n{}^n\mathbf{J}_n(\mathbf{q})\,\dot{\mathbf{q}} \qquad [6.5]$$

where $\mathbf{J_s} = \mathbf{L_s}{}^c\mathbf{V}_n{}^n\mathbf{J}_n$ is the Jacobian of the visual features and where:

– ${}^n\mathbf{J}_n(\mathbf{q})$ is the robot's Jacobian expressed in the end-effector's frame R_n [KHAL 02];

– ${}^c\mathbf{V}_n$ is the kinematic screw's transformation matrix from the camera's frame R_c to frame R_n. This matrix, which remains constant if the camera is rigidly attached to the robot's end-effector, is given by [KHAL 02]:

$${}^c\mathbf{V}_n = \begin{bmatrix} {}^c\mathbf{R}_n & [{}^c\mathbf{t}_n]_\times{}^c\mathbf{R}_n \\ \mathbf{0}_3 & {}^c\mathbf{R}_n \end{bmatrix} \qquad [6.6]$$

where ${}^c\mathbf{R}_n$ and ${}^c\mathbf{t}_n$ are, respectively, the rotation matrix and the translation vector from frame R_c to frame R_n. The elements of the transformation matrix from the camera's frame to the end-effector's frame can be estimated accurately by using hand-eye calibration methods [TSA 89, HOR 95]. Note that visual servoing techniques are usually rather robust in admitting important modeling errors, both in this transformation matrix [ESP 93, MAL 02] and in the robot's Jacobian.

More generally, if the camera is observing a moving object, the differential of s is given by:

$$\dot{\mathbf{s}} = \mathbf{L_s}\,{}^c\mathbf{V}_n\,{}^n\mathbf{J}_n(\mathbf{q})\,\dot{\mathbf{q}} + \frac{\partial \mathbf{s}}{\partial t} \qquad [6.7]$$

where the term $\frac{\partial \mathbf{s}}{\partial t}$ represents the variation of s due to the object's own motion (which is usually not known). In the highly unlikely event that the object's motion is known, and given for example by the kinematic screw vector $\mathbf{v_o}$ in R_c, we get:

$$\dot{\mathbf{s}} = \mathbf{L_s}\,{}^c\mathbf{V}_n\,{}^n\mathbf{J}_n(\mathbf{q})\,\dot{\mathbf{q}} - \mathbf{L_s}\mathbf{v_o} \qquad [6.8]$$

6.2.3. *Eye-to-hand configuration*

Likewise, if we now consider a camera in the scene observing the end-effector of a robot arm, the variation of the visual features rigidly attached to this end-effector is expressed according to the speed of the joint coordinates:

$$\dot{\mathbf{s}} = -\mathbf{L_s}\,{}^c\mathbf{V}_n\,{}^n\mathbf{J}_n(\mathbf{q})\,\dot{\mathbf{q}} + \frac{\partial \mathbf{s}}{\partial t} \qquad [6.9]$$

where $\frac{\partial \mathbf{s}}{\partial t}$ now describes the variations of s due to a possible movement of the camera.

COMMENT.– Notice the difference in signs between Equations [6.5] and [6.9]. This difference is of course due to the configuration change of the sensor with respect to the control variables (see Figure 6.2).

Whether the camera is fixed or mobile, the matrix ${}^c\mathbf{V}_n$ is now variable and has to be estimated at each iteration, which is usually done using a 3-D localization technique (see section 6.2.5.1). If the camera is static, it is therefore more convenient to use one of the following relations:

$$\dot{\mathbf{s}} = -\mathbf{L_s}\,{}^c\mathbf{V}_\emptyset\,{}^\emptyset\mathbf{V}_n\,{}^n\mathbf{J}_n(\mathbf{q})\,\dot{\mathbf{q}} \qquad [6.10]$$

$$= -\mathbf{L_s}\,{}^c\mathbf{V}_\emptyset \begin{bmatrix} \mathbb{I}_3 & [{}^\emptyset\mathbf{t}_n]_\times \\ \mathbf{0}_3 & \mathbb{I}_3 \end{bmatrix} {}^\emptyset\mathbf{J}_n(\mathbf{q})\,\dot{\mathbf{q}} \qquad [6.11]$$

where ${}^{\emptyset}\mathbf{J}_n(\mathbf{q})$ is the robot's Jacobian expressed in its basic frame of reference and where the values of ${}^{\emptyset}\mathbf{V}_n$ and ${}^{\emptyset}\mathbf{t}_n$ are provided by the robot's direct geometric model. This is interesting because the transformation matrix ${}^{c}\mathbf{V}_{\emptyset}$ is then constant and only has to be estimated once beforehand, usually coarsely.

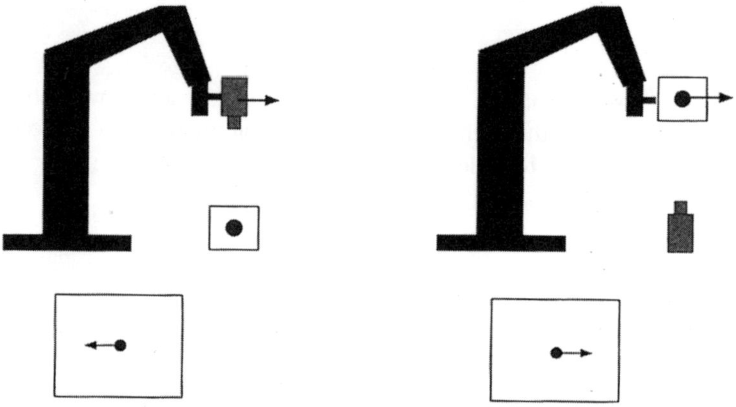

Figure 6.2. *Eye-in-hand configuration (left); eye-to-hand configuration (right)*

In the literature [HAS 93a, HUT 96], most studies focus on eye-in-hand configuration. We can however cite [ALL 93, NEL 94a, HAG 95, KEL 96, CIP 97, HOR 98, RUF 99] in which one or several cameras are used in eye-to-hand configurations.

In any case, the interaction matrix plays an essential role and we will now give its analytical form for a set of visual features. From now on, all the necessary quantities (coordinates and speeds of points, kinematic screw, etc.) are expressed in the camera's frame shown in Figure 6.3.

6.2.4. *Interaction matrix*

6.2.4.1. *Interaction matrix of a 2-D point*

The typical mathematical model for a camera is defined by a perspective projection, such that any point M with coordinates $\mathbf{X} = (X, Y, Z)$ is projected onto the image plane in a point m with coordinates $\mathbf{x} = (x, y)$ with:

$$x = X/Z \ , \ y = Y/Z \qquad [6.12]$$

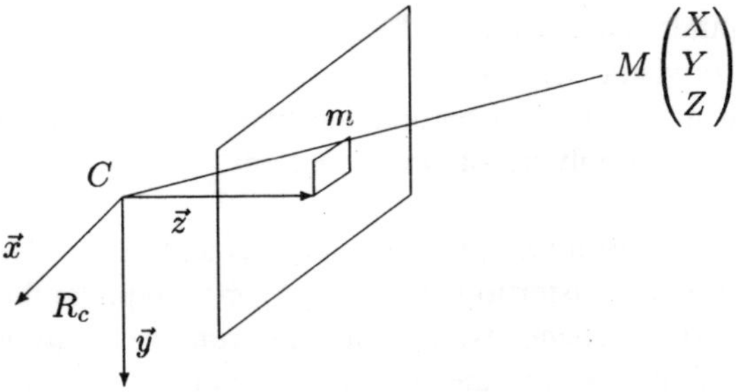

Figure 6.3. *Camera model*

By differentiating this equation, we get the variations in the image of the coordinates x and y of m with respect to the speed $\dot{\mathbf{X}}$ of the coordinates of point M:

$$\dot{\mathbf{x}} = \begin{bmatrix} 1/Z & 0 & -X/Z^2 \\ 0 & 1/Z & -Y/Z^2 \end{bmatrix} \dot{\mathbf{X}} \qquad [6.13]$$

Whatever configuration is chosen (eye-in-hand or eye-to-hand, static or mobile point M), the speed $\dot{\mathbf{X}}$ of M according to the kinematic screw $\mathbf{v}$ between the camera and its environment is given by the fundamental kinematics equation:

$$\dot{\mathbf{X}} = -\boldsymbol{v} - \boldsymbol{\omega} \times \mathbf{X} = -\boldsymbol{v} + [\mathbf{X}]_\times \boldsymbol{\omega} = \begin{bmatrix} -\mathbb{I}_3 & [\mathbf{X}]_\times \end{bmatrix} \mathbf{v} \qquad [6.14]$$

Equation [6.13] can then be simplified using Equation [6.12], written in the form:

$$\dot{\mathbf{x}} = \mathbf{L}_{\mathbf{x}}(\mathbf{x}, Z)\, \mathbf{v} \qquad [6.15]$$

where:

$$\mathbf{L}_{\mathbf{x}}(\mathbf{x}, Z) = \begin{bmatrix} -1/Z & 0 & x/Z & xy & -(1+x^2) & y \\ 0 & -1/Z & y/Z & 1+y^2 & -xy & -x \end{bmatrix} \qquad [6.16]$$

Notice that the terms induced by angular motions only depend on the measurements of x and y in the image. On the other hand, terms induced by translational motions are inversely proportional to the depth of the 3-D point. This effect occurs for all the visual features that can be defined in the image (and describes the classic ambiguity in computer vision between the amplitude of a translational motion and the depth of objects). In visual servoing, it is therefore necessary to insert a 3-D knowledge, even though it is unknown beforehand, whenever trying to control a robot's degrees of freedom that imply translational motions.

Image processing algorithms provide measurements expressed in pixels. If we ignore strongly non-linear distortion effects, due for example to the use of short focal length lenses, the variable change when switching from the coordinates $\mathbf{x}_p = (x_p, y_p)$ of a point, expressed in pixels, to the coordinates $\mathbf{x}$ of this same point, but expressed in meters, is given by:

$$x = (x_p - x_c)/f_x \;\;,\;\; y = (y_p - y_c)/f_y \qquad [6.17]$$

where (x_c, y_c) represents the principal point's coordinates in the image and where $f_x = f/l_x$ and $f_y = f/l_y$ are the ratios between the focal length f of the lens and the dimensions l_x and l_y of a pixel. These parameters, referred to as the intrinsic parameters of the camera, can be estimated beforehand, during a calibration step [TSA 87, BEY 92, ZHA 00], but as with the elements of the hand-eye matrix, coarse approximations are usually sufficient to maintain the stability of visual servoing systems [ESP 93, MAL 99, MAL 02, DEN 02].

It is possible to calculate the interaction matrix related to the coordinates of a point directly expressed in pixels. Using the variable change reciprocal to [6.17], given by:

$$x_p = x_c + f_x x \;\;,\;\; y_p = y_c + f_y y \qquad [6.18]$$

we immediately get:

$$\mathbf{L}_{\mathbf{x}_p} = \begin{bmatrix} f_x & 0 \\ 0 & f_y \end{bmatrix} \mathbf{L}_{\mathbf{x}} \qquad [6.19]$$

where the set of terms contained in $\mathbf{L}_{\mathbf{x}}$, except of course for the depth Z, can be expressed as functions of the intrinsic parameters and coordinates $\mathbf{x}_p$ using [6.17]. If required, the same can be done for the visual features defined later on, working with features expressed in pixels. The main advantage of having an analytical form of the interaction matrix that explicitly depends on the intrinsic parameters, is that it then becomes possible to study how sensitive visual servoing systems are to errors made in the estimation or approximation of these parameters.

Finally, we mention the studies in projective geometry described in [RUF 99] which led to a direct modeling of the Jacobian matrix $\mathbf{J_s}$ such that $\dot{\mathbf{s}} = \mathbf{J_s}\dot{\mathbf{q}}$, in the case where s is comprised of the coordinates of a point located on the end-effector and observed by two external cameras: $\mathbf{s} = (x_g, y_g, x_d, y_d)$. The advantage of such an approach is that it is no longer necessary to know the Jacobian, and hence the geometric model, of the robot being used.

If we now consider a camera equipped with a controllable zoom, thus providing the system with an additional degree of freedom, we get just as simply, from [6.18]:

$$\begin{bmatrix} \dot{x_p} \\ \dot{y_p} \end{bmatrix} = \mathbf{L}_{\mathbf{x}_p}\, \mathbf{v} + \begin{bmatrix} (x_p - x_c)/f \\ (y_p - y_c)/f \end{bmatrix} \dot{f} \qquad [6.20]$$

For purely technological reasons (because for most zooms, position can be controlled, not speed), few studies have used this function, even though it provides an interesting redundancy with respect to the translational motion along the optical axis. We can still mention [HOS 95a, BEN 03].

6.2.4.2. *Interaction matrix of a 2-D geometric primitive*

It is also possible to calculate the interaction matrix related to visual features constructed from geometric primitives [ESP 92]. This is done simply by defining the equations that represent:

– the primitive's nature and configuration in the scene:

$$\mathbf{h}(X, Y, Z, P_1, \ldots, P_n) = 0 \qquad [6.21]$$

– its projection onto the image plane:

$$\mathbf{g}(x, y, p_1, \ldots, p_m) = 0 \qquad [6.22]$$

– the relation between the 3-D primitive and its image (referred to as the limbo surface in the case of a volumetric primitive, see Figure 6.4):

$$1/Z = \mu(x, y, P_1, \ldots, P_l) = 0 \qquad [6.23]$$

As an example, if a straight line in space is represented by the intersection of the two following planes:

$$\mathbf{h}(X, Y, Z, A_1, \ldots, C_2) = \begin{cases} h_1 = A_1X + B_1Y + C_1Z + D_1 = 0 \\ h_2 = A_2X + B_2Y + C_2Z = 0 \end{cases} \qquad [6.24]$$

we immediately obtain, using the equations of perspective projection [6.12]:

– the function μ from h_1:

$$1/Z = Ax + By + C \qquad [6.25]$$

with $A = -A_1/D_1$, $B = -B_1/D_1$ and $C = -C_1/D_1$;

– the equation of the 2-D line, denoted by $\mathcal{D}$, resulting from the projection onto the image of the 3-D line, from h_2:

$$ax + by + c = 0 \text{ with } a = A_2,\ b = B_2,\ c = C_2 \qquad [6.26]$$

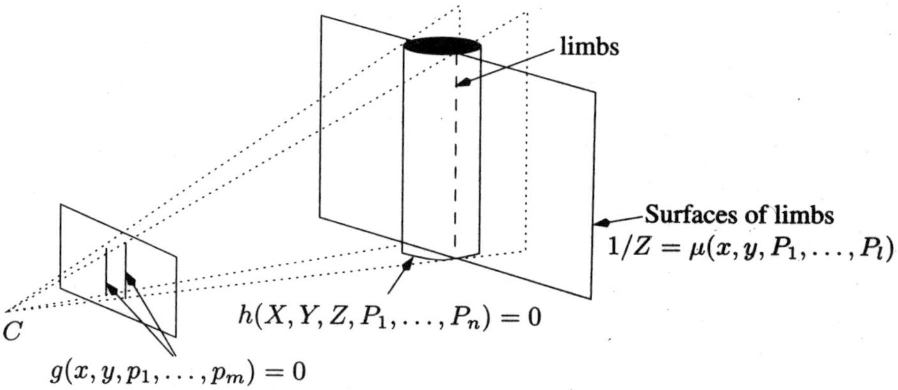

Figure 6.4. *Projection of the primitive onto the image and limb surface in the case of the cylinder*

Because the choice of parameters (a, b, c) is not minimal, it is preferable to choose the (ρ, θ) representation defined by:

$$g(x, y, \rho, \theta) = x\cos\theta + y\sin\theta - \rho = 0 \qquad [6.27]$$

where $\theta = \arctan(b/a)$ and $\rho = -c/\sqrt{a^2 + b^2}$ (see Figure 6.5).

If we differentiate Equation [6.27], which corresponds to the hypothesis that the image of a straight line remains a straight line whatever the camera's motion, we get:

$$\dot{\rho} + (x\sin\theta - y\cos\theta)\,\dot{\theta} = \dot{x}\cos\theta + \dot{y}\sin\theta \ , \quad \forall(x, y) \in \mathcal{D} \qquad [6.28]$$

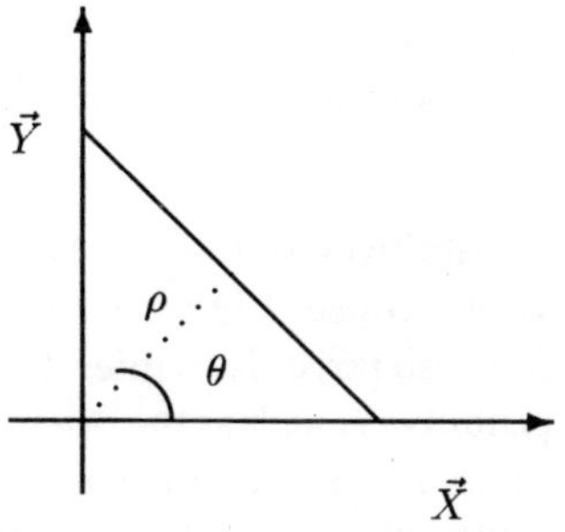

Figure 6.5. *(ρ, θ) representation of the 2-D lines*

Based on Equation [6.27], x is written according to y if $\cos\theta \neq 0$ (or y according to x if that is not the case) and Equation [6.28] can then be written, using [6.15] and [6.25]:

$$(\dot{\rho} + \rho \tan\theta \, \dot{\theta}) + y \, (-\dot{\theta}/\cos\theta) = \mathbf{K_1} \, \mathbf{v} + y \, \mathbf{K_2} \, \mathbf{v} \ , \ \ \forall y \in \mathbb{R} \qquad [6.29]$$

with:

$$\begin{array}{l} \mathbf{K_1} = [\ \lambda_1 \cos\theta \quad \lambda_1 \sin\theta \quad -\lambda_1 \rho \quad \sin\theta \quad -\cos\theta - \rho^2/\cos\theta \quad -\rho\tan\theta\ ] \\ \mathbf{K_2} = [\ \lambda_2 \cos\theta \quad \lambda_2 \sin\theta \quad -\lambda_2 \rho \quad \rho \quad \rho\tan\theta \quad 1/\cos\theta\ ] \end{array}$$

where $\lambda_1 = -A\rho/\cos\theta - C$ and $\lambda_2 = A\tan\theta - B$.

Immediately, we infer that:

$$\left\{ \begin{array}{lcl} \dot{\rho} & = & (\mathbf{K_1} + \rho \sin\theta \, \mathbf{K_2}) \, \mathbf{v} \\ \dot{\theta} & = & -\cos\theta \, \mathbf{K_2} \, \mathbf{v} \end{array} \right. \qquad [6.30]$$

hence:

$$\begin{array}{l} \mathbf{L}_\rho = [\ \lambda_\rho \cos\theta \quad \lambda_\rho \sin\theta \quad -\lambda_\rho \rho \quad (1+\rho^2)\sin\theta \quad -(1+\rho^2)\cos\theta \quad 0\ ] \\ \mathbf{L}_\theta = [\ \lambda_\theta \cos\theta \quad \lambda_\theta \sin\theta \quad -\lambda_\theta \rho \quad -\rho\cos\theta \quad -\rho\sin\theta \quad -1\ ] \end{array} \qquad [6.31]$$

with $\lambda_\rho = -A\rho\cos\theta - B\rho\sin\theta - C$ and $\lambda_\theta = -A\sin\theta + B\cos\theta$.

The same result can be obtained by applying Equation [6.28] to two points of $\mathcal{D}$, for example those with coordinates $(\rho\cos\theta, \rho\sin\theta)$ and $(\rho\cos\theta + \sin\theta, \rho\sin\theta - \cos\theta)$.

Results for more complex primitives (circles, spheres, and cylinders) are given in [CHA 93a], making it possible to use 2-D visual features associated with these primitives in visual servoing. It is also possible to infer the interaction matrix related to features defined from several primitives (such as the orientation between two segments or the distance from a point to a line, for example). The drawback, however, is that it is only possible to work on environments where such geometric primitives exist (hence the more frequent use of characteristic points).

6.2.4.3. *Interaction matrix for complex 2-D shapes*

Recent studies have made it possible to establish the analytical form of the interaction matrix related to visual features representing the projection onto the image of objects with more complex shapes. In [COLO 99, DRU 99], the six terms that correspond to the affine part of the transformation between the image of a planar object in its current position and the image of the same object in the desired position are considered. More precisely, if (x, y) and (x^*, y^*) are the coordinates of a given point on the object in the current image and the desired image, respectively, then we assume that there exists a set of parameters $\boldsymbol{\theta} = (a_1, b_1, c_1, a_2, b_2, c_2)$ such that the relation:

$$\begin{cases} x &= a_1\, x^* + b_1\, y^* + c_1 \\ y &= a_2\, x^* + b_2\, y^* + c_2 \end{cases} \qquad [6.32]$$

is valid for all points of the object. This hypothesis is unfortunately not verified for a camera described by a perspective projection model. Additionally, the interaction matrix related to $\boldsymbol{\theta}$ shows a loss in rank (from 6 to 4) when the object's plane is parallel to the image plane.

Furthermore, if we calculate the Fourier series expansion for the polar signature $\rho(\theta)$ of the contour points of an object in the image (defined such that the coordinates x and y of a contour point are written: $x = x_g + \rho(\theta)\cos\theta$, $y = y_g + \rho(\theta)\sin\theta$ where x_g and y_g are the coordinates of the object's center of gravity), it is possible to calculate the interaction matrix related to the terms of that series [COL 00]. The resulting analytical form, however, is very complex and difficult to understand from a geometrical point of view.

Another possibility is to calculate the interaction matrix related to the moments m_{ij} of an object [CHA 04]. Moments are defined by:

$$m_{ij} = \int\int_{\mathcal{D}} x^i y^j \, dx\, dy \qquad [6.33]$$

where $\mathcal{D}$ is the area occupied by the object in the image and where $i + j$ is the order of the moment. If we assume that the object considered is planar or has a planar limb surface with equation $1/Z = Ax + By + C$, we obtain, for the area a $(= m_{00})$ and the coordinates x_g $(= m_{10}/m_{00})$ and y_g $(= m_{01}/m_{00})$ of the object's center of gravity:

$$\begin{array}{lllllll} \mathbf{L}_a = [& -aA & -aB & a(3/Z_g - C) & 3ay_g & -3ax_g & 0\ [\\ \mathbf{L}_{x_g} = [& -1/Z_g & 0 & x_g/Z_g + \epsilon_1 & x_g y_g + 4n_{11} & -(1 + x_g^2 + 4n_{20}) & y_g\ ] \\ \mathbf{L}_{y_g} = [& 0 & -1/Z_g & y_g/Z_g + \epsilon_2 & 1 + y_g^2 + 4n_{02} & -x_g y_g - 4n_{11} & -x_g\ ] \end{array} \quad [6.34]$$

with $1/Z_g = Ax_g + By_g + C$, $\epsilon_1 = 4(An_{20} + Bn_{11})$, $\epsilon_2 = 4(An_{11} + Bn_{02})$ and where n_{20}, n_{02} and n_{11} are the second order normalized centered moments defined by:

$$n_{ij} = \mu_{ij}/a \text{ with } \left\{ \begin{array}{l} \mu_{20} = m_{20} - ax_g^2 \\ \mu_{02} = m_{02} - ay_g^2 \\ \mu_{11} = m_{11} - ax_g y_g \end{array} \right. \quad [6.35]$$

Note that the area speed $\dot{a}$ is equal to zero for any motion other than the expected translational motion along the camera's optical axis if the object is centered and parallel to the image plane ($A = B = x_g = y_g = 0$). This makes area particularly interesting for controlling this degree of freedom, because of its relative decoupling compared to the other degrees of freedom.

Notice also that the results obtained for the coordinates of the object's center of gravity encompass those given in [6.15] for a purely punctual object, since for a point, we have $n_{20} = n_{11} = n_{02} = 0$ and we can set $A = B = 0$ in [6.34] to again obtain exactly [6.15].

More generally, the interaction matrix related to a moment m_{ij} is given by:

$$\mathbf{L}_{m_{ij}} = \begin{bmatrix} m_{vx} & m_{vy} & m_{vz} & m_{wx} & m_{wy} & m_{wz} \end{bmatrix} \quad [6.36]$$

where:

$$\left\{ \begin{array}{l} m_{vx} = -i(Am_{ij} + Bm_{i-1,j+1} + Cm_{i-1,j}) - Am_{ij} \\ m_{vy} = -j(Am_{i+1,j-1} + Bm_{ij} + Cm_{i,j-1}) - Bm_{ij} \\ m_{vz} = (i + j + 3)(Am_{i+1,j} + Bm_{i,j+1} + Cm_{ij}) - Cm_{ij} \\ m_{wx} = (i + j + 3)m_{i,j+1} + jm_{i,j-1} \\ m_{wy} = -(i + j + 3)m_{i+1,j} - im_{i-1,j} \\ m_{wz} = im_{i-1,j+1} - jm_{i+1,j-1} \end{array} \right.$$

For centered moments defined by:

$$\mu_{ij} = \iint_{\mathcal{D}} (x - x_g)^i (y - y_g)^j \, dx \, dy \qquad [6.37]$$

we get:

$$\mathbf{L}_{\mu_{ij}} = \begin{bmatrix} \mu_{vx} & \mu_{vy} & \mu_{vz} & \mu_{wx} & \mu_{wy} & \mu_{wz} \end{bmatrix} \qquad [6.38]$$

with:

$$\left\{ \begin{array}{ll} \mu_{vx} & = -(i+1)A\mu_{ij} - iB\mu_{i-1,j+1} \\ \mu_{vy} & = -jA\mu_{i+1,j-1} - (j+1)B\mu_{ij} \\ \mu_{vz} & = -A\mu_{wy} + B\mu_{wx} + (i+j+2)C\mu_{ij} \\ \mu_{wx} & = (i+j+3)\mu_{i,j+1} + ix_g\mu_{i-1,j+1} \\ & \quad +(i+2j+3)y_g\mu_{ij} - 4in_{11}\mu_{i-1,j} - 4jn_{02}\mu_{i,j-1} \\ \mu_{wy} & = -(i+j+3)\mu_{i+1,j} - (2i+j+3)x_g\mu_{ij} \\ & \quad -jy_g\mu_{i+1,j-1} + 4in_{20}\mu_{i-1,j} + 4jn_{11}\mu_{i,j-1} \\ \mu_{wz} & = i\mu_{i-1,j+1} - j\mu_{i+1,j-1} \end{array} \right.$$

The numerical value of the interaction matrix related to a moment of order $i + j$ can thus be calculated from the measurement of moments with orders at most $i+j+1$, which is convenient in practice. The values A, B, C characterizing the plane's configuration must also be available (or at least an approximation of these values) in order to calculate the translational terms. As we have already said, this property is true for any visual feature defined in the image.

Based on the moments, it is possible to determine relevant geometric information, such as, as we have seen before, the area and the center of gravity of an object. Furthermore, the main orientation is obtained from the second order centered moments:

$$\alpha = \frac{1}{2} \arctan \left(\frac{2\mu_{11}}{\mu_{20} - \mu_{02}} \right) \qquad [6.39]$$

and we easily get, using [6.38]:

$$\mathbf{L}_{\alpha} = \begin{bmatrix} \alpha_{vx} & \alpha_{vy} & \alpha_{vz} & \alpha_{wx} & \alpha_{wy} & -1 \end{bmatrix} \qquad [6.40]$$

where:

$$\left\{\begin{array}{lcl}\alpha_{vx} & = & aA + bB \\ \alpha_{vy} & = & -cA - aB \\ \alpha_{vz} & = & -A\alpha_{wy} + B\alpha_{wx} \\ \alpha_{wx} & = & -bx_g + ay_g + d \\ \alpha_{wy} & = & ax_g - cy_g + e\end{array}\right.$$

and:

$$\left\{\begin{array}{lcl}a & = & \mu_{11}(\mu_{20} + \mu_{02})/\Delta \\ b & = & [2\mu_{11}^2 + \mu_{02}(\mu_{02} - \mu_{20})]/\Delta \\ c & = & [2\mu_{11}^2 + \mu_{20}(\mu_{20} - \mu_{02})]/\Delta \\ d & = & 5[\mu_{12}(\mu_{20} - \mu_{02}) + \mu_{11}(\mu_{03} - \mu_{21})]/\Delta \\ e & = & 5[\mu_{21}(\mu_{02} - \mu_{20}) + \mu_{11}(\mu_{30} - \mu_{12})]/\Delta \\ \Delta & = & (\mu_{20} - \mu_{02})^2 + 4\mu_{11}^2\end{array}\right.$$

We should point out that translational motions leave α invariant when the object's plane is parallel to the image plane ($\alpha_{vx} = \alpha_{vy} = \alpha_{vz} = 0$ if $A = B = 0$). Note also the direct relation between the variation of α and the angular motion around the optical axis ω_z, an indication, as we could have expected, that α is a good visual feature for controlling this degree of freedom.

One of the different possible strategies in visual servoing consists of directly using all of the measurements available in the image. We then have redundant visual features (that is, more than the number of degrees of freedom that we wish to control), and as we will see in section 6.3.2.2, servoing stability can only be demonstrated in the neighborhood of the convergence position. Another, more promising strategy consists of determining complementary visual features, by building or selection [COR 01, IWA 05, TAH 05], or even by finding a different way of expressing the perspective projection model (for example a spherical projection [HAM 02]). The case of an object's area and orientation discussed earlier are simple and natural examples of such a determination. However, much remains to be done in this field.

6.2.4.4. *Interaction matrix by learning or estimation*

The use of the polar signature or of the moments allows us to consider objects with truly complex shapes, but requires a spatial segmentation phase in the image processing part that can turn out to be extremely difficult in textured environments. To avoid this segmentation phase and be able to process any kind of image, it is possible to conduct a principal component analysis of the desired image and select the principal eigenvectors [NAY 96, DEG 97]. The coefficients of this decomposition form the set s of the visual features. The analytical form of the associated interaction matrix is then unknown (since it is too difficult to obtain) and the servoing is based on a purely numerical estimate provided by a learning technique. This technique consists of generating movements for the different degrees of freedom available and to measure the corresponding variation observed in the image.

Techniques to estimate the interaction matrix have also been used for geometric visual features such as those described in the previous sections. They are all based on the same idea and are performed either offline, by learning [WEI 84, RUF 99, LAP 04], possibly by using a neural network [SUH 93, WEL 96], or online during the servoing [KIN 94, HOS 94, CHA 96, JAG 97, PIE 04]. These studies fall into two categories, those based on purely numerical estimates of the terms of the interaction matrix [WEI 84, SUH 93, WEL 96, HOS 94, JAG 97, PIE 04] or of its pseudoinverse directly [LAP 04], and those that estimate the unknown parameters occurring in this matrix, such as for example the structure of objects or the camera's intrinsic parameters [KIN 94, CHA 96, RUF 99]. The first case is very attractive in practice since it allows us to avoid any modeling phase. The resulting drawback is that it is impossible to demonstrate the system's stability in the presence of inevitable estimation errors. The second option is therefore more satisfactory theoretically speaking, but since it requires an analytical determination of the interaction matrix beforehand, it cannot be applied today to servoing schemes based on visual features as complex as those resulting from a principal component analysis of the image.

6.2.5. *Interaction matrix related to 3-D visual features*

As has been mentioned before, it is also possible to choose visual features no longer expressed directly in the image, but resulting from a reconstruction phase or a 3-D localization phase [WIL 96, MART 97]. These 3-D features are obtained either by a simple triangulation if a calibrated stereoscopic vision system is available, or, in the case of a monocular sensor, by dynamic vision or with a pose estimation method. Dynamic vision techniques rely on the measurement of the camera's motions and of the resulting motion in the image. They are usually rather sensitive to measurement errors [SMI 94, CHA 96]. We will now briefly describe pose estimation techniques, because they are the most commonly used in 3-D visual servoing.

6.2.5.1. *Pose estimation*

There are many methods for estimating a camera's pose with respect to an object using an image of this object. They rely on prior knowledge of the 3-D model of the object and of the camera's calibration parameters. More precisely, for an image acquired at instant t, they provide an estimate $\hat{\mathbf{p}}(t)$ of the real pose $\mathbf{p}(t)$ between the camera's frame and the object's frame based on the measurements $\mathbf{x}(t)$ extracted from the image, the camera's intrinsic parameters and the object's 3-D model, represented for example by the set $\mathbf{X}$ of the 3-D coordinates of the points that constitute it:

$$\hat{\mathbf{p}}(t) = \hat{\mathbf{p}}(\mathbf{x}(t), x_c, y_c, f_x, f_y, \mathbf{X}) \qquad [6.41]$$

Most of the time, the measurements $\mathbf{x}(t)$ are image points [HOR 89, HAR 89, DEM 95], segments [LOW 87, DHO 89], even conics [SAF 92, MA 93], or also cylindrical objects [DHO 90]. But very few methods combine different kinds of primitives (see however [PHO 95] for the combined use of points and lines).

The methods described in the literature are either purely geometric [HOR 89, DHO 89], based on a numerical and iterative linear estimation [DEM 95] or based on non-linear estimation [LOW 87]. Except for very peculiar cases [HOR 89], no analytical solution to this inverse problem is available.

We should point out that in the case of an error in the calibration parameters or in the object's model, the estimate $\hat{\mathbf{p}}(t)$ will be biased and, because of the absence of an analytical solution, it is unfortunately impossible to determine the value of this bias. The same goes for finding the interaction matrix associated with any features built from $\hat{\mathbf{p}}(t)$. This is because, based on [6.41]:

$$\dot{\hat{\mathbf{p}}}(t) = \frac{\partial \hat{\mathbf{p}}}{\partial \mathbf{x}} \dot{\mathbf{x}} = \frac{\partial \hat{\mathbf{p}}}{\partial \mathbf{x}} \mathbf{L}_{\mathbf{x}}\, \mathbf{v} \qquad [6.42]$$

hence:

$$\mathbf{L}_{\hat{\mathbf{p}}} = \frac{\partial \hat{\mathbf{p}}}{\partial \mathbf{x}} \mathbf{L}_{\mathbf{x}} \qquad [6.43]$$

The second term of this matrix product is nothing but the interaction matrix related to $\mathbf{x}$, and is therefore known if $\mathbf{x}$ is comprised of geometric primitives such as points or line segments. On the other hand, the first term, $\frac{\partial \hat{\mathbf{p}}}{\partial \mathbf{x}}$, which represents the variation of the estimate of $\hat{\mathbf{p}}$ according to a variation of the measurements $\mathbf{x}$ in the image, is unknown. We can only note that it is directly related to the estimation method and depends once again on the camera's intrinsic parameters and the object's 3-D model. This is why we will assume from now on that the estimate of $\hat{\mathbf{p}}(t)$ is perfect, which is the case under the (strong) hypotheses that the camera is perfectly calibrated, that the 3-D model of the object is perfectly well known, that the measurements $\mathbf{x}(t)$ are not tainted with any errors, and that the estimation method is free of any numerical instability.

The strongest hypothesis involves the estimation's stability in regards to measurement errors, because if we consider for example four coplanar points, theoretically there exists only one solution to the localization problem [HOR 89]; however, a very small variation of the positions of the four points in the image can cause a very large variation in the estimate of $\hat{\mathbf{p}}$ (hence the matrix $(\frac{\partial \hat{\mathbf{p}}}{\partial \mathbf{x}})$ is very poorly conditioned). Such an effect is illustrated by Figure 6.6. In practice, this effect decreases when considering a large number of points, or non-coplanar points, but there are currently no theoretical results available on the sensitivity of the estimation methods and the measurements to choose, regarding what kind to use, but also how they are arranged in the image and the 3-D space.

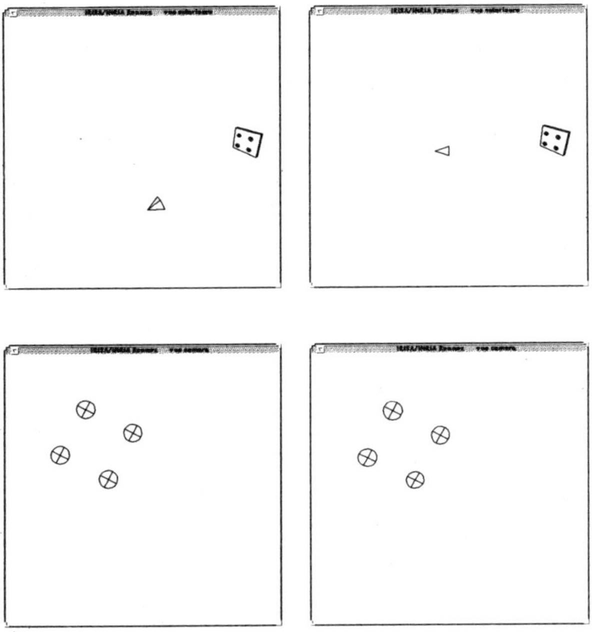

Figure 6.6. *Example of two distinct poses of the camera with respect to the object (top) that provide similar images of this object (bottom)*

Based on $\hat{\mathbf{p}}(t)$, and under the hypotheses mentioned previously, that is, assuming a perfect estimate for $\hat{\mathbf{p}}(t)$ ($\hat{\mathbf{p}}(t) = \mathbf{p}(t)$), we have at our disposal the rotation ${}^c\mathbf{R}_o$ between the camera's frame in its current position R_c and the object's frame R_o attached to the object, as well as the translation ${}^c\mathbf{t}_o$ between theses two frames. We can then infer the position in R_c of any object's point. If, additionally, in the context of an eye-in-hand system, the pose between the camera's frame at its desired position R_{c^*} and the object's frame is known, then we can also infer the displacement necessary to go from R_c to R_{c^*}. With an eye-to-hand system, the same is true of course for an object attached to the robot's end-effector between its current position and its desired position.

We will now give the interaction matrix related to the minimal representation $\theta\mathbf{u}$ of an arbitrary rotation with angle θ about an axis $\mathbf{u}$, then the one associated with the coordinates of a 3-D point.

6.2.5.2. *Interaction matrix related to* $\theta\mathbf{u}$

Remember, first of all, that the $\theta\mathbf{u}$ representation is obtained in a unique manner from the coefficients $r_{ij\ (i=1\cdots 3, j=1\cdots 3)}$ of a rotation matrix $\mathbf{R}$ using the following equation [KHAL 02]:

$$\theta\mathbf{u} = \frac{1}{2\ \mathrm{sinc}\theta}\begin{bmatrix} r_{32} - r_{23} \\ r_{13} - r_{31} \\ r_{21} - r_{12} \end{bmatrix} \qquad [6.44]$$

where $\theta = \arccos((r_{11} + r_{22} + r_{33} - 1)/2)$ and where the sine cardinal $\mathrm{sinc}\theta$, defined by $\sin\theta = \theta\,\mathrm{sinc}\theta$, is a function C^∞ equal to zero in $(2n+1)\pi$, $\forall n \in \mathbb{Z}$. For $\theta = \pi$, the only case not taken into account by [6.44], $\mathbf{u}$ is the eigenvector of $\mathbf{R}$ associated with the eigenvalue 1.

In the case of an eye-in-hand system, it is possible to use the vector $\theta\mathbf{u}$ to represent the rotation ${}^{c^*}\mathbf{R}_c$ between R_{c^*} and R_c. If the matrices ${}^{c^*}\mathbf{R}_{n^*}$ and ${}^{c}\mathbf{R}_n$ are identical, which is usually the case, we can also consider the vector $\theta\mathbf{u}$ associated with the rotation ${}^{n^*}\mathbf{R}_n$. Likewise, with an eye-to-hand system, the vector $\theta\mathbf{u}$ can be used to represent either the rotation ${}^{o^*}\mathbf{R}_o$ between the desired frame and the current frame of the object mounted on the effector, or the rotation ${}^{n^*}\mathbf{R}_n$ if the matrices ${}^{o^*}\mathbf{R}_{n^*}$ and ${}^{o}\mathbf{R}_n$ are identical (which is also usually the case).

In all of the cases mentioned above, the interaction matrix related to $\theta\mathbf{u}$ is given by [MAL 99]:

$$\mathbf{L}_{\theta\mathbf{u}} = \begin{bmatrix} \mathbf{0}_3 & \mathbf{L}_\omega \end{bmatrix} \qquad [6.45]$$

with:

$$\mathbf{L}_\omega = \mathbb{I}_3 - \frac{\theta}{2}[\mathbf{u}]_\times + \left(1 - \frac{\mathrm{sinc}\theta}{\mathrm{sinc}^2\frac{\theta}{2}}\right)[\mathbf{u}]_\times^2 \qquad [6.46]$$

The $\theta\mathbf{u}$ representation is therefore particularly interesting since $\mathbf{L}_\omega$ is singular only for $\theta = 2\pi$. Furthermore, we have:

$$\mathbf{L}_\omega^{-1} = \mathbb{I}_3 + \frac{\theta}{2}\,\mathrm{sinc}^2\frac{\theta}{2}\,[\mathbf{u}]_\times + (1 - \mathrm{sinc}\theta)[\mathbf{u}]_\times^2 \qquad [6.47]$$

which guarantees the following, rather convenient property:

$$\mathbf{L}_\omega^{-1}\,\theta\mathbf{u} = \theta\mathbf{u} \qquad [6.48]$$

If it would be preferable to consider the rotations ${}^c\mathbf{R}_{c^*}$, ${}^n\mathbf{R}_{n^*}$ or ${}^o\mathbf{R}_{o^*}$, we immediately infer from [6.45] that:

$$\mathbf{L}_{\theta\mathbf{u}} = \left[\begin{array}{cc} \mathbf{0}_3 & -\mathbf{L}_\omega \end{array}\right] \qquad [6.49]$$

and we now have:

$$\mathbf{L}_\omega^{-1}\,\theta\mathbf{u} = -\theta\mathbf{u} \qquad [6.50]$$

Note that it is not wise to directly take into account the vector $\theta\mathbf{u}$ associated with the rotation ${}^c\mathbf{R}_o$ and to use the difference between $\theta\mathbf{u}$ and $\theta^*\mathbf{u}^*$ (where $\theta^*\mathbf{u}^*$ represents the desired rotation ${}^{c^*}\mathbf{R}_o$). This is because $\theta\mathbf{u} - \theta^*\mathbf{u}^*$ does not represent a distance in the space SO_3 of rotations [SAM 91].

6.2.5.3. *Interaction matrix related to a 3-D point*

Using the fundamental kinematics equation given in [6.14], we immediately get for any point of the object with coordinates $\mathbf{X}$ connected to the object:

$$\mathbf{L}_{\mathbf{X}} = \left[\begin{array}{cc} -\mathbb{I}_3 & [\mathbf{X}]_\times \end{array}\right] \qquad [6.51]$$

The points taken into account can be characteristic points of the object [MART 96, SCH 04], or also the origin of R_o (we then have $\mathbf{X} = {}^c\mathbf{t}_o$).

Thus, with an eye-in-hand system, if we are interested in the displacement it must achieve, we can also consider the origin of R_{c^*} (we then have $\mathbf{X} = {}^c\mathbf{t}_{c^*}$ and $\mathbf{X}^* = \mathbf{0}$) [MART 97]. In that case, it is even better to consider the position of the origin of the camera's frame expressed in a rigidly fixed frame, such as R_o, or even R_{c^*} or $R_\emptyset$ if the object is static (see Figure 6.7) [WIL 96].

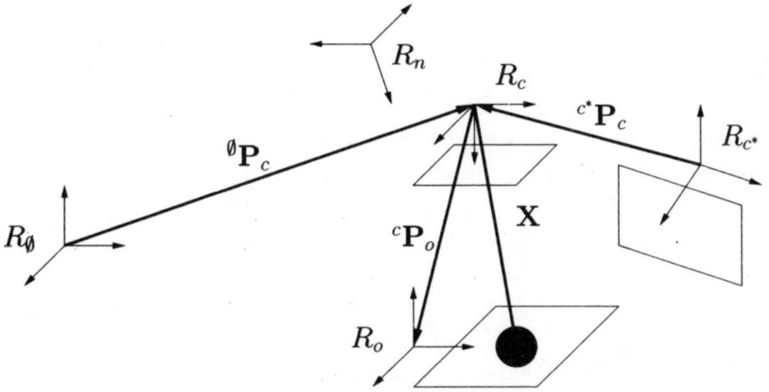

Figure 6.7. *Possible 3-D points with an eye-in-hand system*

For example, if we choose R_o, we have:

$$ {}^o\mathbf{t}_c = -{}^c\mathbf{R}_o^\top {}^c\mathbf{t}_o = -{}^o\mathbf{R}_c {}^c\mathbf{t}_o \qquad [6.52] $$

By differentiating this equation, we get:

$$ \begin{aligned} {}^o\dot{\mathbf{t}}_c &= -{}^o\dot{\mathbf{R}}_c {}^c\mathbf{t}_o - {}^o\mathbf{R}_c {}^c\dot{\mathbf{t}}_o \\ &= -{}^o\mathbf{R}_c \left({}^c\mathbf{R}_o {}^o\dot{\mathbf{R}}_c {}^c\mathbf{t}_o + {}^c\dot{\mathbf{t}}_o \right) \end{aligned} $$

meaning that, using [6.4] and [6.51]:

$$ \begin{aligned} {}^o\dot{\mathbf{t}}_c &= -{}^o\mathbf{R}_c \left([\boldsymbol{\omega}]_\times {}^c\mathbf{t}_o - \boldsymbol{v} + [{}^c\mathbf{t}_o]_\times \boldsymbol{\omega} \right) \\ &= {}^o\mathbf{R}_c \boldsymbol{v} \end{aligned} $$

We therefore have:

$$ \mathbf{L}_{{}^o\mathbf{t}_c} = \left[\begin{array}{cc} {}^o\mathbf{R}_c & \mathbf{0}_3 \end{array} \right] \qquad [6.53] $$

which is independent of the camera's rotational movements. Likewise, if we choose ${}^{c^*}\mathbf{t}_c$, we get:

$$\mathbf{L}_{{}^{c^*}\mathbf{t}_c} = \left[\begin{array}{cc} {}^{c^*}\mathbf{R}_c & 0_3 \end{array} \right] \quad [6.54]$$

and we will then have ${}^{c^*}\mathbf{t}_c{}^* = \mathbf{0}$.

With an eye-to-hand system (see Figure 6.8), and for the same decoupling properties, it is better to consider the position of the origin of either the frame R_o or R_n, and to express the kinematic screw in this origin. This is because if we choose for example ${}^c\mathbf{t}_o$, then, using [6.51] and [6.6], we have:

$$\mathbf{L}_{{}^c\mathbf{t}_o} {}^c\mathbf{V}_o = \left[\begin{array}{cc} -\mathbb{I}_3 & [{}^c\mathbf{t}_o]_\times \end{array} \right] \left[\begin{array}{cc} {}^c\mathbf{R}_o & [{}^c\mathbf{t}_o]_\times {}^c\mathbf{R}_o \\ 0_3 & {}^c\mathbf{R}_o \end{array} \right] \quad [6.55]$$

hence:

$$\mathbf{L}_{{}^c\mathbf{t}_o} {}^c\mathbf{V}_o = \left[\begin{array}{cc} -{}^c\mathbf{R}_o & 0_3 \end{array} \right] \quad [6.56]$$

We can of course express the position of the origin of R_o in any frame. If the robot's reference frame $R_\emptyset$ is chosen, we simply obtain:

$${}^\emptyset\dot{\mathbf{t}}_o = \left[\begin{array}{cc} \mathbb{I}_3 & 0_3 \end{array} \right] {}^\emptyset\mathbf{v}_o \quad [6.57]$$

where ${}^\emptyset\mathbf{v}_o$ is the object's kinematic screw expressed in $R_\emptyset$ and in the origin of R_o. The same result is of course achieved when considering ${}^\emptyset\mathbf{t}_n$ and ${}^\emptyset\mathbf{v}_n$.

6.2.5.4. *Interaction matrix related to a 3-D plane*

Finally, we can also determine the interaction matrix related to 3-D geometric primitives such as line-segments, planes, spheres, etc. For example, in the case of a plane represented by its unit normal $\mathbf{u}$ and its distance to the origin D, we get:

$$\mathbf{L}_{(\mathbf{u},D)} = \left[\begin{array}{cc} 0_3 & [\mathbf{u}]_\times \\ \mathbf{u}^\top & 0 \end{array} \right] \quad [6.58]$$

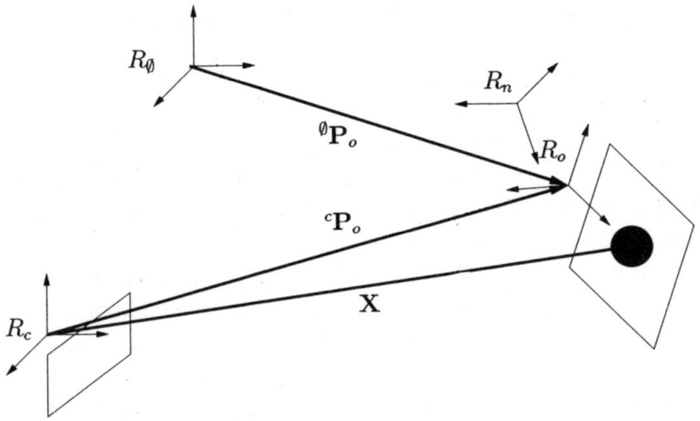

Figure 6.8. *Possible 3-D points with an eye-to-hand system*

6.3. Task function and control scheme

Achieving a robotic task by visual servoing requires the selection of the appropriate visual features and the design of a closed-loop control law. The first phase amounts to defining a task function with properties that ensure that the chosen task will be achieved [SAM 91], the second to regulating this task function. We will first consider the case where we wish to control the 6 degrees of freedom of the robot, in other words to bring the end-effector's frame to a unique desired pose.

If we use a set of k visual features $\mathbf{s}$, the general form of the task function $\mathbf{e}$ is:

$$\mathbf{e}(\mathbf{p}(t)) = \mathbf{C}\,(\mathbf{s}(\mathbf{p}(t)) - \mathbf{s}^*) \qquad [6.59]$$

where:

– $\mathbf{s}(\mathbf{p}(t))$ is the current value of the selected visual features;

– $\mathbf{s}^*$ is the value that $\mathbf{s}$ must reach for the task to be achieved;

– $\mathbf{C}$ is a full-rank $6 \times k$ matrix, referred to as the combination matrix, such that the 6 components of $\mathbf{e}$ are independent and control the robot's 6 degrees of freedom.

6.3.1. *Obtaining the desired value* $\mathbf{s}^*$

Whatever the nature of the visual features that were chosen, the value $\mathbf{s}^*$ is usually obtained, either by defining beforehand the pose that must be achieved between the robot and the object in question, or by learning:

– In the first case, if s includes 2-D features, their desired value can easily be obtained if a 3-D model of the object is available, simply by applying the perspective projection equations to calculate the object's position in the image. Additionally, it is also possible to specify the pose that has to be achieved between the end-effector and the object of interest (for a grasping task for example): the calculation of the visual features (2-D or 3-D) is then immediately obtained if the transformation matrix between the end-effector frame and the camera frame is known. However, in any case, any modeling error in the camera's calibration parameters, in the model of the object (and possibly in the end-effector-camera transform matrix) will have as a result that when the value of s is equal to s^*, the pose actually reached will be different from the one that was specified, because of the bias introduced by the modeling errors.

– Obtaining s^* by learning, though less convenient to achieve in practice, is therefore preferable to ensure that the task is well achieved. It consists in a prior phase of bringing the robot to a desired position with respect to the object, then acquiring the corresponding image, and calculating the value of s^* exactly in the same way as for the future calculations of $s(t)$. In the presence of modeling errors, we find ourselves in the paradoxical situation of having biased desired and current values of visual features, but a pose after convergence that is accurate aside from the measurement errors.

– A third, more elegant solution consists of managing to have the camera observe the end-effector and the object of interest simultaneously. The calculation of s^* can then be achieved automatically [HOR 98]. This solution has rarely been implemented, because, although it seems natural for eye-to-hand systems, it poses significant problems regarding where the camera is placed in the case of eye-in-hand systems.

We will now give in detail the different possible choices for the combination matrix $\mathbf{C}$ by following a (simple) analysis of the system's stability.

6.3.2. *Regulating the task function*

As we saw in the beginning of this section, developing a control law to regulate the task function is separate from defining this function. In the literature, many types of control laws have been suggested: non-linear control laws [HAS 93b, REY 98], LQ or LQG optimal control [PAP 93, HAS 96], based on a GPC controller [GAN 02, GIN 05], even robust H_∞ [KHA 98] or by return of a non-stationary continuous return state feedback [TSAK 98] in the case of mobile robots with nonholonomic constraints. We will simply focus on achieving a decoupled exponential decrease of the task function, that is:

$$\dot{\mathbf{e}} = -\lambda\, \mathbf{e} \qquad [6.60]$$

Using [6.59] and [6.2], if the matrix $\mathbf{C}$ is chosen constant, the differential of $\mathbf{e}$ is given by:

$$\dot{\mathbf{e}} = \mathbf{C}\,\dot{\mathbf{s}} = \mathbf{C}\,\mathbf{L_s}\,\mathbf{v} \qquad [6.61]$$

We saw in sections 6.2.2 and 6.2.3 how to pass from the kinematic screw $\mathbf{v}$ to the joint variables $\dot{\mathbf{q}}$. For simpler notations, we will assume from now on that the control quantity is simply the controllable part of $\mathbf{v}$, denoted by $\mathbf{v_q}$, that is to say $\mathbf{v_q} = \mathbf{v_c}$ in the case of an eye-in-hand system and $\mathbf{v_q} = -\mathbf{v_o}$ in the case of an eye-to-hand system (hence we will not be considering the problems caused by singularities of the robot and its joint limits. Furthermore, we will not be considering the case of a robot with less than six degrees of freedom. We will just point out that, in that case, we must of course work directly in the joint space using [6.7] or [6.11], and not proceed in two steps with $\mathbf{v_q}$ then $\dot{\mathbf{q}}$. We therefore write:

$$\dot{\mathbf{e}} = \mathbf{C}\,\mathbf{L_s}\,\mathbf{v_q} + \frac{\partial \mathbf{e}}{\partial t} \qquad [6.62]$$

where $\frac{\partial \mathbf{e}}{\partial t}$ represents the variations of $\mathbf{e}$ caused either by the object's motion (if an eye-in-hand system is used), or by the camera's motion (if an eye-to-hand system is used). To control the robot's 6 degrees of freedom, it is at least necessary to select $\mathbf{s}$ such that $\mathbf{L_s}$ has rank 6 and we obtain as an ideal control law:

$$\mathbf{v_q} = (\mathbf{C}\,\mathbf{L_s})^{-1}\left(-\lambda\mathbf{e} - \frac{\partial \mathbf{e}}{\partial t}\right) \qquad [6.63]$$

In the case where the visual features are expressed in the image, we saw that the interaction matrix depends on the values of these visual features and on the depth between the camera and the object in question. In the case of 3-D visual features, only some rather strong hypotheses make it possible to obtain the analytical form of this matrix. In any case, measurement and estimation errors are inevitable and the exact value of $\mathbf{L_s}$ is unknown. Only an approximation $\widehat{\mathbf{L_s}}$ can therefore be considered in the control law. Also, the term $\frac{\partial \mathbf{e}}{\partial t}$ is usually unknown. Hence the control law used in practice is:

$$\mathbf{v_q} = \left(\mathbf{C}\,\widehat{\mathbf{L_s}}\right)^{-1}\left(-\lambda\mathbf{e} - \widehat{\frac{\partial \mathbf{e}}{\partial t}}\right) \qquad [6.64]$$

If we assume that this velocity is perfectly achieved, the use of [6.64] in [6.62] leads to:

$$\dot{\mathbf{e}} = -\lambda \, \mathbf{C} \, \mathbf{L_s} \left(\mathbf{C} \, \widehat{\mathbf{L_s}} \right)^{-1} \mathbf{e} - \mathbf{C} \, \mathbf{L_s} \left(\mathbf{C} \, \widehat{\mathbf{L_s}} \right)^{-1} \widehat{\frac{\partial \mathbf{e}}{\partial t}} + \frac{\partial \mathbf{e}}{\partial t} \qquad [6.65]$$

If we assume that $\frac{\partial \mathbf{e}}{\partial t} = \widehat{\frac{\partial \mathbf{e}}{\partial t}} = 0$, then we notice that the positivity condition:

$$\mathbf{C} \, \mathbf{L_s} \left(\mathbf{C} \, \widehat{\mathbf{L_s}} \right)^{-1} > 0 \qquad [6.66]$$

is sufficient to ensure the decrease of $||\mathbf{e}||$ and therefore the system's global asymptotic stability ($||\mathbf{e}||$ is then a Lyapunov function). Also, the resulting behavior will be the same as the one specified in [6.60] under the unique condition that $\widehat{\mathbf{L_s}} = \mathbf{L_s}$ and that $\widehat{\frac{\partial \mathbf{e}}{\partial t}} = \frac{\partial \mathbf{e}}{\partial t}$. We will see in section 6.3.4 how we can estimate $\frac{\partial \mathbf{e}}{\partial t}$, which then makes it possible to reduce tracking errors. We will now focus on different possible choices of $\mathbf{C}$ and $\widehat{\mathbf{L_s}}$. Therefore we will assume from now on that $\frac{\partial \mathbf{e}}{\partial t} = \widehat{\frac{\partial \mathbf{e}}{\partial t}} = 0$ so as not to complicate the notations too much.

6.3.2.1. *Case where the dimension of* $\mathbf{s}$ *is* 6 *(*$k = 6$*)*

If the dimension of $\mathbf{s}$ is 6, it is much more convenient to choose $\mathbf{C} = \mathbb{I}_6$, because the behavior of $\mathbf{s}$ will then be the same as that of $\mathbf{e}$ (meaning that, in the ideal case, all components of $\mathbf{s}$ will have a decoupled exponential decrease). In that case, we get the control law:

$$\mathbf{v_q} = -\lambda \, \widehat{\mathbf{L_s}}^{-1} \mathbf{e} = -\lambda \, \widehat{\mathbf{L_s}}^{-1} (\mathbf{s} - \mathbf{s}^*) \qquad [6.67]$$

and the sufficient stability condition:

$$\mathbf{L_s} \, \widehat{\mathbf{L_s}}^{-1} > 0 \qquad [6.68]$$

If we are able to properly measure the current value of $\mathbf{L_s}$ at each iteration of the control law, taking this estimation into account makes it possible to come closest to the ideal behavior $\dot{\mathbf{s}} = -\lambda \, (\mathbf{s} - \mathbf{s}^*)$.

6.3.2.1.1. 2-D visual features

When considering 2-D visual features, it is unfortunately extremely difficult (up to now) to end up in this type of situation. The main difficulty involves not the estimate of the current value of the interaction matrix, but the selection of the six visual features.

Consider for example the case where s is comprised of the projection coordinates of three points in the image. The interaction matrix $\mathbf{L_s}$ is then a 6×6 matrix and, most of the time, is a full rank matrix. But we can show [MIC 93, PAP 95] that some configurations lead to a loss of rank of $\mathbf{L_s}$. In this case, the singularities are such that the three points are aligned in the image or that the optical center C of the camera belongs to the surface of the cylinder defined by the circumcircle of these three points (see Figure 6.9). It is therefore difficult to ensure that, for any chosen initial position, the robot's motion will avoid going through an isolated singularity (where of course the stability condition [6.68] is no longer satisfied).

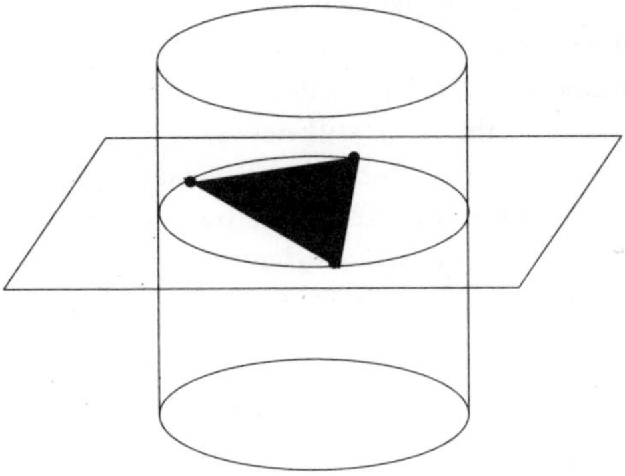

Figure 6.9. *Singularity cylinder*

Also, there are usually four distinct poses between the camera and the scene such that the image of three points is the same [DHO 89]. Minimizing $\|\mathbf{s}-\mathbf{s}^*\|$ can therefore bring the robot to one of the four global minima such that $\|\mathbf{s}-\mathbf{s}^*\| = 0$. Thus, in this case, it is possible to have $\mathbf{s}-\mathbf{s}^* = 0$ even if the pose that was reached is not the specified pose.

When considering visual features of different kinds (such as for example the three straight lines that can be defined from three non-aligned points), the same potential problems arise. A convenient solution consists of restricting the workspace to areas close to the desired pose, that include no isolated singularities, and where minimizing $\|\mathbf{s}-\mathbf{s}^*\|$ draws the robot's end-effector to its desired pose. However, determining the size of these areas is a difficult problem.

Furthermore, if the six terms describing the affine deformation of an object are used (see section 6.2.4.3), a loss of rank of the interaction matrix occurs when the considered object is parallel to the image plane [COLO 99, DRU 99], which renders servoing impossible in the vicinity of this configuration.

Recent studies based on the search of moment combinations have allowed to determine sets of six visual features with very interesting properties (invariant with respect to certain motions, directly related to others) [TAH 05]. However, these results are not yet definitive and the absence of isolated singularities or local minima has not yet been demonstrated.

Because of the different reasons mentioned above, it is very common to use redundant 2-D visual features. We then have k greater than 6, a situation described in section 6.3.2.2.

6.3.2.1.2. 3-D visual features

The use of 3-D visual features makes it possible to avoid the problems mentioned earlier since three parameters $\theta\mathbf{u}$ are now available to represent the orientation and only three position parameters have to be chosen among those given in 6.2.5.3 to have $k = 6$. Remember, however, that it is still necessary to be in the ideal situation for which the different measurements, calibration and estimation errors are negligible, to be able to express the interaction matrix. In the rest of this section, we will therefore assume that we are in this ideal case which (theoretically) ensures the specified behavior $\dot{\mathbf{s}} = -\lambda\,(\mathbf{s} - \mathbf{s}^*)$ and the stability condition [6.68] in the entire workspace. Indeed, we then have:

$$\mathbf{L_s}\,\widehat{\mathbf{L_s}}^{-1} = \mathbf{L_s}\,\mathbf{L_s}^{-1} = \mathbb{I}_6 > 0 \qquad [6.69]$$

If we choose to use $\theta\mathbf{u}$ to represent the rotation ${}^{c^*}\mathbf{R}_c$ and the coordinates $\mathbf{X}$ of an object's point expressed in the camera's current frame R_c, the global interaction matrix related to $\mathbf{s} = (\mathbf{X}, \theta\mathbf{u})$ is given by:

$$\mathbf{L_s} = \begin{bmatrix} -\mathbb{I}_3 & [\mathbf{X}]_\times \\ \mathbf{0}_3 & \mathbf{L}_\omega \end{bmatrix} \qquad [6.70]$$

Notice how appealing this matrix is (block-triangular and not singular except in $\theta = 2\pi$), giving the system an interesting behavior, since, ideally, the trajectory of the considered point is a straight line in the image. If this point is in the camera's field of view in its initial position and in its desired position, then it will be so permanently. By properly selecting this point (in the object's center of gravity for example), we can then minimize the risk of losing a large part of the object during the servoing (without ensuring however that a sufficient number of points necessary to the pose estimation remain visible). At each iteration we can also select the 3-D point that corresponds to the 2-D point closest to the borders of the image plane. But this choice implies a discontinuity in the control's translational components of the control law every time the point is changed. Additionally, without a higher level strategy (such as planning the trajectories in the image), it is always possible to exhibit particular cases that will either cause a part of the object to fall out of view, or lead to perverse effects on the control scheme (if two points are close to opposing edges of the image for example). Finally, the trajectory followed by the camera is a straight line, but only in the camera's *mobile* frame. Hence it will not be an actual straight line if an orientation change is necessary.

To illustrate the behavior of this control law, we will consider a positioning task with respect to four points that form a square. As Figure 6.10 shows, the desired pose of the camera is such that it is parallel and centered with respect to the square, with the image of the four points forming a centered square with its sides parallel to the axes of the image planes. The initial pose chosen corresponds to a significant displacement, particularly in rotation. The results obtained by simulation in ideal conditions (that is to say without introducing measurement errors, calibration errors, or pose estimation errors) are shown in Figure 6.11. We considered, as the coordinates $\mathbf{X}$ in $\mathbf{s}$, those of the origin O in the object's frame of reference, located in the square's center. Notice how, as expected, the trajectory in the image of the projection of O, given as an illustration, forms a perfectly straight line. On the other hand, the camera's trajectory is not a straight line at all. Finally, the components of the camera's velocity show a nice exponential decrease, due to the strong decoupling of this control law.

(a) (b) (c)

Figure 6.10. *Example of a positioning task: (a) desired pose for the camera, (b) initial pose, (c) image of the object for the initial pose and the desired pose*

With an eye-in-hand system, we also get a block-triangular interaction matrix if we consider the vector $\mathbf{X} = {}^{c}\mathbf{t}_{c^*}$ to control the camera's position. But this is of little interest in practice. On the other hand, if we choose to consider the position of the origin of R_c expressed in a frame rigidly linked to the object, for example ${}^{c^*}\mathbf{t}_c$, we then have (see [6.54]):

$$\mathbf{L_s} = \begin{bmatrix} {}^{c^*}\mathbf{R}_c & \mathbf{0}_3 \\ \mathbf{0}_3 & \mathbf{L}_\omega \end{bmatrix} \qquad [6.71]$$

which is block-diagonal and therefore ensures a complete decoupling between the translational motions and the rotational motions. Additionally, the trajectory of the camera will be an actual straight line, something of significant practical interest. To have the robot's end-effector follow a straight line, we simply have to consider ${}^{n^*}\mathbf{t}_n$ in $\mathbf{s}$ instead of ${}^{c^*}\mathbf{t}_c$. Unfortunately, in both cases, there is no longer any control over the object's trajectory in the image, and if the camera's initial position is far from its desired position, there is no guarantee that the object will remain in the camera's field of view during the servoing.

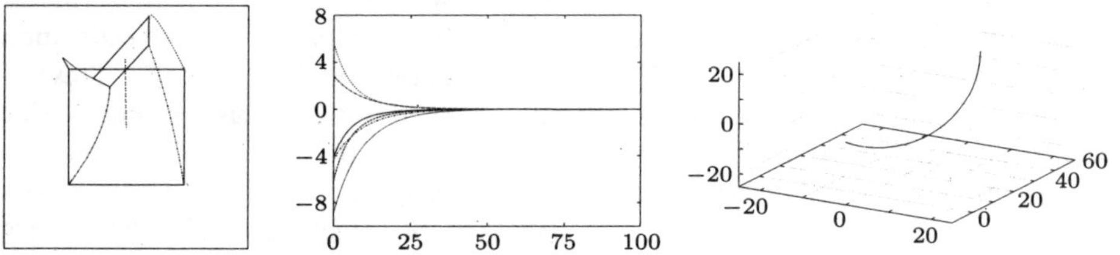

(a) (b) (c)

Figure 6.11. *Servoing results when choosing* $\mathbf{s} = ({}^{c}\mathbf{t}_o, \theta\mathbf{u})$*: (a) trajectories of points in the image, (b) components of* $\mathbf{v}_c$ *(in cm/s and deg/s) calculated at each iteration, (c) trajectory of the origin of the camera's frame in the frame* $\mathcal{R}_{c^*}$ *(in cm)*

The simulation results for this control law, obtained in the exact same conditions as before, are shown in Figure 6.12. They bring support to the comments stated above. Notice also that the decrease of the translational components of $\mathbf{v}_c$ are not as good, because of the coupling of these components induced by the very strong rotation that has to be performed.

Similar choices are of course also possible using an eye-to-hand system (see section 6.2.5.3). As an example, if we select in s the translation ${}^{\emptyset}\mathbf{t}_o$ and the vector $\theta\mathbf{u}$ associated with the rotation ${}^{o^*}\mathbf{R}_o$, we end up, by combining Equations [6.67], [6.57] and [6.48], with the following control law:

$$\begin{cases} {}^{\emptyset}\boldsymbol{v}_o = -\lambda \left({}^{\emptyset}\mathbf{t}_o - {}^{\emptyset}\mathbf{t}_{o^*}\right) \\ {}^{o}\boldsymbol{\omega}_o = -\lambda\, \theta\, \mathbf{u} \end{cases} \qquad [6.72]$$

where ${}^{\emptyset}\boldsymbol{v}_o$ is the translational velocity of R_o expressed in $R_{\emptyset}$ and where ${}^{o}\boldsymbol{\omega}_o$ is its angular velocity expressed in R_o. This control law shows ideal decoupling properties. Additionally, it ensures a straight line trajectory for the origin of R_o both in the 3-D space and in the image. We should however point out that if there are modeling errors present in the robot's Jacobian or a calibration error in the transformation matrix from R_n to R_o, the trajectory actually performed in practice will be different from what is expected. But the closed loop that is used is robust when it comes to these calibration errors and it is possible to quantify this robustness by the analysis of the stability condition [6.68] by reasoning in the joint space.

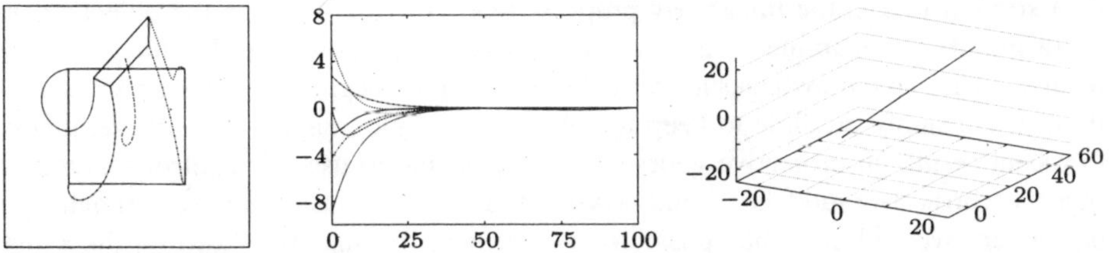

Figure 6.12. *Servoing results when choosing* $\mathbf{s} = ({}^{c^*}\mathbf{t}_c, \theta\mathbf{u})$

6.3.2.1.3. 2 ½-D visual features

As we have already said a number of times, taking 3-D visual features into account is based on the hypothesis that these features can be measured reliably. In practice, they are more sensitive to measurement errors than 2-D visual features, since they are obtained from 2-D measurements and from a pose estimation without any particular smoothing properties. It is therefore a good idea to combine 2-D and 3-D visual features to increase the robustness to measurement errors while maintaining good decoupling properties. In [MAL 99], the task function is defined as follows:

$$\mathbf{e} = \left(\; x - x^* , \quad y - y^* , \quad \log(Z/Z^*), \quad \theta\mathbf{u} \; \right) \qquad [6.73]$$

where:

– (x, y) and (x^*, y^*) are the current coordinates and the desired coordinates, respectively, of a characteristic point in the image;

– Z/Z^* is the ratio of the current depth to the desired depth of this point;

– $\theta\mathbf{u}$ represents the rotation ${}^{c^*}\mathbf{R}_c$ that is to be achieved.

In this case, we obtain the following control law:

$$\mathbf{v_q} = -\lambda \begin{bmatrix} Z\mathbf{L}_{e_v}^{-1} & -Z\mathbf{L}_{e_v}^{-1}\mathbf{L}_{e_{v\omega}} \\ \mathbf{0}_3 & \mathbb{I}_3 \end{bmatrix} \mathbf{e} \qquad [6.74]$$

where (see [6.16] and [6.51]):

$$\mathbf{L}_{e_v} = \begin{bmatrix} -1 & 0 & x \\ 0 & -1 & y \\ 0 & 0 & -1 \end{bmatrix} \quad \text{and} \quad \mathbf{L}_{e_{v\omega}} = \begin{bmatrix} xy & -(1+x^2) & y \\ (1+y^2) & -xy & -x \\ -y & x & 0 \end{bmatrix}$$

Therefore, the resulting decoupling is satisfactory since the control matrix is triangular. Additionally, the trajectory of the characteristic point that was chosen will be a straight line in the image. By properly selecting this point (in the object's center of gravity for example, or as close as possible to the borders of the image, which results in the same drawbacks as those described at the beginning of the previous section), it is usually possible to keep the object inside the image. Furthermore, thanks to recent results in projective geometry, it is possible to use this control scheme on objects whose 3-D models are unknown [MAL 00]. Because the pose estimation is no longer involved, it is then possible to determine the analytical form of the actual interaction matrix (meaning a form that does not rely on the strong hypotheses used before) and get it to display the camera's calibration errors. Thanks to the triangular form of the matrix, it is then possible to determine the analytical conditions that ensure the system's local and global asymptotic stabilities [MAL 99, MAL 02].

In our example, the behavior resulting from the control law [6.74] and from choosing the object's center of gravity as the characteristic point is shown in Figure 6.13. Notice the straight line trajectory of this point in the image and the fact that this behavior is very similar to the one obtained when choosing $\mathbf{s} = ({}^c\mathbf{t}_o, \theta\mathbf{u})$ (go back to Figure 6.11).

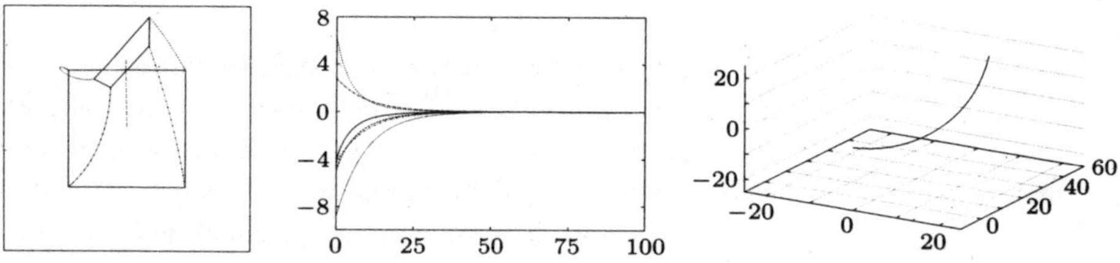

Figure 6.13. *Servoing results when choosing* $\mathbf{s} = (\mathbf{x}_g, \log(Z_g/Z_g^*), \theta\mathbf{u})$

Another version of this technique is described in [MOR 00]. The only difference involves the third component of $\mathbf{e}$ which explicitly takes into account the fact that all of the object's points must remain, as much as possible, inside the image. However, the triangular form of $\mathbf{L}_{e_v}$ is then lost, making it difficult to determine analytical stability conditions when calibration errors are present.

A second 2 $^1/_2$-D visual servoing technique is described in [CHA 00] in the case of an eye-in-hand system. The task function is given by:

$$\mathbf{e} = \left(\ {}^{c^*}\mathbf{t}_c, \quad x - x^*, \quad y - y^*, \quad \theta u_z \ \right) \qquad [6.75]$$

where (x, y) and (x^*, y^*) are again the current and the desired coordinates of a characteristic point in the image, and where u_z is the third component of the rotation axis $\mathbf{u}$ between R_{c^*} and R_c. Using [6.54], [6.16] and [6.45], we infer the expression of the associated control law:

$$\mathbf{v_c} = -\lambda \begin{bmatrix} {}^c\mathbf{R}_{c^*} & \mathbf{0}_3 \\ -\frac{1}{Z}\mathbf{L}_{e_\omega}^{-1}\mathbf{L}_{e_{\omega v}}\,{}^c\mathbf{R}_{c^*} & \mathbf{L}_{e_\omega}^{-1} \end{bmatrix} \mathbf{e} \qquad [6.76]$$

where:

$$\mathbf{L}_{e_{\omega v}} = \begin{bmatrix} -1 & 0 & x \\ 0 & -1 & y \\ 0 & 0 & 0 \end{bmatrix} \quad \text{and} \quad \mathbf{L}_{e_\omega} = \begin{bmatrix} xy & -(1+x^2) & y \\ (1+y^2) & -xy & -x \\ l_1 & l_2 & l_3 \end{bmatrix}$$

$[l_1 \ l_2 \ l_3]$ being the third line of the matrix $\mathbf{L}_\omega$ given in [6.46].

Compared to the previous case, the camera will follow a straight line trajectory, its orientation controlled so that the trajectory of the characteristic point follows a straight line in the image (see Figure 6.14). This control law is therefore extremely useful in practice. If we select, as our characteristic point, the point of the object closest to the limits of the image, the discontinuity of the control law when changing the point will now involve the components of the angular velocity. On the other hand, note that the control matrix is no longer block-triangular, which again makes it difficult to determine analytical conditions ensuring the system's stability when calibration errors are present. The same is true if θu_z is replaced by the orientation of a segment, of a straight line, or of an object in the image (see sections 6.2.4.2 and 6.2.4.3). More 2-D visual features are then used and the only change in the design of the control law consists of replacing the coefficients on the last lines of $\mathbf{L}_{e_{\omega v}}$ and $\mathbf{L}_{e_\omega}$ with their new values.

Finally, in [AND 02], the Plücker coordinates of straight lines are used, which also leads to a 2 $^1/_2$-D visual servoing approach. However the resulting trajectories in 3-D space and in the image are not as satisfactory as in the two cases described previously.

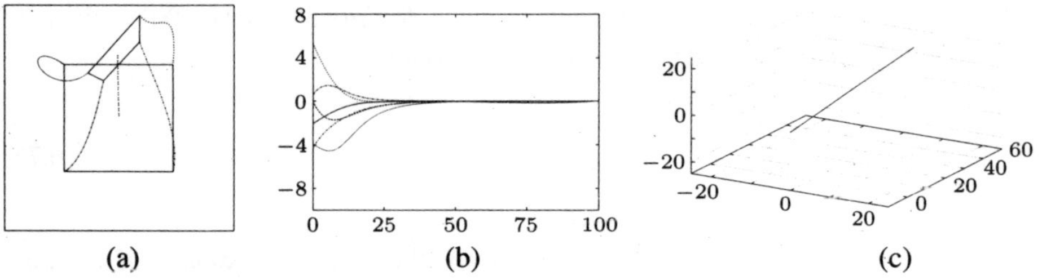

Figure 6.14. *Servoing results when choosing* $\mathbf{s} = ({}^{c^*}\mathbf{t}_c, \mathbf{x}_g, \theta u_z)$

6.3.2.2. *Case where the dimension of* $\mathbf{s}$ *is greater than* 6 *(*$k > 6$*)*

We will now describe the different possible choices for $\mathbf{C}$ and $\widehat{\mathbf{L}_\mathbf{s}}$ (see [6.64]) when the visual features that are chosen are redundant ($k > 6$). Aside for the studies described in [MART 96, SCH 04] where the coordinates of several 3-D points are taken into account, this case only involves the choice of 2-D visual features since selecting six independent visual features is then a difficult problem.

Remember that $\mathbf{C}$ has to be a $6 \times k$ constant matrix with rank 6. The simplest choice consists of choosing as $\mathbf{C}$ the pseudoinverse of an approximate value of the interaction matrix in the desired position:

$$\mathbf{C} = \widehat{\mathbf{L}_\mathbf{s}}^+_{|\mathbf{s}=\mathbf{s}^*} \qquad [6.77]$$

As we saw in the first part of this chapter, the interaction matrix depends on the value of the visual features that are chosen and the depth between the camera and the object. The calculation of $\mathbf{C}$ requires the value of $\mathbf{s}^*$ to be known, as well as the depth parameters in the desired position. If the 3-D model of the object is available, these parameters can easily be calculated by a pose estimation using the desired image. Otherwise, they are usually determined during the task specification itself.

Using [6.77], the control law [6.64] is expressed:

$$\mathbf{v_q} = -\lambda \left(\widehat{\mathbf{L}_\mathbf{s}}^+_{|\mathbf{s}=\mathbf{s}^*} \widehat{\mathbf{L}_\mathbf{s}} \right)^{-1} \mathbf{e} \qquad [6.78]$$

and, by choosing $\widehat{\mathbf{L}_\mathbf{s}}_{|\mathbf{s}=\mathbf{s}^*}$ to approximate $\widehat{\mathbf{L}_\mathbf{s}}$, we get:

$$\mathbf{v_q} = -\lambda\, \mathbf{e} = -\lambda\, \widehat{\mathbf{L}_\mathbf{s}}^+_{|\mathbf{s}=\mathbf{s}^*} (\mathbf{s} - \mathbf{s}^*) \qquad [6.79]$$

We should point that, even if $\mathbf{e}$ is perfectly regulated (i.e. $\mathbf{e} = 0$), it does not necessarily imply that the visual task is achieved (i.e. $\mathbf{s} = \mathbf{s}^*$), because the set of configurations such that:

$$(\mathbf{s} - \mathbf{s}^*) \in \operatorname{Ker} \mathbf{C} \qquad [6.80]$$

leads to $\mathbf{e}$ being equal to zero without $(\mathbf{s}-\mathbf{s}^*)$ being equal to zero. Hence it is important to make sure during the selection of the visual features not to create local minima in the workspace. As an example, consider a centered square parallel to the image plane. It is possible to show that by choosing, as the visual features, the coordinates in the image of the square's four corners, the configurations corresponding to local minima are such that the camera observes the square on its side. The four points are then aligned in the image and what we have is a degenerate case, outside the workspace of course.

Additionally, the stability condition is now written as:

$$\widehat{\mathbf{L}_{\mathbf{s}}}^{+}_{|\mathbf{s}=\mathbf{s}^*} \, \mathbf{L}_{\mathbf{s}} > 0 \qquad [6.81]$$

Even with a perfect estimate of $\widehat{\mathbf{L}_{\mathbf{s}}}_{|\mathbf{s}=\mathbf{s}^*}$, this positivity condition is only ensured in a neighborhood around the desired position. Usually, only local asymptotic stability can thus be demonstrated. Likewise, the decoupled exponential behavior of $\mathbf{e}$ will only be ensured around this desired position. It is therefore possible, if the camera's initial pose is far away from the desired pose, that the resulting trajectories in the image turn out to be poorly satisfactory, or do not even lead to a convergence of the system (see Figure 6.15a). In practice, this only occurs if considerable rotational motions are required [CHA 98].

In our example, note the convergence of the servoing in Figure 6.16, despite the very significant rotation that has to be achieved. However, the components of $\mathbf{v}_c$ do not behave well, except near the convergence.

These problems are commonly solved by directly choosing for $\mathbf{C}$ the pseudo-inverse of an estimated value of the current interaction matrix, instead of a constant matrix:

$$\mathbf{C} = \widehat{\mathbf{L}_{\mathbf{s}}}^{+} \qquad [6.82]$$

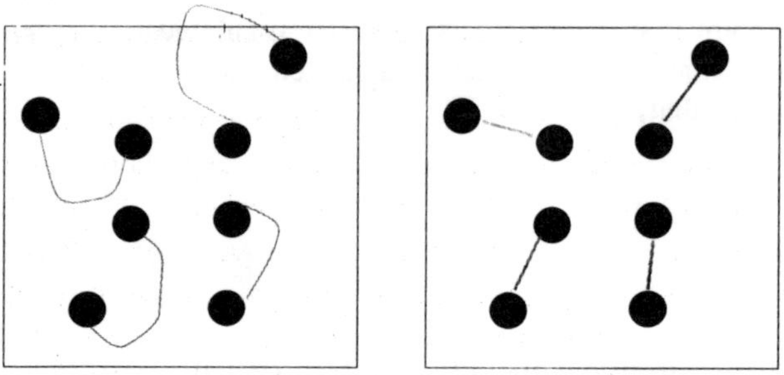

(a) (b)

Figure 6.15. *(a) possible trajectory in the image when choosing* $\mathbf{C} = \widehat{\mathbf{L_s}}^{+}_{|\mathbf{s}=\mathbf{s}^*}$*, (b) expected trajectory in the image when choosing* $\mathbf{C} = \widehat{\mathbf{L_s}}^{+}$

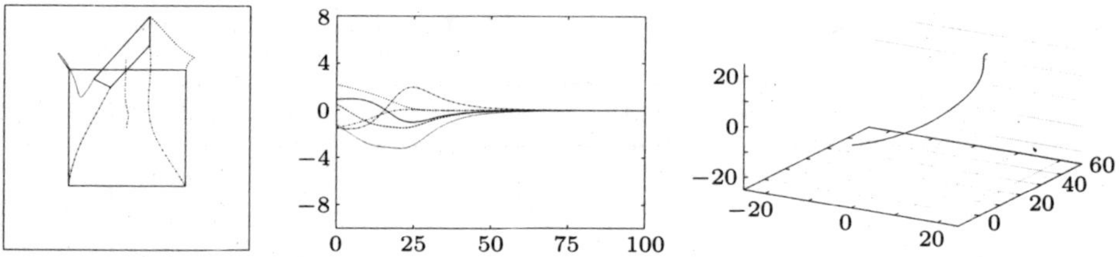

(a) (b) (c)

Figure 6.16. *Servoing results when choosing* $\mathbf{s} = (x_1, y_1, \ldots, x_4, y_4)$ *and* $\mathbf{C} = \mathbf{L_s^+}_{|\mathbf{s}=\mathbf{s}^*}$

This leads to:

$$\mathbf{v_q} = -\lambda\, \widehat{\mathbf{L_s}}^{+}\, (\mathbf{s} - \mathbf{s}^*) \qquad [6.83]$$

It is now necessary at each iteration of the control law to estimate either the 3-D parameters involved in the interaction matrix, or to perform an online numerical estimation of the elements of this matrix (see section 6.2.4.4). In the absence of time smoothing in the calculation of this matrix, the system's behavior will therefore be less stable than in the previous case.

Also, the convergence condition [6.66] no longer applies since the calculation of $\dot{\mathbf{e}}$ would have to take into account the variations of $\mathbf{C}$ (see [6.61]), leading to virtually unfeasible calculations. Once again, only the local asymptotic stability can be demonstrated. Considering the behavior of s, we get:

$$\dot{\mathbf{s}} = -\lambda \mathbf{L_s}\, \widehat{\mathbf{L_s}}^{+}\, (\mathbf{s} - \mathbf{s}^*) \qquad [6.84]$$

As in the previous case, it is therefore impossible to ensure the strict decrease of $\|\mathbf{s} - \mathbf{s}^*\|$ at each iteration since the $k \times k$ matrix $\mathbf{L_s}\ \widehat{\mathbf{L_s}}^+$ only has rank 6. All configurations such that:

$$(\mathbf{s} - \mathbf{s}^*) \in \text{Ker}\ \widehat{\mathbf{L_s}}^+ \qquad [6.85]$$

correspond to attractive local minima, the existence of which is demonstrated in the very simple case of a square [CHA 98].

The drawback of this method is that it attempts to ensure that $\dot{\mathbf{s}} = -\lambda(\mathbf{s} - \mathbf{s}^*)$ (directly providing the control law [6.83]), which implies k constraints when only 6 degrees of freedom are available. In other words, the "task function" $(\mathbf{s} - \mathbf{s}^*)$ is no longer ρ-admissible [SAM 91]. On the other hand, the advantage of this method is that, when it succeeds, it provides very nice trajectories in the image. If, for example, s is comprised of the coordinates of points in the image, the expected trajectories of these points will be straight lines (see Figure 6.15b). In practice, the actual trajectories will not necessarily be as good (since the actual behavior is given by [6.84]). Also, the robot's trajectory needed to achieve these trajectories in the image is not necessarily an advisable one [CHA 98].

These properties are summed-up in Figure 6.17: the trajectory of the points in the image is no longer a true straight line. Additionally, the camera's movement is not ideal given the components of $\mathbf{v}_c$ and the trajectory of the camera's optical center.

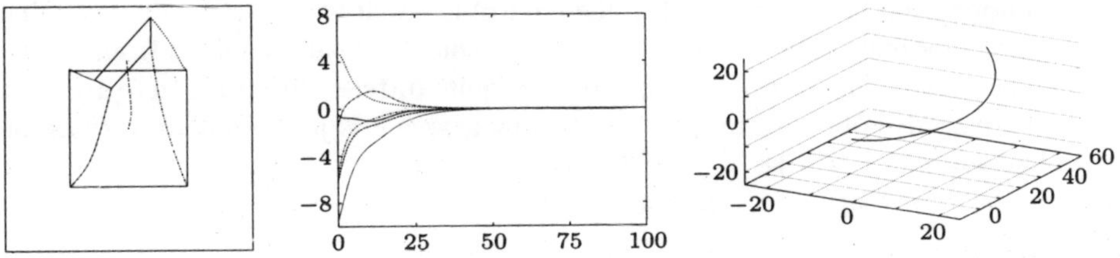

(a) (b) (c)

Figure 6.17. *Servoing results when choosing* $\mathbf{s} = (x_1, y_1, \ldots, x_4, y_4)$ *and* $\mathbf{C} = \mathbf{L_s^+}$

Finally, other methods can be found in the literature. For example, $\widehat{\mathbf{L_s}}^\top$ can be used in the control law (instead of $\widehat{\mathbf{L_s}}^+_{|\mathbf{s}=\mathbf{s}^*}$ or $\widehat{\mathbf{L_s}}^+$) [HAS 93b]. However, the advantages of this method compared to those described before are not obvious, since they do not show good decoupling properties. On the other hand, as was recently suggested [MAL 04], the choice of:

$$\mathbf{C} = \left(\frac{1}{2}\widehat{\mathbf{L_s}} + \frac{1}{2}\widehat{\mathbf{L_s}}_{|\mathbf{s}=\mathbf{s}^*}\right)^+ \quad [6.86]$$

leads to satisfactory results in practice, as Figure 6.18 shows, even if the camera's trajectory is not a straight line.

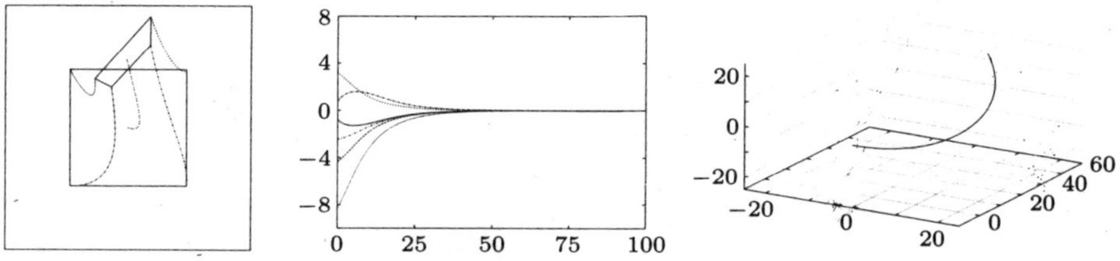

(a) (b) (c)

Figure 6.18. *Servoing results when choosing* $\mathbf{s} = (x_1, y_1, \ldots, x_4, y_4)$ *and* $\mathbf{C} = (\frac{1}{2}\mathbf{L_s} + \frac{1}{2}\mathbf{L_s}_{|\mathbf{s}=\mathbf{s}^*})^+$

Again, we insist on the importance of the choice of the visual features inside the control law. As an illustration, Figure 6.19 shows the results obtained in our example when choosing in s the parameters (ρ_i, θ_i) which represents the four straight lines forming the sides of a square (see section 6.2.4.2). Note that these parameters can of course be directly calculated from the position of the four points in the image. No additional information is necessary. Also, the control law that was chosen is the one that uses a constant matrix for $\mathbf{C}$, namely $\mathbf{L_s^+}_{|\mathbf{s}=\mathbf{s}^*}$ which is easily obtained from [6.31]. As can be seen in Figure 6.19, the behavior is quite different from Figure 6.16, even though it was also obtained using a constant matrix, and is just as satisfactory as the one obtained in Figure 6.18.

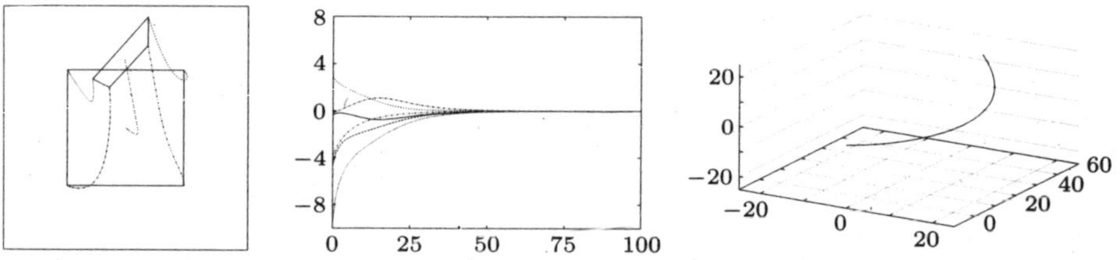

(a) (b) (c)

Figure 6.19. *Servoing results when choosing* $\mathbf{s} = (\rho_1, \theta_1, \ldots, \rho_4, \theta_4)$ *and* $\mathbf{C} = \mathbf{L_s^+}_{|\mathbf{s}=\mathbf{s}^*}$

To conclude this section, we should point out the importance of the conditioning of the interaction matrix and of the combination matrix in the system's behavior [FED 89, NEL 95, SHA 97]. A good conditioning of the former leads to a good sensitivity for the system, whereas a good conditioning of the latter leads to a good robustness for the control law with respect to measurement errors. Even if all of the methods discussed above provide very satisfactory practical results, there is still much work to be done to determine which visual features are the most relevant.

6.3.3. *Hybrid tasks*

We will now consider the case where the k visual features that are chosen do not constrain all of the robot's degrees of freedom. The visual tasks associated with the k constraints can then be categorized depending on the virtual link, an extension of the concept of links between solids, between the sensor and its environment.

6.3.3.1. *Virtual links*

The constraints $\mathbf{s}(\mathbf{p}(t)) - \mathbf{s}^* = 0$ induced by the visual features define, when they are achieved, a *virtual link* between the robot and its environment. Because $\dot{\mathbf{s}} = 0$ is an immediate consequence of $\mathbf{s}(\mathbf{p}(t)) = \mathbf{s}^*$, the set $\mathcal{S}^*$ of motions that leave s unchanged, that is:

$$\mathcal{S}^* = \text{Ker } \mathbf{L_s} \qquad [6.87]$$

enables us to fully characterize this virtual link.

For a pose $\mathbf{p}$ where these constraints are satisfied, the dimension N of $\mathcal{S}^*$ is called the *class* of the virtual link in $\mathbf{p}$. Let $m = 6 - N$. When $m = k$, the k constraints resulting from the visual features are independent. Intuitively, the size of s then corresponds to the number m of degrees of freedom that we can and wish to control using s. As we have already seen previously, it is also possible to come across the case where the visual features are redundant ($k > m$).

Figure 6.20 lists the most common frictionless mechanical links. The class of each link and the number of unconstrained degrees of freedom (in translation (T) or in rotation (R)) that allow such a categorization are also indicated. With a vision sensor, it is possible to achieve all of these links. For example, the case of the rigid link (which constrains the robot's six degrees of freedom) has already been studied in detail in the previous section. For the other links, examples of designs, constructed from the most common geometric primitives (points, straight lines, cylinders, etc.) are given in [CHA 93a].

Name of the link	Class	T	R	Geometric symbol
Rigid	0	0	0	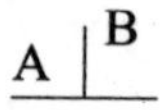
Prismatic	1	1	0	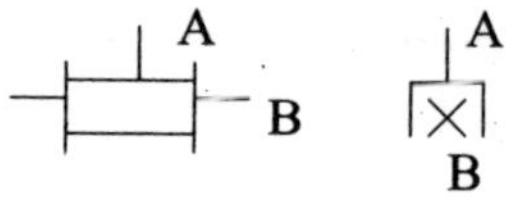
Rotary	1	0	1	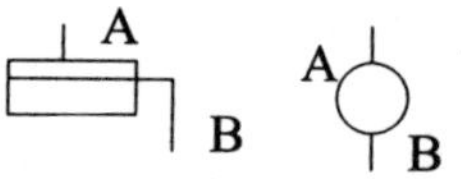
Sliding pivot	2	1	1	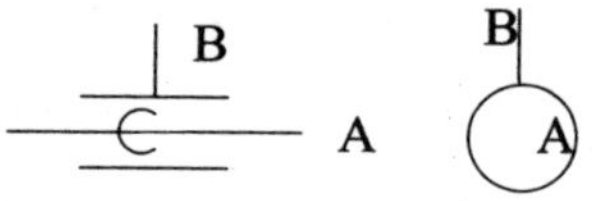
Plane-to-plane contact	3	2	1	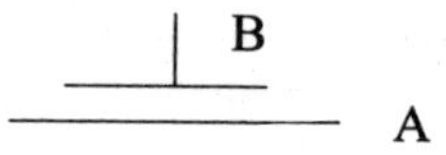
Bearing	3	0	3	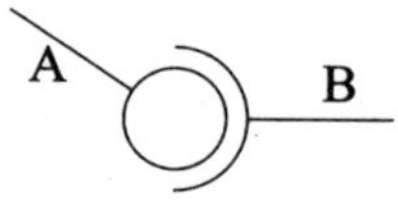
Rectilinear	4	2	2	
Linear annular	4	1	3	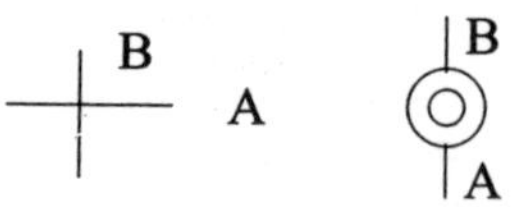
Point contact	5	2	3	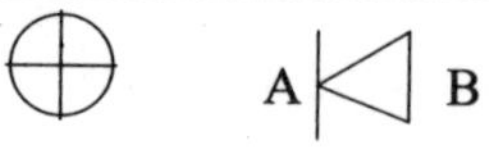

Figure 6.20. *Mechanical links*

As an example, we will explain the case of the bearing link. It can be achieved if the camera is observing a sphere the center of which is located on the optical axis, and therefore has the coordinates $\mathbf{X}_0 = (0, 0, Z_0)$. The image of the sphere is then a centered circle, and if we select in s the area a of the circle and the coordinates (x_g, y_g) of its center of gravity ($\mathbf{s} = (a, x_g, y_g)$), the interaction matrix has the value (see [6.34]):

$$\mathbf{L_s} = \begin{bmatrix} 0 & 0 & 2a/Z_g & 0 & 0 & 0 \\ -1/Z_g & 0 & 0 & 0 & -(1+r^2) & 0 \\ 0 & -1/Z_g & 0 & 1+r^2 & 0 & 0 \end{bmatrix} \qquad [6.88]$$

where $1/Z_g = Z_0/(Z_0^2 - R^2)$ and $r^2 = R^2/(Z_0^2 - R^2)$, r being the circle's radius in the image, and R the sphere's radius. By expressing the interaction matrix in the sphere's center using the screw transformation matrix given in [6.6], we get:

$$\begin{aligned} \mathbf{L_s} &= \begin{bmatrix} 0 & 0 & 2a/Z_g & 0 & 0 & 0 \\ -1/Z_g & 0 & 0 & 0 & -(1+r^2) & 0 \\ 0 & -1/Z_g & 0 & 1+r^2 & 0 & 0 \end{bmatrix} \begin{bmatrix} \mathbb{I}_3 & [\mathbf{X}_0]_\times \\ \mathbf{0}_3 & \mathbb{I}_3 \end{bmatrix} \\ &= \begin{bmatrix} 0 & 0 & 2a/Z_g & 0 & 0 & 0 \\ -1/Z_g & 0 & 0 & 0 & 0 & 0 \\ 0 & -1/Z_g & 0 & 0 & 0 & 0 \end{bmatrix} \end{aligned}$$

which gives the form we wanted for $\mathcal{S}^*$:

$$\mathcal{S}^* = \begin{bmatrix} 0 & 0 & 0 \\ 0 & 0 & 0 \\ 0 & 0 & 0 \\ 1 & 0 & 0 \\ 0 & 1 & 0 \\ 0 & 0 & 1 \end{bmatrix} \qquad [6.89]$$

It is of course possible to select 3-D visual features to achieve this link (if 3-D measurements are available). In this case, we simply have to choose the three coordinates of the sphere's center in $\mathbf{s}$.

We will now discuss in detail how to define a hybrid task combining a visual task controlling $m(< 6)$ degrees of freedom and a secondary task.

6.3.3.2. *Hybrid task function*

Very often, regulating a visual task is not the only objective and this task must be combined with another, such as for example tracking a trajectory or avoiding limits of the joints of the robot, since, aside from the case where the visual task consists of achieving a rigid link, other tasks achieve virtual links with a non-zero class, that is to say with degrees of freedom that are not constrained by the link. For example, a translational motion along the axis of a prismatic link leaves this link unchanged, and it may be advisable to put this available degree of freedom to use on a second objective.

Combining these two objectives can sometimes be done with a simple summation [NEL 94b]. The control law then reaches a compromise that can lead to none of the initial objectives being achieved. Here is another, more elegant approach, described in [SAM 91], among others. It consists of considering the visual task as the priority and to express the second objective as a cost function to minimize under the constraint that the visual task be achieved. The use of this approach is starting to be common in visual servoing. [CHA 94, COS 95, BER 00] give examples where the secondary task consists of performing trajectory tracking. This can be useful for inspection and conformity control applications, but also to perform the 3-D reconstruction of the considered objects [CHA 96] or to ensure that a system is properly positioned [COL 02]. Avoiding limits of the joints and singularities is discussed in [MAR 96]. Secondary tasks can also be visual tasks, to try to avoid occlusions for example [MAR 98], or they can be built from measurements provided by exteroceptive sensors. For example, combining a positioning task by visual servoing with an obstacle avoidance task using a laser type proximity sensor was studied in [CAD 00] in the context of mobile robotics.

Let $\mathbf{e}_1$ be the visual task function and h_s the cost function to minimize, the gradient of which is $\mathbf{e}_2^\top$. The function $\mathbf{e}_1$ has a dimension of $m = n - N \leq k$ where n is the system's number of degrees of freedom, N the class of the desired virtual link, and k the number of visual features used in $\mathbf{s}$. This task function is always written:

$$\mathbf{e}_1 = \mathbf{C}\,(\mathbf{s} - \mathbf{s}^*) \qquad [6.90]$$

where $\mathbf{C}$ is now an $m \times k$ matrix that has full rank m in order for the m components of $\mathbf{e}_1$ to be independent and to control the m desired degrees of freedom. If $\mathbf{C}$ is chosen constant, the interaction matrix of $\mathbf{e}_1$, with a size of $m \times n$ and a full rank equal to m, is given by:

$$\mathbf{L}_{\mathbf{e}_1} = \mathbf{C}\mathbf{L}_{\mathbf{s}} \qquad [6.91]$$

and notice that Ker $\mathbf{L}_{\mathbf{e}_1}$ = Ker $\mathbf{L}_{\mathbf{s}}$.

A task function $\mathbf{e}$ that minimizes h_s under the constraint $\mathbf{e}_1 = 0$ can be expressed in the form [SAM 91]:

$$\mathbf{e} = \mathbf{W}^+\mathbf{e}_1 + (\mathbb{I}_n - \mathbf{W}^+\mathbf{W})\,\mathbf{e}_2 \qquad [6.92]$$

where $\mathbf{W}$ is an $m \times n$ matrix with a full rank equal to m and such that Ker $\mathbf{W}$ = Ker $\mathbf{L_s}$. Hence the matrix $(\mathbb{I}_n - \mathbf{W}^+\mathbf{W})$ is therefore an orthogonal projection operator onto the kernel of $\mathbf{L_s}$ meaning that:

$$(\mathbb{I}_n - \mathbf{W}^+\mathbf{W})\,\mathbf{x} \in \text{Ker}\,\mathbf{L_s}\ ,\ \forall \mathbf{x} \in \mathbb{R}^n \qquad [6.93]$$

Hence whatever the function to minimize, and therefore the secondary task $\mathbf{e}_2$, only movements that do not disturb $\mathbf{e}_1$ will be applied, which also implies that h_s will not necessarily reach its minimum value.

However, because the exact value of the interaction matrix $\mathbf{L_s}$ is usually unknown, $\mathbf{W}$ has to be constructed based on an approximation or an estimate $\widehat{\mathbf{L_s}}$. If the kernel of $\mathbf{W}$ is different from the kernel of $\mathbf{L_s}$, then the secondary task can lead to perturbations in the achievement of $\mathbf{e}_1$. In practice, these perturbations turn out to be not too harmful, unless if the estimate of the interaction matrix is completely erroneous.

To construct the matrix $\mathbf{W}$, the simplest case occurs when $\mathbf{L_s}$ is a full rank matrix with a rank equal to $m = k$. We can then directly take $\mathbf{W} = \widehat{\mathbf{L_s}}$. Otherwise, the m lines of $\mathbf{W}$ can be comprised of the m basis vectors of the subspace generated by $\widehat{\mathbf{L_s}}$. Note that if $\mathbf{L_s}$ has a rank equal to n (in other words if the visual task constrains the system's n degrees of freedom), we can choose $\mathbf{W} = \mathbb{I}_n$. It is then impossible of course to take into account a secondary task, since in that case we have $\mathbf{e} = \mathbf{e}_1$. This also indicates that the task function defined in [6.92] is a generalization of the previous one in [6.59].

By performing the same analysis as the one described in the beginning of section 6.3.2, we get the following control law:

$$\mathbf{v_q} = \widehat{\mathbf{L_e}}^{-1}\left(-\lambda\,\mathbf{e} - \widehat{\frac{\partial \mathbf{e}}{\partial t}}\right) \qquad [6.94]$$

and the sufficient stability condition:

$$\mathbf{L_e}\widehat{\mathbf{L_e}}^{-1} > 0 \qquad [6.95]$$

Given the difficulty of calculating $\mathbf{L_e}$ in practice, it is possible to show [SAM 91] that the condition [6.95] is usually satisfied if we have:

$$\mathbf{L_{e_1}} \mathbf{W}^+ > 0 \qquad [6.96]$$

We can then set $\widehat{\mathbf{L_e}} = \mathbb{I}_n$, which leads to the following control law:

$$\mathbf{v_q} = -\lambda\, \mathbf{e} - \widehat{\frac{\partial \mathbf{e}}{\partial t}} \qquad [6.97]$$

As with the rigid link seen in 6.3.2, it is much better to choose $\mathbf{C} = \mathbb{I}_m$ if the visual features are not redundant ($k = m$). We should point out by the way that it is much less difficult to select non-redundant 2-D visual features for links other than the rigid link (see for example the case of the bearing link described in the previous section). By choosing $\mathbf{C} = \mathbb{I}_m$, Condition [6.96] can be expressed simply as $\mathbf{L_s} \mathbf{W}^+ > 0$, which is respected if the estimate used to build $\mathbf{W}$ is not too coarse. We will also have an exponential decrease for each component of s. In the case where the visual features are redundant ($k > m$), we can choose:

$$\mathbf{C} = \mathbf{W} \widehat{\mathbf{L_s}}^+_{|\mathbf{s}=\mathbf{s}^*} \qquad [6.98]$$

If $\mathbf{W}$ is also built from $\widehat{\mathbf{L_s}}_{|\mathbf{s}=\mathbf{s}^*}$ (and therefore constant), Condition [6.96] is expressed as:

$$\mathbf{W} \widehat{\mathbf{L_s}}^+_{|\mathbf{s}=\mathbf{s}^*} \mathbf{L_s} \mathbf{W}^+ > 0 \qquad [6.99]$$

which at best can only be satisfied in the neighborhood of the configurations such that $\mathbf{s} = \mathbf{s}^*$. In practice, it is also possible to consider matrices $\mathbf{W}$ and $\mathbf{C}$ calculated at each iteration from an estimate of the current value of the interaction matrix, but this choice no longer allows the analytical form of $\mathbf{L_{e_1}}$ to be easily determined.

Finally, if the secondary task makes it possible to know $\frac{\partial \mathbf{e_2}}{\partial t}$, we can choose:

$$\widehat{\frac{\partial \mathbf{e}}{\partial t}} = \mathbf{W}^+ \widehat{\frac{\partial \mathbf{e_1}}{\partial t}} + (\mathbb{I}_n - \mathbf{W}^+\mathbf{W}) \frac{\partial \mathbf{e_2}}{\partial t} \qquad [6.100]$$

where the term $\widehat{\frac{\partial \mathbf{e}_1}{\partial t}}$, if it is properly estimated, allows possible tracking errors to be reduced if the object being observed is moving.

Figure 6.21 illustrates an example of hybrid tasks. We considered the case of positioning a camera with respect to a sphere, thus creating a bearing link as we saw in section 6.3.3.1. The secondary task corresponds to a movement at a constant speed of the three components of the angular velocity. Notice in Figure 6.21e the exponential decrease of the three components of $\mathbf{s}$ (the matrices $\mathbf{W}$ and $\mathbf{C}$ are calculated at each iteration of the control law). Notice also in Figure 6.21f that the secondary task is only properly achieved after the convergence of $\mathbf{s}$ to $\mathbf{s}^*$ (except for the ω_z component which is not involved in the convergence). Finally, note on this same figure that the projection operator $\mathbb{I}_n - \mathbf{W}^+\mathbf{W}$ induces translational velocities v_x and v_y to compensate for the angular velocities ω_y and ω_x caused by the secondary task, and thus preserves the sphere's image as a centered circle.

6.3.4. *Target tracking*

In this short section, we consider the case of an eye-in-hand system tracking a mobile object. However, the principles described below are also valid for an eye-to-hand system where the camera is moving.

A significant part of the studies conducted in visual servoing that deal with target tracking consider the object's motion as a perturbation, that must be eliminated as quickly and efficiently as possible [PAP 93, GAN 02]. Other studies use prior information on the trajectory or the type of movement of the object [ALL 93, HAS 95, RIZ 96, GIN 05]. Furthermore, the use of an integrator is very common in control theory to eliminate tracking errors. Let $\mathbf{I}_k$ be the estimate of $\frac{\partial \mathbf{e}_1}{\partial t}$ at the iteration k. We then have:

$$\begin{aligned} \mathbf{I}_{k+1} &= \mathbf{I}_k + \mu\, \mathbf{e}_{1k} \;\text{ with }\; \mathbf{I}_0 = 0 \\ &= \mu \sum_{j=0}^{k} \mathbf{e}_{1j} \end{aligned} \qquad [6.101]$$

where μ is the gain of the integrator. This technique can only work properly in cases where the object has a constant speed, since we have $\mathbf{I}_{k+1} = \mathbf{I}_k$ if and only if $\mathbf{e}_{1k} = 0$. This implies that the tracking errors are not completely eliminated if the object's motion is more complex.

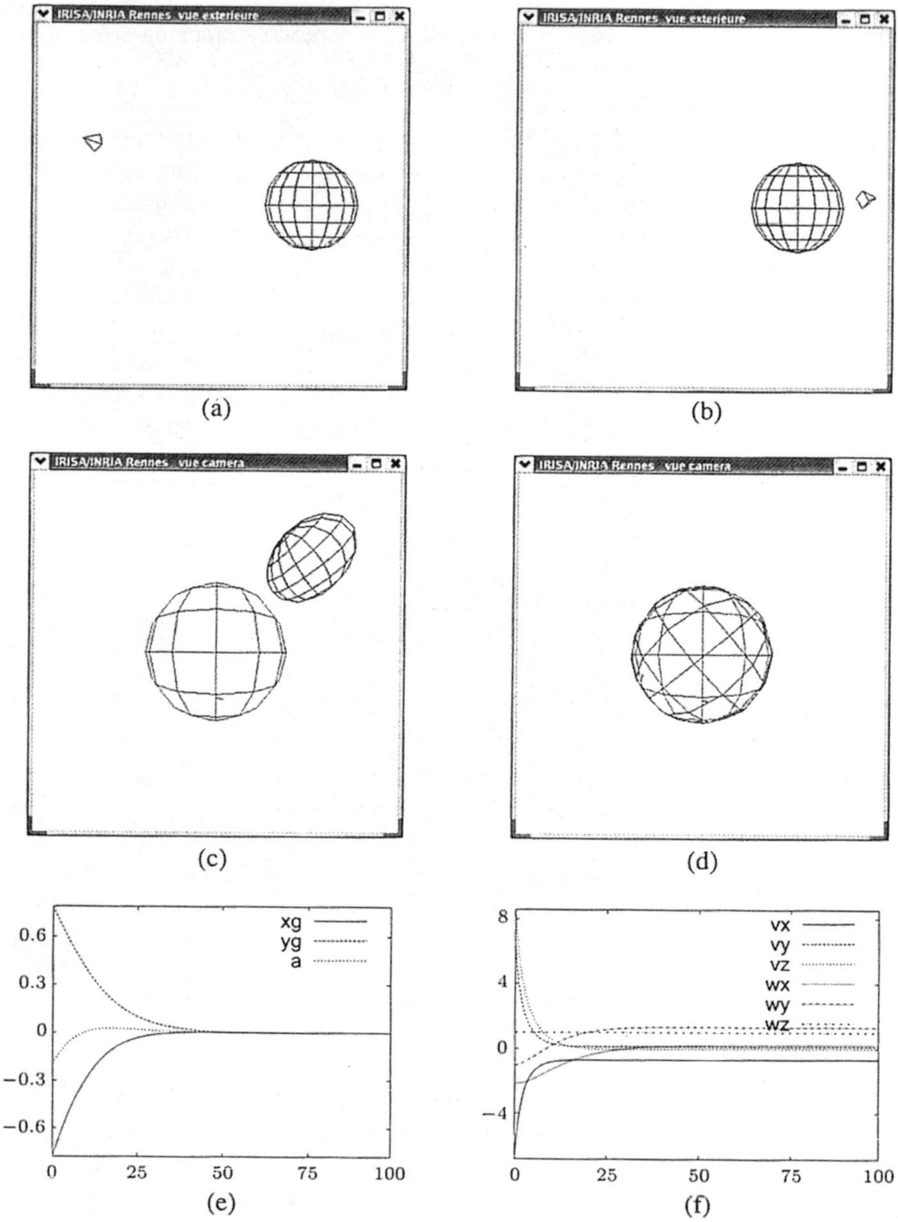

Figure 6.21. *Positioning with respect to a sphere then turning around this sphere: (a) camera's initial pose with respect to the sphere, (b) camera's final pose, (c) superposition of the initial image and of the desired image, (d) superposition of the final image and of the desired image, (e) components of* $\mathbf{s}$ *calculated at each iteration, (f) components of* $\mathbf{v}_c$ *calculated at each iteration*

Other approaches consist of estimating as reliably as possible the object's speed in the image. Indeed, if it is possible to measure the camera's velocity, an estimate of the object's speed is given by:

$$\widehat{\frac{\partial \mathbf{e}_1}{\partial t}} = \widehat{\dot{\mathbf{e}}_1} - \widehat{\mathbf{L}_{\mathbf{e}_1}} \mathbf{v_c} \qquad [6.102]$$

where $\widehat{\dot{\mathbf{e}}_1}$ is for example measured at the iteration k by $\widehat{\dot{\mathbf{e}}_1}_k = (\mathbf{e}_{1k} - \mathbf{e}_{1k-1})/\Delta t$, Δt being the system's sampling period. This leads us to an indirect adaptive control scheme and we can then use a Kalman filter (for example) to smooth this estimate. In [COR 93], such a filter based on a simple state model at a constant speed is presented. In [CHA 93b], a constant acceleration and correlated noise model was chosen. Finally, what is referred to as the GLR algorithm (Generalized Likelihood Ratio) is used in [BENS 95], to detect, estimate and compensate for possible abrupt changes in the object's motion.

6.4. Other exteroceptive sensors

All of the rules described in this chapter are valid for any exteroceptive sensor. The only characteristic involves the modeling of the interaction matrix between the considered sensor and its environment [SAM 91].

Consider for example a narrow field proximity sensor that provides the distance Z between this sensor and the closest object in the sensor's direction (see Figure 6.22). If we assume that the object's surface is perpendicular to the sensor's axis, the interaction matrix related to Z is given by:

$$\mathbf{L}_Z = \begin{bmatrix} 0 & 0 & -1 & 0 & 0 & 0 \end{bmatrix} \qquad [6.103]$$

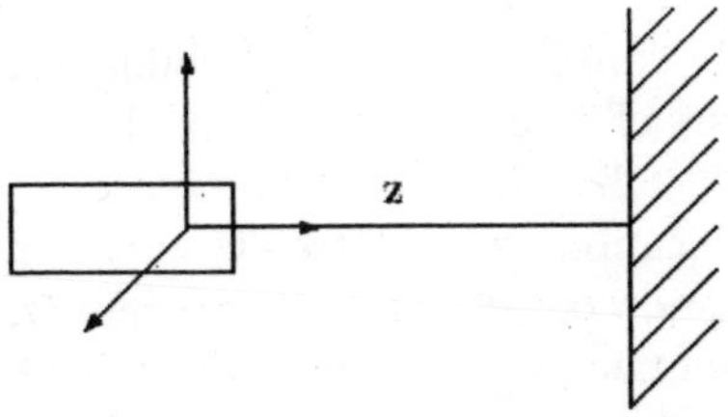

Figure 6.22. *Modeling of a proximity sensor*

Various robotic tasks can then be performed, such as obstacle avoidance or most of the virtual links that we saw previously, by selecting the appropriate number and direction of sensors. For example, a possible choice for creating a plane-to-plane contact is the configuration shown in Figure 6.23. Expressed in the frame of reference $R_O = (O, \vec{x}, \vec{y}, \vec{z})$, the interaction matrix $\mathbf{L}_{Z_i}$ related to each sensor S_i is given by (see [6.6]):

$$\mathbf{L}_{Z_i} = \begin{bmatrix} 0 & 0 & -1 & 0 & 0 & 0 \end{bmatrix} \begin{bmatrix} \mathbb{I}_3 & -[\mathbf{X}_i]_\times \\ 0 & \mathbb{I}_3 \end{bmatrix} \quad [6.104]$$

so in the end:

$$\mathbf{L}_{Z_i} = \begin{bmatrix} 0 & 0 & -1 & -Y_i & X_i & 0 \end{bmatrix} \quad [6.105]$$

where $\mathbf{X}_i = (X_i, Y_i, 0)$ are the coordinates of S_i in R_O. This shows that the interaction matrix that integrates the four sensors has rank 3 and its kernel is:

$$\mathcal{S}^* = \begin{bmatrix} 1 & 0 & 0 \\ 0 & 1 & 0 \\ 0 & 0 & 0 \\ 0 & 0 & 0 \\ 0 & 0 & 0 \\ 0 & 0 & 1 \end{bmatrix} \quad [6.106]$$

The use of force sensors has also been the subject of many studies [SAM 91]. Finally, we can mention [KHA 96] where the considered sensor is comprised of a camera rigidly attached to laser planes.

6.5. Conclusion

In this chapter, we only considered the most common case, that is controlling a system that has six degrees of freedom, for which the selection of relevant visual features to perform the task is the most difficult problem, especially if the rigid link is the objective. Many simplifications occur if the system has fewer degrees of freedom. As an example, the orientation control of a camera mounted on a pan-tilt head, for a centering or mobile object tracking task, presents no modeling difficulties. No 3-D features are even necessary. The main difficulty involves the image processing aspects in order to deal with real, complex objects.

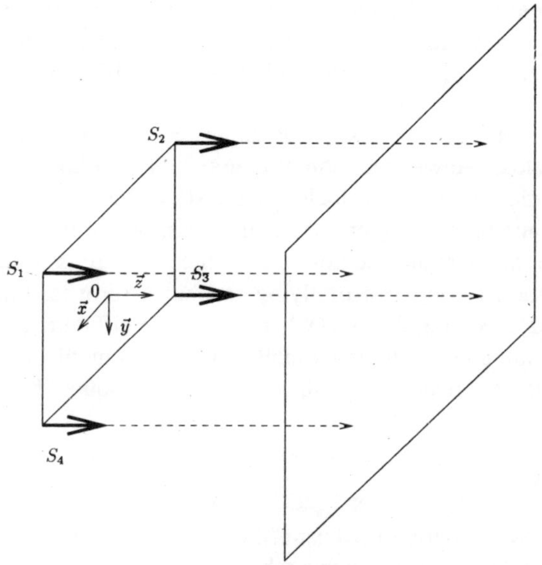

Figure 6.23. *Planar rest link*

For many years, visual servoing techniques could actually only be applied to extremely simple objects (usually producing binary images), because of the slow time processing for complex images. Advances in algorithms, particularly in the field of image motion analysis, but most of all the enormous gains in computing power, have made it possible to consider real, video-rate applications [CRE 00, CRE 01b, TON 97, VIN 00, COM 04]. We should also mention studies where the visual features are no longer geometric, but instead describe a motion in the image sequence [GRO 95, CRE 01a].

The major advances to come in visual servoing involve significant progress needed in determining the optimal visual features and taking into account unknown objects, that do not require prior 3-D knowledge. The studies in the field of projective geometry have already shown promising results [MAL 00, RUF 99]. In practice, if the learning of the desired image is impossible, obtaining the desired value to be reached when working with a coarsely calibrated system can be quite difficult. Furthermore, aspects of trajectory planning in the sensor's space [HOS 95b] are currently the subject of successful studies [MEZ 02, COW 02, ZAN 04]. With the same idea, visual task sequencing would deserve to be developed further. It would then be possible to broaden the very local aspect of the current techniques. Likewise, merging measurements provided by several exteroceptive sensors, which may or may not be of different types, inside the control scheme should make it possible to deal with many new applications [CAD 00, MAL 01].

Also, the development of omni-directional vision sensors [BAK 99, GAS 00] paves the way for many studies in the field of visual features modeling, given how peculiar and complex its projection model is [BAR 02, MEZ 04].

Finally, the use of visual servoing has expanded considerably over the past few years in several fields. We will mention of course mobile robotics, particularly applications in automatic driving of vehicles [DIC 91, PIS 95, KHA 98, TSAK 98]. In addition to the image processing problems, the main difficulty lies in designing control laws that take into account the non-holonomic constraints of this type of robots. We can also mention the control of flying machines (drones, blimps, helicopters) [ZHA 99, HAM 02, ZWA 02, RIV 04] where one of the difficulties come from the fact that these aircrafts are under-actuated. Finally, the field of medical robotics is showing great promise for the application of visual servoing techniques [KRU 03, GIN 05].

6.6. Bibliography

[ALL 93] ALLEN P.K., YOSHIMI B., TIMCENKO A., MICHELMAN P., "Automated tracking and grasping of a moving object with a robotic hand-eye system", *IEEE Transactions on Robotics and Automation*, vol. 9, no. 2, p. 152-165, 1993.

[AND 02] ANDREFF N., ESPIAU B., HORAUD R., "Visual servoing from lines", *International Journal of Robotics Research*, vol. 21, no. 8, p. 679-700, 2002.

[BAK 99] BAKER S., NAYAR S.K., "A theory of single-viewpoint catadioptric image formation", *International Journal of Computer Vision*, vol. 35, no. 2, p. 1-22, 1999.

[BAR 02] BARRETO J.P., MARTIN F., HORAUD R., "Visual servoing/tracking using central catadioptric images", *Eighth International Symposium on Experimental Robotics (ISER'02)*, Bombay, p. 863-869, 2002.

[BEN 03] BENHIMANE S., MALIS E., "Vision-based control with respect to planar and non-planar objects using a zooming camera", *IEEE International Conference on Advanced Robotics (ICAR'03)*, Coimbra, p. 991-996, 2003.

[BENS 95] BENSALAH F., CHAUMETTE F., "Compensation of abrupt motion changes in target tracking by visual servoing", *IEEE/RSJ International Conference on Intelligent Robots and Systems (IROS'95)*, Pittsburgh, p. 181-187, 1995.

[BER 00] BERRY F., MARTINET P, GALLICE J., "Turning around an unknown object using visual servoing", *IEEE/RSJ International Conference on Intelligent Robots and Systems (IROS'00)*, Takamatsu, p. 257-262, 2000.

[BEY 92] BEYER H., "Accurate calibration of CCD cameras", *IEEE International Conference on Computer Vision and Pattern Recognition (CVPR'92)*, Urbana Champaign, p. 96-101, 1992.

[CAD 00] CADENAT V., SOUERES P., COURDESSES M., "Using redundancy to perform a vision-based task amidst obstacles", *Sixth International IFAC Symposium on Robot Control (SYROCO'00)*, Vienna, p .205-210, 2000.

[CHA 93a] CHAUMETTE F., RIVES P., ESPIAU B., "Classification and realization of the different vision-based tasks", In [HAS 93a], p. 199-228, 1993.

[CHA 93b] CHAUMETTE F., SANTOS A., "Tracking a moving object by visual servoing", *Twelfth World Congress IFAC*, Sydney, vol. 9, p. 643-648, 1993.

[CHA 94] CHAUMETTE F., "Visual servoing using image features defined upon geometrical primitives", *Thirty-third IEEE Conference on Decision and Control (CDC'94)*, Orlando, vol. 4, p. 3782-3787, 1994.

[CHA 96] CHAUMETTE F., BOUKIR S., BOUTHEMY P., JUVIN D., "Structure from controlled motion", *IEEE Transactions on Pattern Analysis and Machine Intelligence*, vol. 18, no. 5, p. 492-504, 1996.

[CHA 98] CHAUMETTE F., "Potential problems of stability and convergence in image-based and position-based visual servoing", In *The Confluence of Vision and Control*, LNCIS, vol. 237, p. 66-78, Springer-Verlag, 1998.

[CHA 00] CHAUMETTE F., MALIS E., "2 1/2 D visual servoing: A possible solution to improve image-based and position-based visual servoings", *IEEE International Conference on Robotics and Automation (ICRA'00)*, San Francisco, vol. 1, p. 630-635, 2000.

[CHA 04] CHAUMETTE F., "Image moments: A general and useful set of features for visual servoing", *IEEE Transactions on Robotics*, vol. 20, no. 4, p. 713-723, 2004.

[CIP 97] CIPOLLA R., HOLLINGHURST N., "Visually guided grasping in unstructured environment", *Robotics and Autonomous Systems*, vol. 19, p. 337-346, 1997.

[COL 00] COLLEWET C., CHAUMETTE F., "A contour approach for image-based control of objects with complex shape", *IEEE/RSJ International Conference on Intelligent Robots and Systems (IROS'00)*, Takamatsu, p. 751-756, 2000.

[COL 02] COLLEWET C., CHAUMETTE F., "Positioning a camera with respect to planar objects of unknown shape by coupling 2D visual servoing and 3D estimations", *IEEE Transactions on Robotics and Automation*, vol. 18, no. 3, p.322-333, 2002.

[COLO 99] COLOMBO C., ALLOTTA B., "Image-based robot task planning and control using a compact visual representation", *IEEE Transactions on Systems, Man and Cybernetics*, vol. A-29, no. 1, p. 92-100, 1999.

[COM 04] COMPORT A., MARCHAND E., CHAUMETTE F., "Robust model-based tracking for robot vision", *IEEE/RSJ International Conference on Intelligent Robots and Systems (IROS'04)*, Sendai, vol. 1, p. 692-697, 2004.

[COR 93] CORKE P., GOODS M., "Controller design for high performance visual servoing", *Twelfth World Congress IFAC*, Sydney, vol. 9, p. 395-398, 1993.

[COR 01] CORKE P., HUTCHINSON S., "A new partitioned approach to image-based visual servo control", *IEEE Transactions on Robotics and Automation*, vol. 17, no. 4, p. 507-515, 2001.

[COS 95] COSTE-MANIÈRE E., COUVIGNOU P., KHOSLA P.K., "Visual servoing in the task-function framework: A contour following task", *Journal of Intelligent and Robotic Systems*, vol. 12, p. 1-21, 1995.

[COW 02] COWAN N., WEINGARTEN J., KODITSCHEK D., "Visual servoing via navigation functions", *IEEE Transactions on Robotics and Automation*, vol. 18, no. 4, p. 521-533, 2002.

[CRE 00] CRÉTUAL A., CHAUMETTE F., "Dynamic stabilization of a pan and tilt camera for sub-marine image visualization", *Computer Vision and Image Understanding*, vol. 79, no. 1, p. 47-65, 2000.

[CRE 01a] CRÉTUAL A., CHAUMETTE F., "Visual servoing based on image motion", *International Journal of Robotics Research*, vol. 20, no. 11, p. 857-877, 2001.

[CRE 01b] CRÉTUAL A., CHAUMETTE F., "Application of motion-based visual servoing to target tracking", *International Journal of Robotics Research*, vol. 20, no. 11, p. 878-890, 2001.

[DEG 97] DEGUCHI K., "Direct interpretation of dynamic images and camera motion for visual servoing without image feature correspondence", *Journal of Robotics and Mechatronics*, vol. 9, no. 2, p. 104-110, 1997.

[DEM 95] DEMENTHON D., DAVIS L.S., "Model-based object pose in 25 lines of code", *International Journal of Computer Vision*, vol. 15, no. 1/2, p. 123-141, 1995.

[DEN 02] DENG L., JANABI-SHARIFI F., WILSON W., "Stability and robustness of visual servoing methods", *IEEE International Conference on Robotics and Automation (ICRA'02)*, Washington DC, p. 1604-1609, 2002.

[DHO 89] DHOME M., RICHETIN M., LAPRESTE J.T., RIVES G., "Determination of the attitude of 3D objects from a single perspective view", *IEEE Transactions on Pattern Analysis and Machine Intelligence*, vol. 11, no. 12, p. 1265-1278, 1989.

[DHO 90] DHOME M., LAPRESTE J.T., RIVES G., RICHETIN M., "Spatial localization of modelled objects of revolution in monocular perspective vision", *European Conference on Computer Vision (ECCV'90)*, Antibes, LNCS, vol. 427, p. 475-485, Springer-Verlag, 1990.

[DIC 91] DICKMANNS E., CHRITIANS T., "Relative 3D state estimation for autonomous visual guidance of road vehicles", *Robotics and Autonomous Systems*, vol. 7, no. 2/3, p. 85-98, 1991.

[DRU 99] DRUMMOND T., CIPOLLA R., "Visual tracking and control using Lie algebra", *IEEE International Conference on Computer Vision and Pattern Recognition (CVPR'99)*, Fort Collins, vol. 2, p. 652-657, 1999.

[ESP 92] ESPIAU B., CHAUMETTE F., RIVES P., "A new approach to visual servoing in robotics", *IEEE Transactions on Robotics and Automation*, vol. 8, no. 6, p. 313-326, 1992.

[ESP 93] ESPIAU B., "Effect of camera calibration errors on visual servoing in robotics", *International Symposium on Experimental Robotics (ISER'93)*, Kyoto, 1993.

[FAU 93] FAUGERAS O., *Three-Dimensionnal Computer Vision: A Geometric Viewpoint*, MIT Press, Cambridge, 1993.

[FED 89] FEDDEMA J., LEE C., MITCHELL O., "Automatic selection of image features for visual servoing of a robot manipulator", *IEEE International Conference on Robotics and Automation (ICRA'89)*, Scottsdale, vol. 2, p. 832-837, 1989.

[GAN 02] GANGLOFF J., DE MATHELIN M., "Visual servoing of a 6 DOF manipulator for unknown 3D profile following", *IEEE Transactions on Robotics and Automation*, vol. 18, no. 4, p. 511-520, 2002.

[GAS 00] GASPAR J., WINTERS N., SANTOS-VICTOR J., "Vision-based navigation and environmental representations with an omni-directional camera", *IEEE Transactions on Robotics and Automation*, vol. 16, no. 6, p. 890-898, 2000.

[GIN 05] GINHOUX R., GANGLOFF J., DE MATHELIN M., SOLER L., ARENAS SANCHEZ M., MARESCAUX J., "Active filtering of physiological motion in robotized surgery using predictive control", *IEEE Transactions on Robotics*, vol. 21, no. 1, p. 67-79, 2005.

[GRO 95] GROSSO E., TISTARELLI M., "Active/dynamic stereo vision", *IEEE Transactions on Pattern Analysis and Machine Intelligence*, vol. 17, no. 11, p. 1117-1127, 1995.

[HAG 95] HAGER G., CHANG W., MORSE A.S., "Robot feedback control based on stereo vision: Towards calibration-free hand-eye coordination", *IEEE Control Systems Magazine*, vol. 15, no. 1, p. hbox30-39, 1995.

[HAM 02] HAMEL T., MAHONY R., "Visual servoing of an under-actuated dynamic rigid body system: An image-based approach", *IEEE Transactions on Robotics and Automation*, vol. 18, no. 2, p. 187-198, 2002.

[HAR 89] HARALICK R., JOO H., LEE C.N., ZHUANG X., VAIDYA V., KIM M., "Pose estimation from corresponding point data", *IEEE Transactions on Systems, Man and Cybernetics*, vol. 19, no. 6, p. 1426-1445, 1989.

[HAS 93a] HASHIMOTO K. (ed.), *Visual Servoing: Real Time Control of Robot Manipulators Based on Visual Sensory Feedback*, World Scientific Series in Robotics and Automated Systems, vol. 7, World Scientific Press, Singapore, 1993.

[HAS 93b] HASHIMOTO K., KIMURA H., "LQ Optimal and non linear approaches to visual servoing", In [HAS 93a], p. 165-198, 1993.

[HAS 95] HASHIMOTO K., KIMURA H., "Visual servoing with non linear observer", *IEEE International Conference on Robotics and Automation (ICRA'95)*, Nagoya, p. 484-489, 1995.

[HAS 96] HASHIMOTO K., EBINE T., KIMURA H., "Visual servoing with hand-eye manipulator-optimal control approach", *IEEE Transactions on Robotics and Automation*, vol. 12, no. 5, p. 766-774, 1996.

[HOR 89] HORAUD R., CONIO B., LEBOULLEUX O., LACOLLE B., "An analytic solution for the perspective 4-points problem", *Computer Vision, Graphics, and Image Processing*, vol. 47, no. 1, p. 33-44, 1989.

[HOR 95] HORAUD R., DORNAIKA F., "Hand-eye calibration", *International Journal of Robotics Research*, vol. 14, no. 3, p. 195-210, 1995.

[HOR 98] HORAUD R., DORNAIKA F., ESPIAU B., "Visually guided object grasping", *IEEE Transactions on Robotics and Automation*, vol. 14, no. 4, p. 525-532, 1998.

[HOS 94] HOSODA K., ASADA M., "Versatile visual servoing without knowledge of true jacobian", *IEEE/RSJ International Conference on Intelligent Robots and Systems (IROS'94)*, Munich, 1994.

[HOS 95a] HOSODA K., MORIYAMA H., ASADA M., "Visual servoing utilizing zoom mechanism", *IEEE International Conference on Robotics and Automation (ICRA'95)*, Nagoya, 1995.

[HOS 95b] HOSODA K., SAKAMOTO K., ASADA M., "Trajectory generation for obstacle avoidance of uncalibrated stereo visual servoing without 3D reconstruction", *IEEE/RSJ International Conference on Intelligent Robots and Systems (IROS'95)*, Pittsburgh, vol. 3, p.29-34, 1995.

[HUT 96] HUTCHINSON S., HAGER G., CORKE P., "A tutorial on visual servo control", *IEEE Transactions on Robotics and Automation*, vol. 12, no. 5, p. 651-670, 1996.

[IWA 05] IWATSUKI M., OKIYAMA N., "A new formulation of visual servoing based on cylindrical coordinate system", *IEEE Transactions on Robotics*, vol. 21, no. 2, p. 266-273, 2005.

[JAG 97] JÄGERSAND M., FUENTES O., NELSON R., "Experimental evaluation of uncalibrated visual servoing for precision manipulation", *IEEE International Conference on Robotics and Automation (ICRA'97)*, Albuquerque, vol. 3, p. 2874-2880, 1997.

[KEL 96] KELLY R., "Robust asymptotically stable visual servoing of planar robots", *IEEE Transactions on Robotics and Automation*, vol. 12, no. 5, p. 759-766, 1996.

[KHA 96] KHADRAOUI D., MOTYL G., MARTINET P., GALLICE J., CHAUMETTE F., "Visual servoing in robotics scheme using a camera/laser-stripe sensor", *IEEE Transactions on Robotics and Automation*, vol. 12, no. 5, p. 743-750, 1996.

[KHA 98] KHADRAOUI D., ROUVEURE R., DEBAIN C., MARTINET P., BONTON P., GALLICE J., "Vision-based control in driving assistance of agricultural vehicles", *International Journal of Robotics Research*, vol. 17, no. 10, p. 1040-1054, 1998.

[KHAL 02] KHALIL W., DOMBRE E., *Modeling, identification and control of robots*, Hermes Penton Sciences, London, 2002.

[KIN 94] KINOSHITA K., DEGUCHI K., "Simultaneous determination of camera pose and intrinsic parameters by visual servoing", *International Conference on Pattern Recognition (ICPR'94)*, Jerusalem, vol. 1, p. 285-290, 1994.

[KRA 05] KRAGIC D., CHRISTENSEN H., "Special issue on advances in robot vision", *Robotics and Autonomous System*, vol. 52, no. 1, 2005.

[KRU 03] KRUPA A., GANGLOFF J., DOIGNON C., DE MATHELIN M., MOREL G., LEROY J., SOLER L., MARESCAUX J., "Autonomous 3-D positioning of surgical instruments in robotized laparoscopic surgery using visual servoing", *IEEE Transactions on Robotics and Automation*, vol. 19, no. 5, p. 842-853, 2003.

[LAP 04] LAPRESTÉ J.-T., JURIE F., CHAUMETTE F., "An efficient method to compute the inverse jacobian matrix in visual servoing", *IEEE International Conference on Robotics and Automation (ICRA'04)*, New Orleans, vol. 1, p. 727-732, 2004.

[LOW 87] LOWE D.G., "Three-dimensional object recognition from single two-dimensional images", *Artificial Intelligence*, vol. 31, p. 355-394, 1987.

[MA 93] DE MA S., "Conics-based stereo, motion estimation, and pose determination", *International Journal of Computer Vision*, vol. 10, no. 1, p. 7-25, 1993.

[MA 03] MA Y., SOATTO S., KOSECKA J., SASTRY S., *An Invitation to 3-D Vision: From Images to Models*, Springer-Verlag, 2003.

[MAL 99] MALIS E., CHAUMETTE F., BOUDET S., "2 1/2 D visual servoing", *IEEE Transactions on Robotics and Automation*, vol. 15, no. 2, p. 238-250, 1999.

[MAL 00] MALIS E., CHAUMETTE F., "2 1/2 D visual servoing with respect to unknown objects through a new estimation scheme of camera displacement", *International Journal of Computer Vision*, vol. 37, no. 1, p. 79-97, 2000.

[MAL 01] MALIS E., MOREL G., CHAUMETTE F., "Robot control from disparate multiple sensors", *International Journal of Robotics Research*, vol. 20, no. 5, p. 364-378, 2001.

[MAL 02] MALIS E., CHAUMETTE F., "Theoretical improvements in the stability analysis of a new class of model-free visual servoing methods", *IEEE Transactions on Robotics and Automation*, vol. 18, no. 2, p. 176-186, 2002.

[MAL 04] MALIS E., "Improving vision-based control using efficient second-order minimization techniques", *IEEE International Conference on Robotics and Automation (ICRA'04)*, New Orleans, p. 1843-1848, 2004.

[MAR 96] MARCHAND E., CHAUMETTE F., RIZZO A., "Using the task function approach to avoid robot joint limits and kinematic singularities in visual servoing", *IEEE/RSJ International Conference on Intelligent Robots and Systems (IROS'96)*, Osaka, vol. 3, p. 1083-1090, 1996.

[MAR 98] MARCHAND E., HAGER G., "Dynamic sensor planning in visual servoing", *IEEE International Conference on Robotics and Automation (ICRA'98)*, Leuven, vol. 3, p. 1988-1993, 1998.

[MART 96] MARTINET P., GALLICE J., KHADRAOUI D., "Vision-based control law using 3D visual features", *World Automation Congress (WAC'96)*, Montpellier, vol. 3, p. 497-502, 1996.

[MART 97] MARTINET P., DAUCHER N., GALLICE J., DHOME M., "Robot control using monocular pose estimation", *Workshop on New Trends in Image-based Robot Servoing (IROS'97)*, Grenoble, p. 1-12, 1997.

[MEZ 02] MEZOUAR Y., CHAUMETTE F., "Path planning for robust image-based control", *IEEE Transactions on Robotics and Automation*, vol. 18, no. 4, p. 534-549, 2002.

[MEZ 04] MEZOUAR Y., HAJ ABDELKADER H., MARTINET P., CHAUMETTE F., "Central catadioptric visual servoing from 3D straight lines", *IEEE/RSJ International Conference on Intelligent Robots and Systems (IROS'04)*, Sendai, vol. 1, p. 343-349, 2004.

[MIC 93] MICHEL H., RIVES P., Singularities in the determination of the situation of a robot effector from the perspective view of three points, INRIA Research Report, no. 1850, 1993.

[MOR 00] MOREL G., LEIBEZEIT T., SZEWCZYK J., BOUDET S., POT J., "Explicit incorporation of 2D constraints in vision-based control of robot manipulators", *International Symposium on Experimental Robotics (ISER'00)*, Sydney, LNCIS, vol. 250, p. 99-108, Springer-Verlag, 2000.

[NAY 96] NAYAR S.K., NEME S.A., MURASE H., "Subspace methods for robot vision", *IEEE Transactions on Robotics and Automation*, vol. 12, no. 5, p. 750-758, 1996.

[NEL 94a] NELSON B., KHOSLA P.K., "Integrating sensor placement and visual tracking strategies", *IEEE International Conference on Robotics and Automation (ICRA'94)*, p. 1351-1356, 1994.

[NEL 94b] NELSON B., KHOSLA P.K., "The resolvability ellipsoid for visual servoing", *IEEE International Conference on Computer Vision and Pattern Recognition (CVPR'94)*, Seattle, p. 829-832, 1994.

[NEL 95] NELSON B., KHOSLA P.K., "Strategies for increasing the tracking region of an eye-in-hand system by singularity and joint limits avoidance", *International Journal of Robotics Research*, vol. 14, no. 3, p. 255-269, 1995.

[PAP 93] PAPANIKOLOPOULOS N., KHOSLA P.K., KANADE T., "Visual tracking of a moving target by a camera mounted on a robot: A combination of control and vision", *IEEE Transactions on Robotics and Automation*, vol. 9, no. 1, p. 14-35, 1993.

[PAP 95] PAPANIKOLOPOULOS N., "Selection of features and evaluation of visual measurements during robotic visual servoing tasks", *Journal of Intelligent and Robotics Systems*, vol. 13, p. 279-304, 1995.

[PHO 95] PHONG T.Q., HORAUD R., YASSINE A., TAO P.D., "Object pose from 2D to 3D point and line correspondences", *International Journal of Computer Vision*, vol. 15, no. 3, p. 225-243, 1995.

[PIE 04] PIEPMEIER J., MC MURRAY G., LIPKIN H., "Uncalibrated dynamic visual servoing", *IEEE Transactions on Robotics and Automation*, vol. 20, no. 1, p. 143-147, 2004.

[PIS 95] PISSARD-GIBOLLET R., RIVES P., "Applying visual servoing techniques to control a mobile hand-eye system", *IEEE International Conference on Robotics and Automation (ICRA'95)*, Nagoya, 1995.

[REY 98] REYES F., KELLY R., "Experimental evaluation of fixed-camera direct visual controllers on a direct-drive robot", *IEEE International Conference on Robotics and Automation (ICRA'98)*, Leuven, vol. 3, p. 2327-2332, 1998.

[RIV 04] RIVES P., AZINHEIRA J., "Linear structures following by an airship using vanishing points and horizon line in a visual servoing scheme", *IEEE International Conference on Robotics and Automation (ICRA'04)*, New Orleans, vol. 1, p. 255-260, 2004.

[RIZ 96] RIZZI A., KODITSCHEK D., "An active visual estimator for dexterous manipulation", *IEEE Transactions on Robotics and Automation*, vol. 12, no. 5, p. 697-713, 1996.

[RUF 99] RUF A., HORAUD R., "Rigid and articulated motion seen with an uncalibrated stereo rig", *International Conference on Computer Vision (ICCV'99)*, Corfou, 1999.

[SAF 92] SAFAEE-RAD R., TCHOUKANOV I., SMITH K.C., BENHABIB B., "Three-dimensional location estimation of circular features for machine vision", *IEEE Transactions on Robotics and Automation*, vol. 8, no. 5, p. 624-639, 1992.

[SAM 91] SAMSON C., LE BORGNE M., ESPIAU B., *Robot Control: The Task Function Approach*, Clarendon Press, Oxford, 1991.

[SCH 04] SCHRAMM F., MOREL G., MICAELLI A., LOTTIN A., "Extended 2D visual servoing", *IEEE International Conference on Robotics and Automation*, New Orleans, p. 267-273, 2004.

[SHA 97] SHARMA R., HUTCHINSON S., "Motion perceptibility and its application to active vision-based servo control", *IEEE Transactions on Robotics and Automation*, vol. 13, no. 4, p. 607-616, 1997.

[SMI 94] SMITH C., PAPANIKOLOPOULOS N., "Computation of shape through controlled active exploration", *IEEE International Conference on Robotics and Automation (ICRA'94)*, San Diego, vol. 3, p. 2516-2521, 1994.

[SUH 93] SUH I.H., "Visual servoing of robot manipulators by fuzzy membership function based neural networks", In [HAS 93a], p. 285-315, 1993.

[TAH 05] TAHRI O., CHAUMETTE F., "Point-based and region-based image moments for visual servoing of planar objects", *IEEE Transactions on Robotics*, vol. 21, no. 6, 1116-1127, 2005.

[TON 97] TONKO M., SCHÄFER K., HEIMES F., NAGEL H.-H., "Towards visually servoed manipulation of car engine parts", *IEEE International Conference on Robotics and Automation (ICRA'97)*, Albuquerque, vol. 4, p. 3166-3171, 1997.

[TSA 87] TSAI R.Y., "A versatile camera calibration technique for high-accuracy 3D machine vision metrology using off-the-shelf TV cameras and lenses", *IEEE Journal on Robotics and Automation*, vol. 3, no. 4, p. 323-344, 1987.

[TSA 89] TSAI R.Y., LENZ R., "A new technique for fully autonomous and efficient 3D robotics hand-eye calibration", *IEEE Transactions on Robotics and Automation*, vol. 5, p. 345-358, 1989.

[TSAK 98] TSAKIRIS D., RIVES P., SAMSON C., "Extending visual servoing techniques to nonholonomic mobile robots", In *The Confluence of Vision and Control, LNCIS*, vol. 237, p. 106-117, Springer-Verlag, 1998.

[VIN 00] VINCZE M., HAGER G. (eds.), *Robust Vision for Vision-based Control of Motion*, IEEE Press, Piscataway, 2000.

[WEI 84] WEISS L.E., Dynamic Visual Servo Control of Robots. An Adaptive Image-based Approach, PhD Thesis, CMU-RI-TR-84-16, Carnegie Mellon University, 1984.

[WEL 96] WELLS G., VENAILLE C., TORRAS C., "Vision-based robot positioning using neural networks", *Image and Vision Computing*, vol. 14, p. 715-732, 1996.

[WIL 96] WILSON W., HULLS C., BELL G., "Relative end-effector control using cartesian position-based visual servoing", *IEEE Transactions on Robotics and Automation*, vol. 12, no. 5, p. 684-696, 1996.

[ZAN 04] ZANNE P., MOREL G., PLESTAN F., "A robust 3D vision based control and planning", *IEEE International Conference on Robotics and Automation (ICRA'04)*, New Orleans, p. 4423-4428, 2004.

[ZHA 99] ZHANG H., OSTROWSKI J., "Visual servoing with dynamics: Control of an unmanned blimp", *IEEE International Conference on Robotics and Automation (ICRA'99)*, Detroit, p. 618-623, 1999.

[ZHA 00] ZHANG Z. Y., "A flexible new technique for camera calibration", *IEEE Transactions on Pattern Analysis and Machine Intelligence*, vol. 22, no. 11, p. 1330-1334, 2000.

[ZWA 02] VAN DER ZWAAN S., BERNARDINO A., SANTOS-VICTOR J., "Visual station keeping for floating robots in unstructured environments", *Robotics and Autonomous Systems*, vol. 39, no. 3-4, p. 145-155, 2002.

Chapter 7

Modeling and Control of Flexible Robots

7.1. Introduction

The object of this chapter is to give an overview of the current state of the art regarding the modeling and control of flexible robot manipulators. The modeling method suggested in section 7.2 is devoted to control. It particularly leads to reduced complexity models as compared to their finite element competitors, more adapted to analysis than to control. Within this framework, the approach presented leads to the most competitive algorithms in term of calculation time and enables an easy tuning of the model resolution. Section 7.3 presents the control methods of flexible structures for regulation and tracking objectives in the joint space and operational space.

7.2. Modeling of flexible robots

7.2.1. *Introduction*

In the 1980s many studies dealt with the dynamic modeling of flexible multi-body systems and their control.

Chapter written by Frédéric BOYER, Wisama KHALIL, Mouhacine BENOSMAN and Georges LE VEY.

As far as modeling is concerned, among the most advanced developments we quote those of Simo [SIM 86], [SIM 88], Cardona and Géradin [CAR 88], Kim and Haug [KIM 88], Serna and Bayo [SER 89], Sharf and Damaren [SHA 92], Schwertassek *et al.* [SCH 99], etc. In the field of robotics, two types of systems justified these efforts. The first one concerns the production industry. These are the manufacturing robots for which there are efforts to increase their bandwidths in order to accelerate the work pace. In addition, progress made regarding the material and mechanical design fields make it possible now to lighten the structures considerably, so it is obvious that a future generation of high-speed industrial robots has to go through the control of their structural deformations. Nevertheless, lightening and flexibility of the industrial robots are limited by gravity. The second systems concerns the manipulator arms traveling within an orbit, they are by nature very flexible systems because of their extreme length and smoothness and due to the heavy loads that they handle (up to a few tons). To solve the problem caused by the dynamic modeling of these two types of systems, two great theories were born. The first, which is the most widely used [BOO 84], [BOY 96a], [BRE 97], [HUG 89], [MEI 91], [SIN 84], [SHA 90], [SHA 92], etc., is more particularly adapted to the modeling of high-speed industrial robots. Indeed, this theory, known as the "floating frame", regards the deformations of each isolated body as the result of linear disturbances of the overall rigid motions of a reference frame which, as the term "floating" indicates, cannot be attached to anything material [CAN 77].

In this theory, two great formulations distinguish the models, according to whether they are written in the formalism attributed to Lagrange [BOO 84], [CHE 90], [SIN 84] or "Newton-Euler" [BOY 96a], [HUG 89], [MEI 91], [SHA 90], [SHA 92]. It seems established today that the second formalism is the most efficient one, both in the field of flexible manipulators and of their rigid counterparts. Among other advantages of this formulation, we mention the easier refinement of the model's texture (in this respect we quote [BOY 00] which extends the method of the floating frame, until then restricted to the linear deformations, to the domain of quadratic disturbances), its algorithmic efficiency (recursive algorithms in $o(n)$ [BOY 95], [BOY 98], [DEL 92], [FIS 97], [VER 94]), its general nature (the Newton-Euler models lead to the Lagrangian models through the operation known as assembly [BOY 96b], [GLA 99]).

The second theory [CAR 88], [IBR 98], [SIM 86], [SIM 88], [BOY 04], [BOY 05] was initially developed with the aim to model the huge space structures, subjected to finite deformations, but the theory can equally model low amplitude vibratory phenomena. It is based on the use of Galilean references (not mobile like in the previous situation) and takes into account each finite overall motion attached to a beam. This approach belongs to the non-linear method of finite elements and is most often based on a Newmark implicit numeric integration [SIM 91] associated to an exact linearization [HUG 78] coupled to a Newton-Raphson algorithm. Finally, we must note that if the general approach is, from the point of view of the mechanic, "more rigorous" than the floating frame approach, once the fields of deformation are reduced through modal truncation, makes it possible to obtain models which are better adapted to control engineering applications (lower state dimension).

7.2.2. *Generalized Newton-Euler model for a kinematically free elastic body*

The objective of this section is to generalize the Newton-Euler equations for an elastic body denoted *S*, kinematically free and subject to external actions similar to those applied to an isolated body in an open robotic chain. For more details on the developments which will follow we invite the reader to refer to [BOY 94], [BOY 96a]. When such a model is sought, the following question arises: which principle of the dynamic balance is to be applied? The equations of Lagrange or the theorems of Newton and Euler? Actually we will see that neither of these two methods is naturally adapted to our objective, and that we will consequently have to apply a principle of the dynamic balance that unifies both of these two approaches: "the principle of virtual powers". In order to clarify this last point, we will firstly reconsider the fundamental difference between a Newton-Euler model and its Lagrangian counterpart. This difference is based on the concept of motion description formalism.

7.2.2.1. *Definition: formalism of a dynamic model*

The continuous medium mechanics proposes two formalisms to describe the motion of a solid material [GER 86]:

i) Lagrangian formalism: the motion is described by the knowledge of the successive positions occupied by each material particle. The motion is consequently described in terms of trajectories. The material particles are identified by their position in a reference configuration denoted Σ_o and usually chosen as the initial configuration of the motion. In this formalism the basic kinematic unknowns are unknowns of position in an infinite number for the general case (position field). Once certain constraints such as rigidity are considered, or once the reduction operation (modal, finite elements) is carried out, the position field is parameterized only by a finite set of time functions which take the meaning of generalized coordinates of Lagrange mechanics. Consequently, the Lagrangian model of a holonomous system (which is the case of a flexible manipulator) can be written as follows:

$$\mathbf{f}(\ddot{\mathbf{q}},\dot{\mathbf{q}},\mathbf{q},\mathbf{Q}) = \mathbf{0} \qquad [7.1]$$

where $\mathbf{q}$ is the vector of generalized coordinates (Lagrangian kinematic unknowns) and $\mathbf{Q}$ is the vector of external generalized forces applied on the system.

ii) Eulerian formalism: the motion is described knowing the velocity field over the current configuration $\Sigma(t)$ of the body at every instant t of the motion. Knowing this field makes it possible to obtain $\Sigma(t+dt)$ from $\Sigma(t)$ without an explicit reference to the past of the motion and particularly to the reference frame. Such a description can consequently be described as amnesic and it is contrary to the mnesic nature of the Lagrangian description. Attention, the mnesic or amnesic nature refers here only to the geometric condition and not to the internal condition of materials governed by their constitutive laws. In this formalism the basic kinematic unknowns are defined by the velocity field which, once the possible constraints are considered, can be reduced to a finite set of velocities such as, for example, the components of the velocity screw of a rigid body. Consequently, in such formalism, the dynamic model of the material body will be written:

$$\mathbf{f}(\dot{\mathbf{V}},\mathbf{V},\mathbf{\Phi},\mathbf{F})=\mathbf{0} \qquad [7.2]$$

where $\mathbf{V}$ represents the velocity field on the body, while [illegible] represents the dual quantities representing the external loads applied onto the body. The configuration of the system is represented by the symbol $\mathbf{\Phi}$ and is related to the velocity field through the so-called reconstruction relations, such as: $\dot{\mathbf{\Phi}}=\mathbf{h}(\mathbf{\Phi},\mathbf{V})$. We shall note finally that the Newton-Euler model of the rigid body, usually obtained from the theorems of Newton and Euler, is of the type [7.2], where the velocity field degenerates into a simple screw. Thanks to what has just been recalled, we are now ready to decouple the formalism of a dynamic model from the balance principle which generates it. Thus, model [7.1] applied to the rigid body can for example result from equations [7.2], once the unknown fields are replaced by their parameterized expression (by a set of generalized coordinates). Reciprocally, the form [7.2] can result, in the case of a rigid body, from Lagrange equations through the inverse change of kinematic unknowns.

7.2.2.2. *Choice of formalism*

Keeping in mind the previous definition, we notice that the Eulerian formalism is imposed by our initial objective, while the Lagrangian formalism is essential to describe the system's internal deformations (here reduced to an elastic body only). Indeed, the elasticity of the internal state of the body of a chain depends only on its current configuration and on a privileged configuration, called reference configuration, and naturally identified in its elastic static state. Thus, the description of elastic deformations is necessarily mnesic and the formalism adopted to describe it is Lagrangian. This opposition makes us separate the motion of each elastic body into two types of motions (this is the basis of the floating frame approach). The first type does not imply any deformation and consequently the corresponding motions are labeled as rigid. The second type refers to pure deformation motions, which are labeled as elastic.

Such a classification requires a prior selection of the reference of rigid motions. Multiple choices are possible. In this chapter we will deal only with "embedded references". Such a choice is based on the following physical reality. In the vicinity of the joint between a body and its predecessor in the chain, there always is a three-dimensional material area that does not get deformed. We will thus assign a reference frame $R(t)$ to this area, and the mobile configuration, i.e. embedded and denoted $\Sigma_o(t)$, will be defined as the "prolongation" of the rigid geometry of the body which resulted from this area (see Figure 7.1). Note that such a choice will force us to adopt modal bases that verify the condition of "embedding at the origin of the mobile reference frame $R(t)$ of the body". The continuous succession of $\Sigma_o(t)$ in space defines the rigid reference motion. We will call S_o the fictitious rigid body occupying at every moment the embedded configuration. As for the elastic motion, it will be identified with the relative motion of the actual body S compared to S_o. Once the classification of motions is done, we will adopt for each segment:

– the Eulerian formalism to describe its rigid reference motions;

– the Lagrangian formalism to describe its elastic motions.

It follows that the kinematic unknown quantities of the elastic body are obtained by a set of Eulerian unknown quantities (subscripted "*r*" as "rigid"), and by a set of Lagrangian unknown quantities (subscripted "*e*" as "elastic"), and that its dynamic equations are written as follows:

$$\mathbf{f}(\dot{\mathbf{V}}_r, \mathbf{V}_r, \ddot{\mathbf{q}}_e, \dot{\mathbf{q}}_e, \mathbf{\Phi}, \mathbf{F}_r, \mathbf{Q}_e) = \mathbf{0}$$

where $\mathbf{Q}_e$ represents the vector of the elastic generalized forces. Thus, such a model is neither Eulerian, nor Lagrangian, but both at the same time. Moreover, it enables the generalization of the Newton-Euler model for an elastic body because, by removing the elastic unknown quantities, it gives again the form [7.2] of the Newton-Euler model for a rigid body. Consequently, we will call this model "generalized Newton-Euler model". After these general comments, we will establish the kinematic model of the elastic body in this mixed formalism.

7.2.2.3. *Kinematic model of a free elastic body*

In order to lighten the developments, we will adopt the formalism of wrenches and screws, which will be noted by double characters. In this formalism, the Galilean screw field of the body S occupying the current configuration $\Sigma(t)$ is written as:

$$M \in \Sigma(t) \mapsto \underline{\mathbb{V}}(M) = \begin{Bmatrix} \underline{\mathbf{v}}(M) \\ \underline{\boldsymbol{\omega}}(M) \end{Bmatrix}$$

where "_" underlines either a vector of the space $\mathbb{R}^3$, or a vector matrix as a screw or a wrench. The vector $\underline{\mathbf{v}}(M)$ (respectively $\underline{\boldsymbol{\omega}}(M)$) represents the linear Galilean velocity (respectively the Galilean angular velocity) in M (M_o) (point of $\Sigma(t)$ image of $M_o \in \Sigma_o(t)$ (see Figure 7.1)). If we insert the motion of the reference body S_o, we have:

$$M \in \Sigma(t) \mapsto \underline{\mathbf{V}}(M) = \underline{\mathbf{V}}_r(M) + \underline{\mathbf{V}}_e(M) = \begin{Bmatrix} \underline{\mathbf{v}}_r + \underline{\boldsymbol{\omega}}_r \times \underline{\mathbf{r}}(M) \\ \underline{\boldsymbol{\omega}}_r \end{Bmatrix} + \begin{Bmatrix} \underline{\mathbf{v}}_e(M) \\ \underline{\boldsymbol{\omega}}_e(M) \end{Bmatrix} \quad [7.3]$$

where we used the rigidity property of S_o and chose the point A – the origin of the mobile frame $R(t)$ (and the embedding point) – as a reference frame of its velocity field. The vector $\underline{\mathbf{AM}}$ is denoted here $\underline{\mathbf{r}}(M)$. The velocities $\underline{\boldsymbol{\omega}}_r$ and $\underline{\mathbf{v}}_r$ create a set of Eulerian unknown quantities for the rigid component of the motion and are grouped together in the body reference screw:

$$\underline{\mathbf{V}}_r = \begin{Bmatrix} \underline{\mathbf{v}}_r \\ \underline{\boldsymbol{\omega}}_r \end{Bmatrix}$$

The fields of vectors $\underline{\mathbf{v}}_e$ and $\underline{\boldsymbol{\omega}}_e$ are respectively the fields of elastic linear velocity and elastic angular velocity. The hypothesis of elasticity makes it possible to totally parameterize the internal state of the solid by the position field applying S_o on S. This field is later reduced to a truncated basis of Rayleigh-Ritz functions [MEI 89]:

$$\underline{\mathbf{V}}_e(M) = \begin{Bmatrix} \underline{\mathbf{v}}_e(M) \\ \underline{\boldsymbol{\omega}}_e(M) \end{Bmatrix} = \sum_{k=1}^{N} \begin{Bmatrix} \underline{\boldsymbol{\Phi}}_{dk}(M_o) \\ \underline{\boldsymbol{\Phi}}_{rk}(M_o) \end{Bmatrix} \dot{q}_{ek} \quad [7.4]$$

where $\underline{\boldsymbol{\Phi}}_{dk}$ and $\underline{\boldsymbol{\Phi}}_{rk}$ are respectively the kth displacement and rotation form functions, and N is the total number of form functions. We shall note that the vector $\underline{\mathbf{r}}(M)$ appearing in [7.3] can be expanded as follows: $\underline{\mathbf{r}}(M) = \underline{\mathbf{r}}(M_o) + \sum_{k=1}^{N} \underline{\boldsymbol{\Phi}}_{dk}(M_o)\, q_{ek}$. Finally, we shall underline that in the three-dimensional case of a classic medium (non-Cosserat) the rotation form functions are not independent of the displacement form functions, but they are linked by the "curl operator":

$$\underline{\boldsymbol{\Phi}}_{rk}(M_o) = \frac{1}{2} \underline{\nabla} \times \underline{\boldsymbol{\Phi}}_{dk}(M_o)$$

where $\underline{\nabla}$ is the "nabla operator". In agreement with our choice of reference configurations, the form functions are supposed to be embedded in $A = A_o$, i.e. they verify:

$$\underline{\Phi}_{dk}(A) = \underline{\Phi}_{rk}(A) = \underline{0}$$

Equations [7.3] and [7.4] define the kinematic model of elastic body. This model is well parameterized by a set of Eulerian and Lagrangian kinematic unknown quantities: ($\underline{\omega}_r, \underline{V}_r, q_{e1}, q_{e2} .., q_{eN}$).

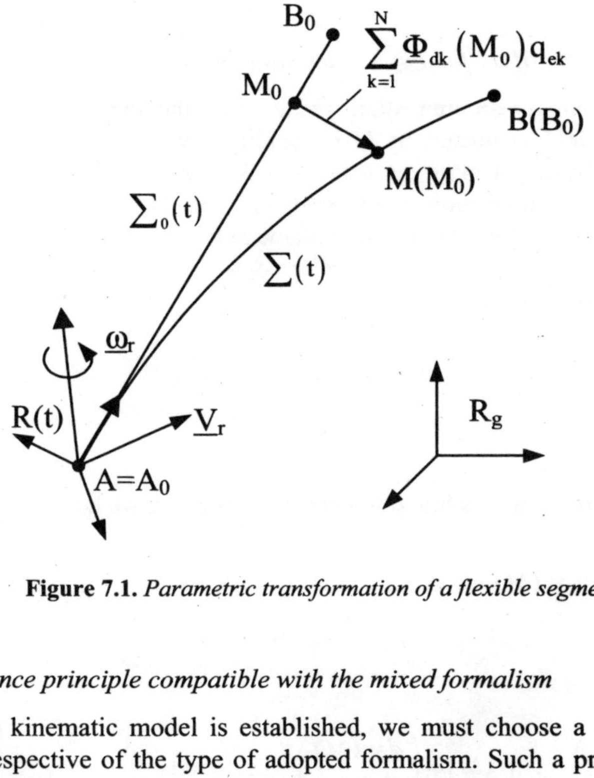

Figure 7.1. *Parametric transformation of a flexible segment*

7.2.2.4. *Balance principle compatible with the mixed formalism*

Once the kinematic model is established, we must choose a dynamic balance principle irrespective of the type of adopted formalism. Such a principle, reuniting the two formalisms previously discussed, is the principle of virtual powers. It is defined as follows:

$$P^*_{acc/g}(S) = P^*_{int}(S) + P^*_{ext/g}(S) \qquad [7.5]$$

Where, from left to right, we have the virtual power of acceleration quantities, the virtual power of constitutive internal forces and that of external load. The index "/g" means "as compared to a Galilean reference denoted R_g". For the virtual balance to be relevant, the virtual velocity field applied to the system "frozen" in time must be compatible with the parametric transformation [7.3] and [7.4]. We will use the symbol "*" in order to distinguish a virtual field from its real counterpart. Under these conditions we will adopt as the test field:

$$\underline{\mathbb{V}}^*(M) = \underline{\mathbb{V}}_r^*(M) + \underline{\mathbb{V}}_e^*(M) = \begin{Bmatrix} \underline{\mathbf{v}}_r^* + \underline{\boldsymbol{\omega}}_r^* \times \underline{\mathbf{r}}(M) \\ \underline{\boldsymbol{\omega}}_r^* \end{Bmatrix} + \sum_{k=1}^{N} \begin{Bmatrix} \underline{\boldsymbol{\Phi}}_{dk}(M_o) \\ \underline{\boldsymbol{\Phi}}_{rk}(M_o) \end{Bmatrix} \dot{q}_{ek}^* \quad [7.6]$$

7.2.2.5. *Virtual power of the field of acceleration quantities*

This contribution is an immediate result once the dynamic wrench field is known. Such a field evaluated in M is identified with the dynamic wrench of material particle having a mass dm located in M at t. It has as dynamic resultant $dm\underline{\boldsymbol{\gamma}}(M)$, and as dynamic momentum $dm\underline{\mathbf{h}}(M)$. This last momentum is zero, the particle having no spin, since there is no volume extension (non-Cosserat medium).

This makes it possible to write:

$$d\underline{\mathbb{T}}_{acc}(M) = dm(M) \begin{Bmatrix} \underline{\boldsymbol{\gamma}}(M) \\ \underline{\mathbf{0}} \end{Bmatrix}$$

Once the motion of the mobile reference S_o is inserted, we have:

$$\underline{\boldsymbol{\gamma}}(M) = \underline{\boldsymbol{\gamma}}_r + \underline{\boldsymbol{\gamma}}_e(M) + 2\,\underline{\boldsymbol{\omega}}_r \times \underline{\mathbf{v}}_e(M) + \underline{\boldsymbol{\omega}}_r \times (\underline{\boldsymbol{\omega}}_r \times \underline{\mathbf{r}}(M)) + \underline{\boldsymbol{\alpha}}_r \times \underline{\mathbf{r}}(M) \quad [7.7]$$

with:

$$\underline{\boldsymbol{\gamma}}_r = \left(\frac{d\underline{\mathbf{v}}_r}{dt}\right)_{/R_g}, \quad \underline{\boldsymbol{\gamma}}_e(M) = \left(\frac{d\underline{\mathbf{v}}_e(M)}{dt}\right)_{/R(t)}, \quad \underline{\boldsymbol{\alpha}}_r = \left(\frac{d\underline{\boldsymbol{\omega}}_r}{dt}\right)_{/R_g}$$

where the symbol "/" means "compared to" and, as we recall, R_g is the Galilean inertial reference frame, while $R(t)$ is the mobile reference frame fixed at the embedding in A.

The virtual power generated by the acceleration quantity of the considered particle is written as the interior product of its virtual distributor and its dynamic wrench, both evaluated in M:

$$dP^*_{acc,g}(M) = \; <\underline{\mathbb{V}}^*(M), d\underline{\mathbb{T}}_{acc}(M)>$$

Let us recall now that the vector space of wrenches and the one of screws are in a relation of duality via the internal product <, >. Later on, we will adopt the following writing convention. The internal product of a distributor $\underline{\mathbb{V}} = \begin{Bmatrix} \underline{\mathbf{v}} \\ \underline{\boldsymbol{\omega}} \end{Bmatrix}$, and of a wrench $\underline{\mathbb{T}} = \begin{Bmatrix} \underline{\mathbf{F}} \\ \underline{\mathbf{C}} \end{Bmatrix}$ will be marked:

$$<\underline{\mathbb{V}}, \underline{\mathbb{T}}> = \underline{\mathbf{F}}.\underline{\mathbf{v}} + \underline{\mathbf{C}}.\underline{\boldsymbol{\omega}}$$

where the symbol "." indicates the contracted product of two tensors (the scalar product of the two vectors being a particular case). We will also use the alternative writing:

$$<\underline{\mathbb{V}}, \underline{\mathbb{T}}> = \underline{\mathbf{F}}.\underline{\mathbf{v}} + \underline{\mathbf{C}}.\underline{\boldsymbol{\omega}} = \underline{\mathbf{v}}^{\mathrm{T}}.\underline{\mathbf{F}} + \underline{\boldsymbol{\omega}}^{\mathrm{T}}.\underline{\mathbf{C}}$$

where $\underline{\mathbf{v}}^{\mathrm{T}}$ indicates the dual vector of vector $\underline{\mathbf{v}}$. Finally, by integrating on the entire body the virtual power of the acceleration quantities of each particle in the virtual field [7.4], we will have:

$$P^*_{acc,g} = \int_{\Sigma(t)} dP^*_{acc,g}(M) = \int_{\Sigma_o(t)} \underline{\boldsymbol{\gamma}}(M)\, dm.\underline{\mathbf{v}}^*_r +$$

$$\int_{\Sigma_o(t)} (\underline{\mathbf{r}}(M_o) + \sum_{k=1}^{N} \underline{\boldsymbol{\Phi}}_{dk}(M_o) q_{ek}) \times \underline{\boldsymbol{\gamma}}(M)) dm.\underline{\boldsymbol{\omega}}^*_r + \sum_{k=1}^{N} \int_{\Sigma_o(t)} \underline{\boldsymbol{\gamma}}(M).\underline{\boldsymbol{\Phi}}_{dk}(M_o) dm\, \dot{q}^*_{ek} \quad [7.8]$$

In order not to complicate the notations, in [7.8] and subsequently, the accelerations depending on M appearing in the integrals on $\Sigma_o(t)$ are considered to be composed functions: $\underline{\gamma}(M) = \underline{\gamma}(M(M_o))$.

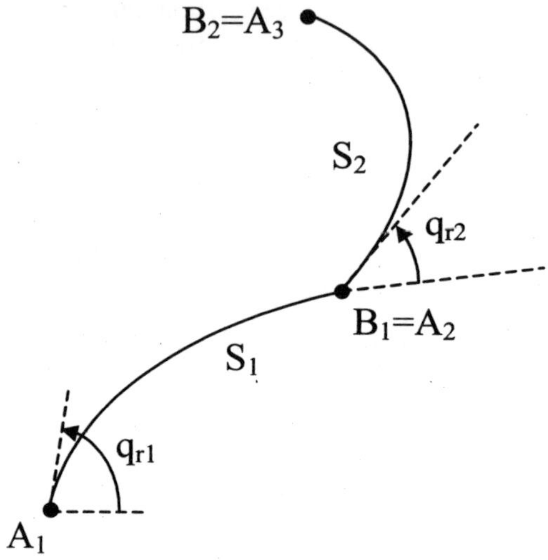

Figure 7.2. *Parametric transformation of the chain*

7.2.2.6. *Virtual power of external forces*

Later on, we will consider the case of an open robotic chain, with point and revolute joints. Under these restrictions, each body is subjected to:

– gravity;

– the control and constraint torques transmitted by the joints to the contact points A and B of the body with the rest of the chain.

We shall note here that since the joints are punctual, B is the embedding point of the next body in the chain (see Figure 7.2).

Virtual power of the gravity field

The gravity wrench is a "force wrench" of the type:

$$d\underline{\mathbb{T}}_g(M) = dm(M)\begin{Bmatrix} \underline{\mathbf{g}} \\ \underline{\mathbf{0}} \end{Bmatrix}$$

where $\underline{\mathbf{g}}$ is the gravitational acceleration.

Consequently, its virtual power is written:

$$P^*_{pes,g} = \int_{\Sigma(t)} < \underline{\mathbb{V}}^*(M), d\underline{\mathbb{T}}_g(M) > \qquad [7.9]$$

If we compare [7.9] with the initial expression of $P^*_{acc,g}$, we realize that to consider the gravity means to replace $\underline{\gamma}(M)$ by $\underline{\gamma}(M) - \underline{\mathbf{g}}$ in [7.7].

Virtual power of constraint and control wrenches

The "reaction-control" wrench transmitted by the joints on the two sides of the body b are written:

$$\underline{\mathbb{T}}_c(A) = \begin{Bmatrix} \underline{\mathbf{F}} \\ \underline{\mathbf{C}} \end{Bmatrix}, \ \underline{\mathbb{T}}_c(B) = \begin{Bmatrix} -\underline{\mathbf{F}}' \\ -\underline{\mathbf{C}}' \end{Bmatrix}$$

where we use a negative sign for the forces and torques applied by a body to its predecessor in the chain. The virtual power of these wrenches is written:

$$P^*_{c,g} = < \underline{\mathbb{V}}^*(A), \underline{\mathbb{T}}_c(A) > + < \underline{\mathbb{V}}^*(B), \underline{\mathbb{T}}_c(B) >,$$

i.e. in terms of the mixed set of kinematic unknowns (Lagrangian-Eulerian):

$$P^*_{c,g} = [\underline{\mathbf{F}} - \underline{\mathbf{F}}'].\underline{\mathbf{v}}^*_r$$

$$+ \left[\underline{\mathbf{C}} - \underline{\mathbf{C}}' - (\underline{\mathbf{r}}(B_o) + \sum_{k=1}^{N} \underline{\mathbf{\Phi}}_{dk}(B_o) q_{ek}) \times \underline{\mathbf{F}}' \right] . \underline{\boldsymbol{\omega}}^*_r +$$

$$+ \sum_{k=1}^{N} \left[-\underline{\mathbf{F}}'.\underline{\mathbf{\Phi}}_{dk}(B_o) - \underline{\mathbf{C}}'.\underline{\mathbf{\Phi}}_{rk}(B_o) \right] \dot{q}^*_{ek} \qquad [7.10]$$

where B_o is the antecedent of B by the current deformation applying $\Sigma_o(t)$ on $\Sigma(t)$ (see Figure 7.1).

7.2.2.7. *Virtual power of elastic cohesion forces*

Since our body is assumed to be elastic, its internal state can be entirely described by a potential of deformation according to the elastic parametric Rayleigh-Ritz transformation at the current moment:

$$U_{int}(S) = U_{int}(q_{ei}).$$

We obtain the virtual power from the internal forces:

$$P^*_{int}(S) = -\sum_{i=1}^{N} \left(\frac{\partial U_{int}}{\partial q_{ei}} \right) \dot{q}^*_{ei}$$

Limiting our developments to linear elasticity, the potential is taken in the quadratic form:

$$U_{int}(q_{ei}) = \frac{1}{2} \sum_{i=1}^{N} \sum_{j=1}^{N} k_{ij} q_{ei} q_{ej}$$

where k_{ij} represents the generalized Rayleigh-Ritz stiffness. Finally, we obtain the following expression of the required power:

$$P^*_{int}(S) = \sum_{i=1}^{N} \sum_{j=1}^{N} -(k_{ij}.q_{ej})\dot{q}^*_{ei} \qquad [7.11]$$

7.2.2.8. *Balance of virtual powers*

We can now state the balance of virtual powers. Considering [7.8], [7.9], [7.10] and [7.11] in [7.5], we obtain the following integral equation:

$$\int_{\Sigma_o(t)} \underline{\gamma}(M)\, dm.\underline{\mathbf{v}}^*_r + \int_{\Sigma_o(t)} (\underline{\mathbf{r}}(M_o) + \sum_{k=1}^{N} \underline{\mathbf{\Phi}}_{dk}(M_o) q_{ek}) \times \underline{\gamma}(M)) dm.\underline{\boldsymbol{\omega}}^*_r$$

$$+ \sum_{i=1}^{N} \int_{\Sigma_o(t)} \underline{\gamma}(M).\underline{\mathbf{\Phi}}_{di}(M_o) dm\, \dot{q}^*_{ei} \;=\; \left[\underline{\mathbf{F}} - \underline{\mathbf{F}}' \right] .\underline{\mathbf{v}}^*_r +$$

$$+ \left[(\underline{\mathbf{C}} - \underline{\mathbf{C}}') - (\underline{\mathbf{r}}(B_o) + \sum_{k=1}^{N} \underline{\mathbf{\Phi}}_{dk}(B_o) q_{ek}) \times \underline{\mathbf{F}}' \right] .\underline{\boldsymbol{\omega}}^*_r +$$

$$+ \sum_{i=1}^{N} \left[-\underline{\mathbf{F}}'.\underline{\mathbf{\Phi}}_{di}(B_o) - \underline{\mathbf{C}}'.\underline{\mathbf{\Phi}}_{ri}(B_o) \right] \dot{q}^*_{ei} + \sum_{i=1}^{N} \sum_{j=1}^{N} -(k_{ij}.q_{ej})\dot{q}^*_{ei} \qquad [7.12]$$

which can be re-written as follows:

$$\underline{\mathbf{A}}.\underline{\mathbf{v}}_r^* + \underline{\mathbf{B}}.\underline{\boldsymbol{\omega}}_r^* + \sum_{i=1}^{N} Q_{ei}.\dot{q}_{ei}^* = 0$$

The Eulerian velocities $\underline{\mathbf{v}}_r^*$ and $\underline{\boldsymbol{\omega}}_r^*$ and their Lagrangian counterparts $\dot{q}_{e1}^*, \dot{q}_{e2}^*, \ldots \dot{q}_{eN}^*$ are independent and arbitrary because "virtual". This enables us to draw from [7.12] three sets of dynamic balances (2 Eulerian and 1 Lagrangian) with the form:

$$\underline{\mathbf{A}} = \underline{\mathbf{0}},\ \underline{\mathbf{B}} = \underline{\mathbf{0}},\ Q_{e1} = 0, \ldots, Q_{eN} = 0$$

7.2.2.9. *Linear rigid balance in integral form*

From "$\underline{\mathbf{A}} = \underline{\mathbf{0}}$" and from [7.7] we obtain:

$$\int_{\Sigma_o(t)} dm\ \underline{\boldsymbol{\gamma}}_r + \underline{\boldsymbol{\alpha}}_r \times \left(\int_{\Sigma_o(t)} \underline{\mathbf{r}}(M_o) + \sum_{k=1}^{N} \underline{\boldsymbol{\Phi}}_{dk}(M_o)\ q_{ek} dm \right) +$$

$$+\underline{\boldsymbol{\omega}}_r \times \left(\underline{\boldsymbol{\omega}}_r \times \left(\int_{\Sigma_o(t)} \underline{\mathbf{r}}(M_o) + \sum_{k=1}^{N} \underline{\boldsymbol{\Phi}}_{dk}(M_o)\ q_{ek} dm \right)\right) +$$

$$+\sum_{k=1}^{N} \int_{\Sigma_o(t)} \underline{\boldsymbol{\Phi}}_{dk}(M_o)\ dm\ \ddot{q}_{ek} + 2\underline{\boldsymbol{\omega}}_r \times \left(\sum_{k=1}^{N} \int_{\Sigma_o(t)} \underline{\boldsymbol{\Phi}}_{dk}(M_o)\ dm\ \dot{q}_{ek}\right) = \underline{\mathbf{F}} - \underline{\mathbf{F}}' + \int_{\Sigma_o(t)} dm\ \underline{\mathbf{g}}$$

[7.13]

7.2.2.10. *Angular rigid balance in integral form*

From "$\underline{\mathbf{B}} = \underline{\mathbf{0}}$" and from [7.7] we obtain:

$$\int_{\Sigma_o(t)} (\underline{\underline{\hat{\mathbf{r}}}}^T(M).\underline{\underline{\hat{\mathbf{r}}}}(M))dm.\underline{\boldsymbol{\alpha}}_r + \left(\int_{\Sigma_o(t)} (\underline{\mathbf{r}}(M_o) + \sum_{k=1}^{N} \underline{\boldsymbol{\Phi}}_{dk}(M_o)\ q_{ek})\ dm \right) \times \underline{\boldsymbol{\gamma}}_r +$$

$$+\sum_{k=1}^{N} \left(\int_{\Sigma_o(t)} \underline{\mathbf{r}}(M_o) \times \underline{\boldsymbol{\Phi}}_{dk}(M_o)\ dm + \sum_{i=1}^{N} \int_{\Sigma_o(t)} \underline{\boldsymbol{\Phi}}_{di}(M_o) \times \underline{\boldsymbol{\Phi}}_{dk}(M_o)\ dm\ q_{ei} \right) \ddot{q}_{ek} +$$

$$+\underline{\boldsymbol{\omega}}_r \times \Big(\int_{\Sigma_o(t)} (\underline{\underline{\hat{\mathbf{r}}}}^T(M).\underline{\underline{\hat{\mathbf{r}}}}(M))dm.\underline{\boldsymbol{\omega}}_r \Big) +$$

$$+2\sum_k \int_{\Sigma_o(t)} \left(\left(\underline{\mathbf{r}}(M_o) + \sum_{i=1}^{N} \underline{\boldsymbol{\Phi}}_{di}(M_o) q_{ei} \right) \times \left(\underline{\boldsymbol{\omega}}_r \times \underline{\boldsymbol{\Phi}}_{dk}(M_o) \right) \right) dm\, \dot{q}_{ek} =$$

$$\underline{\mathbf{C}} - \underline{\mathbf{C}}' - \left(\underline{\mathbf{r}}(B_o) + \sum_{k=1}^{N} \underline{\boldsymbol{\Phi}}_{dk}(B_o)\, q_{ek} \right) \times \underline{\mathbf{F}}'$$

$$+ \left(\int_{\Sigma_o(t)} (\underline{\mathbf{r}}(M_o) + \sum_{k=1}^{N} \underline{\boldsymbol{\Phi}}_{dk}(M_o) q_{ek})\, dm \right) \times \underline{\mathbf{g}} \qquad [7.14]$$

where above as well as further on, "$\underline{\underline{\ }}$" underlines a tensor of $\mathbb{R}^3 \otimes \mathbb{R}^3$ ($\otimes$ is the usual tensorial product), and $\underline{\underline{\hat{\mathbf{w}}}}$ is the antisymmetric tensor associated to vector $\underline{\mathbf{w}}$ such that: $\underline{\underline{\hat{\mathbf{w}}}}.\underline{\mathbf{u}} = \underline{\mathbf{w}} \times \underline{\mathbf{u}}$, $\forall \underline{\mathbf{u}} \in \mathbb{R}^3$.

7.2.2.11. *Elastic balances in integral form*

The ith elastic balance results from "$Q_{ei} = 0$" and from [7.7]:

$$\sum_{k=1}^{N} \int_{\Sigma_o(t)} (\underline{\boldsymbol{\Phi}}_{di}(M_o).\underline{\boldsymbol{\Phi}}_{dk}(M_o))dm\, \ddot{q}_{ek} + \int_{\Sigma_o(t)} \underline{\boldsymbol{\Phi}}_{di}(M_o)dm.\underline{\boldsymbol{\gamma}}_r +$$

$$+ \left(\sum_{k=1}^{N} \int_{\Sigma_o(t)} (\underline{\boldsymbol{\Phi}}_{dk}(M_o) \times \underline{\boldsymbol{\Phi}}_{di}(M_o))dm\, q_{ek} + \int_{\Sigma_o(t)} \underline{\mathbf{r}}(M_o) \times \underline{\boldsymbol{\Phi}}_{di}(M_o)dm \right) .\underline{\boldsymbol{\alpha}}_r$$

$$+ 2\sum_{k=1}^{N} \Big(\int_{\Sigma_o(t)} (\underline{\boldsymbol{\Phi}}_{dk}(M_o) \times \underline{\boldsymbol{\Phi}}_{di}(M_o)).dm \Big).\underline{\boldsymbol{\omega}}_r \dot{q}_{ek} +$$

$$+ \left(\int_{\Sigma_o(t)} (\underline{\boldsymbol{\omega}}_r \times \underline{\mathbf{r}}(M_o)) \times \underline{\boldsymbol{\Phi}}_{di}(M_o)dm + \sum_{k=1}^{N} \int_{\Sigma_o(t)} (\underline{\boldsymbol{\omega}}_r \times \underline{\boldsymbol{\Phi}}_{dk}(M_o) q_{ek}) \times \underline{\boldsymbol{\Phi}}_{di}(M_o)dm \right) .\underline{\boldsymbol{\omega}}_r$$

$$+\sum_{k=1}^{N} k_{ik} q_{ek} = -\underline{\mathbf{F}}' .\underline{\boldsymbol{\Phi}}_{di}(B_o) - \underline{\mathbf{C}}' .\underline{\boldsymbol{\Phi}}_{ri}(B_o) + \int_{\Sigma_o(t)} \underline{\boldsymbol{\Phi}}_{di}(M_o)\, dm.\underline{\mathbf{g}} \qquad [7.15]$$

7.2.2.12. *Linear rigid balance in parametric form*

In order to extract the kinematic unknown quantities from the preceding integrals and to factorize the tensorial inertial parameters independent of time, the following relation is used:

$$\underline{\mathbf{V}}_1 \times (\underline{\mathbf{V}}_2 \times \underline{\mathbf{V}}_3) = -(\hat{\underline{\underline{\mathbf{V}}}}_1^{\mathrm{T}} . \hat{\underline{\underline{\mathbf{V}}}}_2) . \underline{\mathbf{V}}_3$$

Moreover, the integrals appearing in [7.13], [7.16] and [7.15] referring to quantities related to the static geometry of the body, the integration field $\Sigma_o(t)$ can be replaced by Σ_o, the initial configuration of the body.

Hence, balances [7.13], [7.14] and [7.15] are rewritten in a parametric form which we will now develop.

Linear rigid balance in parametric form

$$m\underline{\boldsymbol{\gamma}}_r + \underline{\boldsymbol{\alpha}}_r \times \underline{\mathbf{MS}} + \underline{\boldsymbol{\omega}}_r \times (\underline{\boldsymbol{\omega}}_r \times \underline{\mathbf{MS}}) + \\ + \sum_{k=1}^{N} \underline{\mathbf{b}}_k \ddot{q}_{ek} + 2\underline{\boldsymbol{\omega}}_r \times (\sum_{k=1}^{N} \underline{\mathbf{b}}_k \dot{q}_{ek}) = \underline{\mathbf{F}} - \underline{\mathbf{F}}' + m\,\underline{\mathbf{g}}, \qquad [7.16]$$

where we introduced the following inertial parameters:

– the body mass:

$$\int_{\Sigma_o} dm = m;$$

– the vector of its rigid "first moments" of inertia:

$$\int_{\Sigma_o} \underline{\mathbf{r}}(\overset{\circ}{M}_o) dm = \underline{\mathbf{MS}}_{\mathbf{r}};$$

– its kth elastic counterpart:

$$\int_{\Sigma_o} \underline{\boldsymbol{\Phi}}_{dk}(M_o) dm = \underline{\mathbf{b}}_{\mathbf{k}};$$

– the total vector of its first moments of inertia:

$$\int_{\Sigma_o} \underline{\mathbf{r}}(M_o) + \sum_{k=1}^{N} \underline{\mathbf{\Phi}}_{dk}(M_o) q_{ek} dm = \underline{\mathbf{MS}}_r + \sum_{k=1}^{N} \underline{\mathbf{b}}_k q_{ek} = \underline{\mathbf{MS}}.$$

Angular rigid balance in parametric form

$$\underline{\underline{\mathbf{I}}} . \underline{\boldsymbol{\alpha}}_r + \underline{\mathbf{MS}} \times \underline{\boldsymbol{\gamma}}_r + \sum_{k=1}^{N} \left(\underline{\boldsymbol{\beta}}_k + \sum_{i=1}^{N} \underline{\boldsymbol{\lambda}}_{ki} q_{ei} \right) \ddot{q}_{ek} +$$

$$+ \underline{\boldsymbol{\omega}}_r \times (\underline{\underline{\mathbf{I}}} . \underline{\boldsymbol{\omega}}_r) + 2 \sum_{k=1}^{N} \underline{\underline{\mathbf{I}}}_{re,k} . \underline{\boldsymbol{\omega}}_r \dot{q}_{ek} + 2 \sum_{k,i=1}^{N} \underline{\underline{\mathbf{I}}}_{ee,ik} . \underline{\boldsymbol{\omega}}_r q_{ei} \dot{q}_{ek} =$$

$$+ \underline{\mathbf{C}} - \underline{\mathbf{C}}' - \left(\underline{\mathbf{r}}(B_o) + \sum_{k=1}^{N} \underline{\mathbf{\Phi}}_{dk}(B_o) q_{ek} \right) \times \underline{\mathbf{F}} + \underline{\mathbf{MS}} \times \underline{\mathbf{g}} \qquad [7.17]$$

where we introduced the total tensor of inertia of the body:

$$\underline{\underline{\mathbf{I}}} = \int_{\Sigma_o} (\underline{\underline{\hat{\mathbf{r}}}}^T (M) . \underline{\underline{\hat{\mathbf{r}}}}(M)) . dm = \underline{\underline{\mathbf{I}}}_{rr} + \sum_{j=1}^{N} (\underline{\underline{\mathbf{I}}}_{re,j} + \underline{\underline{\mathbf{I}}}_{re,j}^T) q_{ej} + \sum_{i,j=1}^{N} \underline{\underline{\mathbf{I}}}_{ee,ij} q_{ei} q_{ej} .$$

This tensor depends on the configuration, and its detailed expression means it is necessary to know the following constant tensors:

– the tensor of inertia of the rigid solid:

$$\int_{\Sigma_o} \underline{\underline{\hat{\mathbf{r}}}}^T (M_o) . \underline{\underline{\hat{\mathbf{r}}}}(M_o) \, dm = \underline{\underline{\mathbf{I}}}_{rr} ;$$

– the tensor of inertia of the rigid-elastic:

$$\int_{\Sigma_o} \underline{\underline{\hat{\mathbf{r}}}}^T (M_o) . \underline{\underline{\hat{\mathbf{\Phi}}}}_{di} (M_o) \, dm = \underline{\underline{\mathbf{I}}}_{re,i} ;$$

– the tensor of inertia of the elastic-elastic:

$$\int_{\Sigma_o} \underline{\underline{\hat{\mathbf{\Phi}}}}_{di}^{T}(M_o).\underline{\underline{\hat{\mathbf{\Phi}}}}_{dj}(M_o)\, dm = \underline{\underline{\mathbf{I}}}_{ee,ij}.$$

Angular balances require also knowing the two sets of following vectors:

$$\int_{\Sigma_o} \underline{\mathbf{r}}(M_o)\times\underline{\mathbf{\Phi}}_{dk}(M_o)\, dm = \underline{\boldsymbol{\beta}}_k;$$

$$\int_{\Sigma_o} \underline{\mathbf{\Phi}}_{di}(M_o)\times\underline{\mathbf{\Phi}}_{dk}(M_o)\, dm = \underline{\boldsymbol{\lambda}}_{ik}.$$

i^{th} elastic balance in parametric form

$$\sum_{k=1}^{N} m_{ik}\ddot{q}_{ek} + \underline{\mathbf{b}}_i.\underline{\boldsymbol{\gamma}}_r + (\sum_{k=1}^{N}\underline{\boldsymbol{\lambda}}_{ki}q_{ek} + \underline{\boldsymbol{\beta}}_i).\underline{\boldsymbol{\alpha}}_r +$$

$$+ 2\sum_{k=1}^{N}\underline{\boldsymbol{\lambda}}_{ki}.\underline{\boldsymbol{\omega}}_r.\dot{q}_{ek} - \underline{\boldsymbol{\omega}}_r.\underline{\underline{\mathbf{I}}}_{er,i}.\underline{\boldsymbol{\omega}}_r - \sum_{k=1}^{N}\underline{\boldsymbol{\omega}}_r.\underline{\underline{\mathbf{I}}}_{ee,ik}.\underline{\boldsymbol{\omega}}_r\, q_{ek} +$$

$$+\sum_{k=1}^{N} k_{ik}q_{ek} = -\underline{\mathbf{F}}'.\underline{\mathbf{\Phi}}_{di}(B_o) - \underline{\mathbf{C}}'.\underline{\mathbf{\Phi}}_{ri}(B_o) + \underline{\mathbf{b}}_i.\underline{\mathbf{g}}, \qquad [7.18]$$

which means it is necessary to know, apart from the parameters previously introduced, the generalized Rayleigh-Ritz masses:

$$\int_{\Sigma_o} (\underline{\mathbf{\Phi}}_{di}(M_o).\underline{\mathbf{\Phi}}_{dk}(M_o))dm = m_{ik}.$$

7.2.2.13. *Intrinsic matrix form of the generalized Newton-Euler model*

Finally, we can bring together the parametric balances [7.16], [7.17] and [7.18] as the following intrinsic matrix:

$$\begin{pmatrix} \underline{\underline{\mathbf{M}}} & \underline{\underline{\mathbf{M\hat{S}}}}^T & \underline{\underline{\mathbf{MS}}}_{de} \\ \underline{\underline{\mathbf{M\hat{S}}}} & \underline{\underline{\mathbf{I}}} & \underline{\underline{\mathbf{MS}}}_{re} \\ \underline{\underline{\mathbf{MS}}}_{de}^T & \underline{\underline{\mathbf{MS}}}_{re}^T & \mathbf{m}_{ee} \end{pmatrix} \begin{Bmatrix} \underline{\boldsymbol{\gamma}}_r \\ \underline{\boldsymbol{\alpha}}_r \\ \ddot{\mathbf{q}}_e \end{Bmatrix} + \begin{Bmatrix} \underline{\mathbf{F}}_{in}(\mathbf{q}_e, \dot{\mathbf{q}}_e, \underline{\boldsymbol{\omega}}_r) \\ \underline{\mathbf{C}}_{in}(\mathbf{q}_e, \dot{\mathbf{q}}_e, \underline{\boldsymbol{\omega}}_r) \\ \mathbf{Q}_{in}(\mathbf{q}_e, \dot{\mathbf{q}}_e, \underline{\boldsymbol{\omega}}_r) \end{Bmatrix} + \begin{Bmatrix} \underline{\mathbf{0}} \\ \underline{\mathbf{0}} \\ \mathbf{k}_{ee}\mathbf{q}_e \end{Bmatrix} + \begin{Bmatrix} \underline{\mathbf{F}}_g \\ \underline{\mathbf{C}}_g \\ \mathbf{Q}_g \end{Bmatrix} = \begin{Bmatrix} \Delta\underline{\mathbf{F}}_c \\ \Delta\underline{\mathbf{C}}_c \\ \Delta\mathbf{Q}_c \end{Bmatrix} \qquad [7.19]$$

Where the blocks of [7.19] imply:

– the matrices of generalized coordinates, velocities and Rayleigh-Ritz accelerations:

$$\mathbf{q}_e = (q_{e1}, q_{e2}, \ldots q_{eN})^T \ , \ \dot{\mathbf{q}}_e = (\dot{q}_{e1}, \dot{q}_{e2}, \ldots \dot{q}_{eN})^T \ , \ \ddot{\mathbf{q}}_e = (\ddot{q}_{e1}, \ddot{q}_{e2}, \ldots \ddot{q}_{eN})^T ;$$

– the matrices of generalized mass and stiffness of the structure:

$$\mathbf{m}_{ee} = mat_{i,j=1..N}(m_{ij}) \, , \ \mathbf{k}_{ee} = mat_{i,j=1..N}(k_{ij})$$

where $mat_{i,j=1..N}(a_{ij})$ is the square matrix of generic component a_{ij}, and where i and j are respectively the row and column indices.

– the tensor of linear mass of the body having the mass m:

$$\underline{\underline{\mathbf{M}}} = m \, \underline{\underline{\boldsymbol{\delta}}}$$

where $\underline{\underline{\boldsymbol{\delta}}}$ is the tensor whose components in the mobile base of $R(t)$ are the symbols of Kronecker;

– the antisymmetric tensor of the first moments of inertia:

$$\underline{\underline{\mathbf{M\hat{S}}}} = \underline{\underline{\mathbf{M\hat{S}}}}_r + \sum_{k=1}^{N} \underline{\underline{\hat{\mathbf{b}}}}_k \, q_{ek} .$$

As well as the row matrices of vectors which we underlined as vectors:

$$\underline{\mathbf{MS}}_{de} = \left(\underline{\mathbf{b}}_1, \ldots \underline{\mathbf{b}}_N\right), \ \underline{\mathbf{MS}}_{re} = \left(\underline{\boldsymbol{\beta}}_1 + \sum_{k=1}^{N} \underline{\boldsymbol{\lambda}}_{k1} q_{ek}, \ldots, \underline{\boldsymbol{\beta}}_N + \sum_{k=1}^{N} \underline{\boldsymbol{\lambda}}_{kN} q_{ek}\right)$$

their duals being written by simple matrix transposition and passage to dual vectors:

$$\underline{\mathbf{MS}}_{de}^T = \left(\underline{\mathbf{b}}_1^T, \ldots \underline{\mathbf{b}}_N^T\right)^T, \ \underline{\mathbf{MS}}_{re}^T = \left(\underline{\boldsymbol{\beta}}_1^T + \sum_{k=1}^{N} \underline{\boldsymbol{\lambda}}_{k1}^T q_{ek}, \ldots, \underline{\boldsymbol{\beta}}_N^T + \sum_{k=1}^{N} \underline{\boldsymbol{\lambda}}_{kN}^T q_{ek}\right)^T ;$$

– the resultant of inertial forces:

$$\underline{\mathbf{F}}_{in}(\mathbf{q}_e,\dot{\mathbf{q}}_e,\underline{\boldsymbol{\omega}}_r)=\underline{\boldsymbol{\omega}}_r\times(\underline{\boldsymbol{\omega}}_r\times(\underline{\mathbf{MS}}_r+\sum_{k=1}^{N}\underline{\mathbf{b}}_k q_{ek}))+2\underline{\boldsymbol{\omega}}_r\times(\sum_{k=1}^{N}\underline{\mathbf{b}}_k\dot{q}_{ek});$$

– the resultant of inertial moments:

$$\underline{\mathbf{C}}_{in}(\mathbf{q}_e,\dot{\mathbf{q}}_e,\underline{\boldsymbol{\omega}}_r)=\underline{\boldsymbol{\omega}}_r\times(\underline{\underline{\mathbf{I}}}.\underline{\boldsymbol{\omega}}_r)+2\sum_{k=1}^{N}\underline{\underline{\mathbf{I}}}_{re,k}.\underline{\boldsymbol{\omega}}_r.\dot{q}_{ek}+2\sum_{k,i=1}^{N}\underline{\underline{\mathbf{I}}}_{ee,ki}.\underline{\boldsymbol{\omega}}_r q_{ei}\dot{q}_{ek};$$

– the generalized inertial forces:

$$\mathbf{Q}_{in}(\mathbf{q}_e,\dot{\mathbf{q}}_e,\underline{\boldsymbol{\omega}}_r)=col_{i=1,..N}\left(2\sum_{k=1}^{N}\underline{\boldsymbol{\lambda}}_{ki}.\underline{\boldsymbol{\omega}}_r.\dot{q}_{ek}-\underline{\boldsymbol{\omega}}_r.\underline{\underline{\mathbf{I}}}_{er,i}.\underline{\boldsymbol{\omega}}_r-\sum_{k=1}^{N}\underline{\boldsymbol{\omega}}_r.\underline{\underline{\mathbf{I}}}_{ee,ik}.\underline{\boldsymbol{\omega}}_r q_{ek}\right);$$

– the gravity forces and moments:

$$\underline{\mathbf{F}}_g=m.\underline{\mathbf{g}},\quad \underline{\mathbf{C}}_g=\left(\underline{\mathbf{MS}}_r+\sum_{k=1}^{N}\underline{\mathbf{b}}_k\,q_{ek}\right)\times\underline{\mathbf{g}};$$

– the ith gravity generalized force:

$$Q_{g,i}=\underline{\mathbf{b}}_i.\underline{\mathbf{g}};$$

– the resultants of forces and torques transmitted by the joints:

$$\Delta\underline{\mathbf{F}}_{\mathbf{c}}=\underline{\mathbf{F}}-\underline{\mathbf{F}}',\quad \Delta\underline{\mathbf{C}}_{\mathbf{c}}=\underline{\mathbf{C}}-\underline{\mathbf{C}}'-\underline{\mathbf{r}}(B)\times\underline{\mathbf{F}}';$$

– the ith generalized force transmitted by the joints:

$$\Delta\mathbf{Q}_{c,i}=-\underline{\mathbf{F}}'.\underline{\boldsymbol{\Phi}}_{di}(B_o)-\underline{\mathbf{C}}'.\underline{\boldsymbol{\Phi}}_{ri}(B_o).$$

NOTES.

– We note that the matrix product of [7.19] represents, in turn, the contracted product of a vector by a tensor, the scalar product of the two vectors, the product of a scalar by a vector or a simple product of the two scalars, according to the tensorial nature of the concerned factors.

– Equation [7.19] generalizes the Newton-Euler model of a rigid solid to the case of an elastic solid subjected to small deformations in the expected sense because by removing the elastic unknown quantities and equations in [7.19] we have the usual Newton-Euler equations for a rigid solid:

$$\begin{pmatrix} \underline{\underline{\mathbf{M}}} & \underline{\underline{\hat{\mathbf{MS}}}}_r^T \\ \underline{\underline{\hat{\mathbf{MS}}}}_r & \underline{\underline{\mathbf{I}}}_{rr} \end{pmatrix} \begin{Bmatrix} \underline{\gamma}_r \\ \underline{\alpha}_r \end{Bmatrix} + \begin{Bmatrix} \underline{\omega}_r \times (\underline{\omega}_r \times \underline{\mathbf{MS}}_r) \\ \underline{\omega}_r \times (\underline{\underline{\mathbf{I}}}_{rr} . \underline{\omega}_r) \end{Bmatrix} + \begin{Bmatrix} m.\underline{\mathbf{g}} \\ \underline{\mathbf{MS}}_r \times \underline{\mathbf{g}} \end{Bmatrix} = \begin{Bmatrix} \Delta \underline{\mathbf{F}}_c \\ \Delta \underline{\mathbf{C}}_c \end{Bmatrix}$$

– The equations given above are written in an intrinsic form and can consequently be expressed in any reference frame; nevertheless, the embedding reference frame $R(t)$ leads to the simplest formulation.

7.2.3. *Velocity model of a simple open robotic chain*

We will now consider a simple open robotic chain, each body of which is of the type we have previously studied. The number of bodies is n and they are denoted $S_1,...S_n$, starting from the base towards the end-effector. The base, marked S_0, is assumed to be rigid. The joints are assumed to be point-revolute. The unit vectors, support of joint axes, are denoted starting from the base towards the end-effector: $\underline{\mathbf{a}}_1$, $\underline{\mathbf{a}}_2,...\underline{\mathbf{a}}_n$. Their angular variables are denoted $q_1, q_2,...q_n$. The dynamics of isolated segments results simply from [7.19], by subscripting all the tensorial magnitudes of this generic balance with the index of the body considered. We will now complete the set of dynamic equations thus defined with a recurrence on the reference screws of the bodies, imposed by the joints. This recurrence creates a kinematic model of the chain.

In order to establish the kinematic model of the joint $j+1$, let us consider the two bodies S_j and S_{j+1} connected by this joint. These two bodies are connected to the contact point $B_j = A_{j+1}$, where B is the last point of S_j and A_{j+1} the first point of S_{j+1} (these two points are often aligned because of the assumption of point character of the joint). We then write the velocity screws of S_j and S_{j+1} in B_j and A_{j+1} respectively. We have for the first of these two screws:

$$\underline{\mathbb{V}}_j(B_j) = \underline{\mathbb{V}}_{rj} + \begin{Bmatrix} \underline{\omega}_{rj} \times \underline{\mathbf{r}}_j(B_j) \\ \underline{\mathbf{0}} \end{Bmatrix} + \underline{\mathbb{V}}_{ej}(B_j) \qquad [7.20]$$

as for the second, we have more simply:

$$\underline{\mathbb{V}}_{j+1}(A_{j+1}) = \underline{\mathbb{V}}_{r,j+1} \qquad [7.21]$$

Thus, the relative motion introduced by the joint centered in A_{j+1} being modeled as:

$$\underline{\mathbb{V}}_{j+1}(A_{j+1}) - \underline{\mathbb{V}}_j(A_{j+1}) = \underline{\mathbb{V}}_{j+1}(A_{j+1}) - \underline{\mathbb{V}}_j(B_j) = \dot{q}_{rj+1}\begin{Bmatrix} \underline{\mathbf{0}} \\ \underline{\mathbf{a}}_{j+1} \end{Bmatrix} \qquad [7.22]$$

we obtain, considering [7.20] and [7.21] in [7.22], the recurrence on the reference screws:

$$\underline{\mathbb{V}}_{rj+1} = \underline{\mathbb{V}}_{rj} + \sum_{k=1}^{N_j}\begin{Bmatrix} \underline{\mathbf{\Phi}}_{dj,k}(B_{oj}) \\ \underline{\mathbf{\Phi}}_{rj,k}(B_{oj}) \end{Bmatrix}\dot{q}_{ej,k} + \begin{Bmatrix} \underline{\boldsymbol{\omega}}_{rj} \times \underline{\mathbf{r}}_j(A_{j+1}) \\ \underline{\mathbf{0}} \end{Bmatrix} + \begin{Bmatrix} \underline{\mathbf{0}} \\ \underline{\mathbf{a}}_{j+1} \end{Bmatrix}\dot{q}_{rj+1} \qquad [7.23]$$

where B_{oj} is the antecedent by the current deformation on S_j from the point $B_j = A_{j+1}$.

7.2.4. *Acceleration model of a simple open robotic chain*

In order to generalize the Newton-Euler algorithms developed for the control and simulation of the rigid robot manipulators in the case of a flexible manipulator, it is necessary to establish the recurrence on the accelerations of the segments. This recurrence is obtained from its velocity counterpart by simple Galilean derivation, using the operators:

$$\frac{d.}{dt}_{/R_g} = \frac{d.}{dt}_{/R_j(t)} + \underline{\boldsymbol{\omega}}_{rj} \times ., \; j = 1,..n$$

where $R_j(t)$ is the reference frame (i.e. embedding) of S_j. After some tiresome but simple calculations we obtain the required recurrence.

Recurrence on the accelerations

For $j = 0,..,n-1$:

$$\dot{\underline{\mathbb{V}}}_{rj+1} = \dot{\underline{\mathbb{V}}}_{rj} + \sum_{k=1}^{N_j} \begin{Bmatrix} \underline{\mathbf{\Phi}}_{dj,k}(B_{oj}) \\ \underline{\mathbf{\Phi}}_{rj,k}(B_{oj}) \end{Bmatrix} \ddot{q}_{ej,k} + \begin{Bmatrix} \underline{\boldsymbol{\alpha}}_{rj} \times \underline{\mathbf{r}}_j(A_{j+1}) \\ \underline{\mathbf{0}} \end{Bmatrix} + \begin{Bmatrix} \underline{\mathbf{0}} \\ \underline{\mathbf{a}}_{j+1} \end{Bmatrix} \ddot{q}_{rj+1} + \begin{Bmatrix} \underline{\mathbf{g}}_{linr,j+1} \\ \underline{\mathbf{g}}_{angr,j+1} \end{Bmatrix} \quad [7.24]$$

where we introduced the following notations for $j = 0,..,n$:

$$\underline{\mathbf{g}}_{lin,j+1} = \underline{\boldsymbol{\omega}}_{rj} \times (2\underline{\mathbf{\Phi}}_{dj}(B_{oj})\dot{\mathbf{q}}_{ej}) + \underline{\boldsymbol{\omega}}_{rj} \times (\underline{\boldsymbol{\omega}}_{rj} \times \underline{\mathbf{r}}_j(A_{j+1}))$$

$$\underline{\mathbf{g}}_{ang,j+1} = \dot{q}_{rj+1}(\underline{\boldsymbol{\omega}}_{rj} + \underline{\mathbf{\Phi}}_{rj}(B_{oj})\dot{\mathbf{q}}_{ej}) \times \underline{\mathbf{a}}_{j+1} + (\underline{\boldsymbol{\omega}}_{rj} \times \underline{\mathbf{\Phi}}_{rj}(B_{oj})\dot{\mathbf{q}}_{ej}),$$

$$\frac{d}{dt_{/R_g}} \begin{Bmatrix} \underline{\mathbf{v}}_{rj} \\ \underline{\boldsymbol{\omega}}_{rj} \end{Bmatrix} = \begin{Bmatrix} \underline{\boldsymbol{\gamma}}_{rj} \\ \underline{\boldsymbol{\alpha}}_{rj} \end{Bmatrix} = \dot{\underline{\mathbb{V}}}_{rj},$$

$$\begin{pmatrix} \underline{\mathbf{\Phi}}_{dj}(B_{oj}) \\ \underline{\mathbf{\Phi}}_{rj}(B_{oj}) \end{pmatrix} = \begin{pmatrix} \underline{\mathbf{\Phi}}_{dj,1}(B_{oj}), \underline{\mathbf{\Phi}}_{dj,2}(B_{oj}),.., \underline{\mathbf{\Phi}}_{dj,N_j}(B_{oj}) \\ \underline{\mathbf{\Phi}}_{rj,1}(B_{oj}), \underline{\mathbf{\Phi}}_{rj,2}(B_{oj}).., \underline{\mathbf{\Phi}}_{rj,N_j}(B_{oj}) \end{pmatrix}$$

7.2.5. *Generalized Newton-Euler model for a flexible manipulator*

We are now able to write the generalized Newton-Euler equations for a flexible robot manipulator. This is practically done by associating to the balance equations [7.19] written for each body, the recurrence on the velocities [7.23] and on the accelerations [7.24]. In addition, the "reaction-control wrench" applied by S_j on S_{j+1} via the centered point joint in A_{j+1} is denoted:

$$\underline{\mathbb{T}}_{j+1}(A_{j+1}) = \begin{Bmatrix} \underline{\mathbf{F}}_{j+1} \\ \underline{\mathbf{C}}_{j+1} \end{Bmatrix}$$

with these notations we obtain the algebraic differential system:

Dynamic equations

For $j = 1,..,n$:

$$\begin{pmatrix} \underline{\underline{\mathbf{M}}}_j & \underline{\underline{\mathbf{M\hat{S}}}}_j^T & \underline{\mathbf{MS}}_{dej} \\ \underline{\underline{\mathbf{M\hat{S}}}}_j & \underline{\underline{\mathbf{I}}}_j & \underline{\mathbf{MS}}_{rej} \\ \underline{\mathbf{MS}}_{dej}^{\mathbf{T}} & \underline{\mathbf{MS}}_{rej}^{\mathbf{T}} & \mathbf{m}_{eej} \end{pmatrix} \begin{Bmatrix} \underline{\boldsymbol{\gamma}}_{rj} \\ \underline{\boldsymbol{\alpha}}_{rj} \\ \ddot{\mathbf{q}}_{ej} \end{Bmatrix} + \begin{Bmatrix} \underline{\mathbf{F}}_{inj} + \underline{\mathbf{F}}_{gj} \\ \underline{\mathbf{C}}_{inj} + \underline{\mathbf{C}}_{gj} \\ \mathbf{Q}_{inj} + \mathbf{Q}_{gj} + \mathbf{k}_{eej}.\mathbf{q}_{ej} \end{Bmatrix} =$$

$$\begin{Bmatrix} \underline{\mathbf{F}}_j - \underline{\mathbf{F}}_{j+1} \\ \underline{\mathbf{C}}_j - \underline{\mathbf{C}}_{j+1} - \underline{\mathbf{r}}_j(A_{j+1}) \times \underline{\mathbf{F}}_{j+1} \\ col_{j=1,..n}(col_{k=1,..N_j}(-\underline{\mathbf{F}}_{j+1}.\underline{\boldsymbol{\Phi}}_{dj,k}(B_{oj}) - \underline{\mathbf{C}}_{j+1}.\underline{\boldsymbol{\Phi}}_{rj,k}(B_{oj}))) \end{Bmatrix} \quad [7.25]$$

Recurrence on the velocities

For $j = 0,..,n-1$:

$$\underline{\mathbb{V}}_{rj+1} = \underline{\mathbb{V}}_{rj} + \sum_{k=1}^{N_j} \begin{Bmatrix} \underline{\boldsymbol{\Phi}}_{dj,k}(B_{oj}) \\ \underline{\boldsymbol{\Phi}}_{rj,k}(B_{oj}) \end{Bmatrix} \dot{q}_{ej,k} + \begin{Bmatrix} \underline{\boldsymbol{\omega}}_{rj} \times \underline{\mathbf{r}}_j(A_{j+1}) \\ \underline{\mathbf{0}} \end{Bmatrix} + \begin{Bmatrix} \underline{\mathbf{0}} \\ \underline{\mathbf{a}}_{j+1} \end{Bmatrix} \dot{q}_{rj+1} \quad [7.26]$$

Recurrence on the accelerations

For $j = 0,..,n-1$:

$$\dot{\underline{\mathbb{V}}}_{rj+1} = \dot{\underline{\mathbb{V}}}_{rj} + \sum_{k=1}^{N_j} \begin{Bmatrix} \underline{\boldsymbol{\Phi}}_{dj,k}(B_{oj}) \\ \underline{\boldsymbol{\Phi}}_{rj,k}(B_{oj}) \end{Bmatrix} \ddot{q}_{ej,k} + \begin{Bmatrix} \underline{\boldsymbol{\alpha}}_{rj} \times \underline{\mathbf{r}}_j(A_{j+1}) \\ \underline{\mathbf{0}} \end{Bmatrix} + \begin{Bmatrix} \underline{\mathbf{0}} \\ \underline{\mathbf{a}}_{j+1} \end{Bmatrix} \ddot{q}_{rj+1} + \begin{Bmatrix} \underline{\mathbf{g}}_{linr,j+1} \\ \underline{\mathbf{g}}_{angr,j+1} \end{Bmatrix} \quad [7.27]$$

In order to complete this model it is necessary to add the boundary conditions, in motion for the base and in forces for the end-effector.

7.2.6. *Extrinsic Newton-Euler model for numerical calculus*

Finally, the numerical application of this model for the simulation and control problem requires expressing the preceding intrinsic equations in a set of adapted reference frames. This is carried out in practice by projecting all the tensorial quantities attached to a body S_j in the corresponding reference frame $R_j(t)$. We will write ${}^j\mathbf{v}$, the matrix of components of a vector $\underline{\mathbf{v}}$ or of a tensor $\underline{\underline{\mathbf{v}}}$, in $R_j(t)$. Moreover, the 6 × 6 matrices of corresponding wrenches and screws components will be indicated in double characters. Hence, we obtain, based on [7.25], [7.26] and [7.27], the generalized Newton-Euler model of an open chain with revolute joints and embedded references, which are as follows:

Body balance

For $j = 1,..,n$:

$$\begin{pmatrix} \mathbb{J}_{rrj} & \mathbf{J}_{rej} \\ \mathbf{J}_{erj} & \mathbf{J}_{eej} \end{pmatrix} \begin{Bmatrix} {}^j\dot{\mathbb{V}}_j \\ \ddot{\mathbf{q}}_{ej} \end{Bmatrix} + \begin{Bmatrix} {}^j\mathbb{C}_j \\ \mathbf{c}_j \end{Bmatrix} = \begin{Bmatrix} {}^j\mathbb{F}_j - {}^{j+1}\mathbb{T}_j^T\, {}^{j+1}\mathbb{F}_{j+1} \\ -{}^j\mathbf{\Phi}_j^T\, {}^j\mathbb{R}_{j+1}\, {}^{j+1}\mathbb{F}_{j+1} \end{Bmatrix} \tag{7.28}$$

Recurrence on the bodies

For $j = 1,..,n$

$${}^j\mathbb{V}_j = {}^j\mathbb{T}_{j-1}\, {}^{j-1}\mathbb{V}_{j-1} + {}^j\mathbb{R}_{j-1}\, {}^{j-1}\mathbf{\Phi}_{j-1}\dot{\mathbf{q}}_{ej-1} + \dot{q}_{rj}\, {}^j\mathbb{A}_j \tag{7.29}$$

Recurrence on the accelerations

For $j = 1,..,n$

$${}^j\dot{\mathbb{V}}_j = {}^j\mathbb{T}_{j-1}\, {}^{j-1}\dot{\mathbb{V}}_{j-1} + {}^j\mathbb{R}_{j-1}\, {}^{j-1}\mathbf{\Phi}_{j-1}\, \ddot{\mathbf{q}}_{ej-1} + {}^j\mathbb{H}_j \tag{7.30}$$

where we introduced:

– the 6 × 1 matrices:

$${}^j\mathbb{V}_j = \begin{Bmatrix} {}^j\mathbf{v}_{rj} \\ {}^j\boldsymbol{\omega}_{rj} \end{Bmatrix},\ {}^j\dot{\mathbb{V}}_j = \begin{Bmatrix} {}^j\boldsymbol{\gamma}_{rj} \\ {}^j\boldsymbol{\alpha}_{rj} \end{Bmatrix},\ {}^j\mathbb{A}_j = \begin{Bmatrix} \mathbf{0} \\ {}^j\mathbf{a}_j \end{Bmatrix},\ {}^j\mathbb{C}_j = \begin{Bmatrix} {}^j\mathbf{F}_{in,j} + {}^j\mathbf{F}_{g,j} \\ {}^j\mathbf{C}_{in,j} + {}^j\mathbf{C}_{g,j} \end{Bmatrix},$$

$${}^j\mathbb{F}_j = \begin{Bmatrix} {}^j\mathbf{F}_j \\ {}^j\mathbf{C}_j \end{Bmatrix} + \tau_j.{}^j\mathbb{A}_j,\ {}^j\mathbb{G}_j = \begin{Bmatrix} {}^j\mathbf{g}_{lin,j} \\ {}^j\mathbf{g}_{ang,j} \end{Bmatrix},\ {}^j\mathbb{H}_j = \ddot{q}_{rj}.{}^j\mathbb{A}_j + {}^j\mathbb{G}_j$$

where τ_j is the torque delivered by the actuator of the jth joint;

– the 6×6 matrices:

$$^{j}\mathbb{R}_{j-1} = \begin{pmatrix} ^{j}\mathbf{R}_{j-1} & \mathbf{0}_{3\times3} \\ \mathbf{0}_{3\times3} & ^{j}\mathbf{R}_{j-1} \end{pmatrix}, \; ^{j}\mathbb{T}_{j-1} = \begin{pmatrix} ^{j}\mathbf{R}_{j-1} & -^{j}\mathbf{R}_{j-1}.^{j-1}\hat{\mathbf{P}}_{j} \\ \mathbf{0}_{3\times3} & ^{j}\mathbf{R}_{j-1} \end{pmatrix},$$

$$\mathbb{J}_{rrj} = \begin{pmatrix} m_j \mathbf{I}_{3,3} & ^{j}\mathbf{M}\hat{\mathbf{S}}_j^T \\ ^{j}\mathbf{M}\hat{\mathbf{S}}_j & ^{j}\mathbf{I}_j \end{pmatrix}$$

where $^{j}\mathbf{R}_{j-1}$ is the 3×3 matrix of the axes of R_{j-1} in those of R_j, and $^{j-1}P_j$ is the 3×1 matrix of the position of point A_j in R_{j-1};

– the $6\times N_j$ matrices:

$$^{j}\mathbf{\Phi}_j = \begin{pmatrix} ^{j}\mathbf{\Phi}_{dj1}(B_{oj}),.., ^{j}\mathbf{\Phi}_{djNj}(B_{oj}) \\ ^{j}\mathbf{\Phi}_{rj1}(B_{oj}),.., ^{j}\mathbf{\Phi}_{rjNj}(B_{oj}) \end{pmatrix}, \; \mathbf{J}_{rej} = \mathbf{J}_{erj}^T = \begin{pmatrix} ^{j}\mathbf{MS}_{dej} \\ ^{j}\mathbf{MS}_{rej} \end{pmatrix};$$

– the $N_j \times 1$ matrices:

$$\mathbf{c}_j = \mathbf{Q}_{inj} + \mathbf{Q}_{gj} + \mathbf{k}_{eej}\mathbf{q}_{ej};$$

– the $N_j \times N_j$ matrices:

$$\mathbf{J}_{eej} = \mathbf{m}_{eej}.$$

Finally, we will see later on how it is possible to obtain the inverse and direct dynamics of a flexible robot manipulator based on these recursive equations. Alternatively, these two problems can also be dealt with due to the Lagrangian model [BOO 84], [BOY 94], [CHE 90], [SIN 84]:

$$\begin{pmatrix} \mathbf{M}_{rr} & \mathbf{M}_{re} \\ \mathbf{M}_{er} & \mathbf{M}_{ee} \end{pmatrix} \cdot \begin{Bmatrix} \ddot{\mathbf{q}}_r \\ \ddot{\mathbf{q}}_e \end{Bmatrix} = \begin{Bmatrix} \mathbf{T}_r \\ \mathbf{T}_e \end{Bmatrix} + \begin{Bmatrix} \boldsymbol{\tau} \\ \mathbf{0} \end{Bmatrix} \quad [7.31]$$

where matrices $\mathbf{M}_{rr}$ and $\mathbf{M}_{ee}$ are respectively the rigid and elastic inertial matrices, $\mathbf{M}_{re}$ represents the coupling matrix between the two sub-systems, $\mathbf{T}_r$ and $\mathbf{T}_e$ contain the elastic, centrifugal Coriolis and gravity forces, acting respectively on the rigid and flexible degrees of freedom and τ is the vector of torques generated by the actuators. In order to obtain this model, we will have to give up the direct calculation using the Lagrange equations and use instead the operation known as "assembly" of the Newton-Euler model [BOY 96b], [GLA 99].

7.2.7. *Geometric model of an open chain*

The three sets of equations [7.28], [7.29] and [7.30] must be completed by a geometric model of the robot manipulator, i.e. by a model enabling to pass from the reference frame of any body of the chain to its successor. Let two such bodies be S_{j-1} and S_j. By using the 4×4 homogenous transformation formalism, the transformation applying the reference frame R_{j-1} on R_j is written:

$$ {}^{j-1}\mathbf{T}_j = \begin{pmatrix} {}^{j-1}\mathbf{R}_j & {}^{j-1}\mathbf{P}_j \\ \mathbf{0} & 1 \end{pmatrix}. $$

where ${}^{j-1}\mathbf{R}_j$ and ${}^{j-1}\mathbf{P}_j$ were previously defined. These two matrices are easily calculated due to the relation:

$$ {}^{j-1}\mathbf{T}_j = \begin{pmatrix} {}^{j-1}\mathbf{R}_{ej-1}\,{}^{j-1}\mathbf{R}_{rj} & {}^{j-1}\mathbf{P}_{rj} + {}^{j-1}\mathbf{d}_{ej} \\ \mathbf{0} & 1 \end{pmatrix}. $$

where the following notations were introduced:

– the 4×4 matrix:

$$ \mathbf{T}_{rj} = \begin{pmatrix} \mathbf{R}_{rj} & \mathbf{P}_{rj} \\ \mathbf{0} & 1 \end{pmatrix} $$

is the homogenous transformation applying R_{j-1} on R_j when the robot is considered rigid (the set of these homogenous transformations creates the usual geometric model of the rigid robot);

– the 3×1 matrix:

$$ {}^{j-1}\mathbf{d}_{ej} = \sum_{k=1}^{N_{j-1}} {}^{j-1}\mathbf{\Phi}_{dj-1,k}(B_{oj-1})\, q_{ej-1,k} $$

represents the displacement induced by the deformation of S_{j-1} evaluated at the connection point with S_j;

– the 3×3 matrix:

$$\mathbf{R}_{ej-1} = \sum_{k=1}^{N_{j-1}} {}^{j-1}\hat{\mathbf{\Phi}}_{rj-1,k}(B_{oj-1}).q_{ej-1,k}$$

represents the rotation induced by the deformation of the body S_{j-1} evaluated at the connection point with S_j. It is relevant only under the assumption of small elastic displacements and small deformations of the body, i.e. within the framework of linear elasticity. In order to exceed this limit, we refer the reader to [BOY 99a], [BOY 99b], [BOY 00] where a non-linear kinematic of a Euler-Bernoulli beam as well as generalized Newton-Euler models are detailed, creating a consistent approximation of order one and two of the energy balance (consistent linear and quadratic models). These two models particularly make it possible to model certain non-linear effects occurring in fast dynamics (such as the "dynamic stiffening" [SHA 95]). In addition, the quadratic model can model the dynamics of a manipulator subjected to finite elastic displacements (i.e. the arrows at the end of the body representing half the length of the bodies).

7.2.8. *Recursive calculation of the inverse and direct dynamic models for a flexible robot*

7.2.8.1. *Introduction*

Let us reconsider the form [7.28] of the dynamic model of a flexible robot. The inverse dynamic model, which is used for the control, makes it possible to calculate the vector of actuator torques as well as the vector of elastic accelerations in terms of the state of the system (rigid and elastic positions and velocities) and of the vector of rigid accelerations. At this level, it is important to notice that the elastic accelerations are neither input nor output accelerations of the algorithm, but internal variables constrained by the lower part of equation [7.31]:

$$\ddot{\mathbf{q}}_e = \mathbf{M}_{ee}^{-1}(\mathbf{T}_e - \mathbf{M}_{er}\ddot{\mathbf{q}}_r) \qquad [7.32]$$

By reporting this result in the higher part of equation [7.31], we obtain the reduced joint dynamics:

$$(\mathbf{M}_{rr} - \mathbf{M}_{re}\mathbf{M}_{ee}^{-1}\mathbf{M}_{er})\ddot{\mathbf{q}}_r + (\mathbf{M}_{re}\mathbf{M}_{ee}^{-1}\mathbf{T}_e - \mathbf{T}_r) = \boldsymbol{\tau} \qquad [7.33]$$

In section 7.2.8.2, we will present a recursive and effective algorithm (from the point of view of the number of operations) in order to calculate the inverse dynamic model without explicitly calculating the matrix elements of equation [7.31] (or alternatively of the two equations above). This algorithm creates a generalization of the inverse algorithm for rigid robot dynamics proposed by Luh, Walker and Paul [LUH 80] which is presented in Chapter 1 of this book.

To simulate the behavior of the robot, the direct dynamic model is used. This model makes it possible to calculate rigid and elastic accelerations according to the joint torques and to the system status vector (rigid and elastic position and velocities). The calculation of the direct dynamic model starting from form [7.31] requires the inversion of the total inertial matrix of the robot (or, alternatively, the inversion of the reduced dynamics matrix [7.33]). We present in section 7.2.8.3 a recursive and effective algorithm in order to obtain rigid and elastic accelerations without explicitly calculating the matrix elements of equation [7.31]. This algorithm constitutes a generalization of the algorithm for rigid robots proposed by [ARM 79], [BRA 86], [FEA 83], [KHA 02].

7.2.8.2. *Recursive algorithm of the inverse dynamic model*

The algorithm uses three recurrences:

i) forward recurrence for $j = 1,...n$ in order to calculate ${}^{j}\mathbb{H}_j$, ${}^{j}\mathbb{C}_j, \mathbf{c}_j$, $\mathbb{J}_{rrj}$, $\mathbf{J}_{rej}$, $\mathbf{J}_{erj}$, ${}^{j-1}\mathbb{R}_j$, ${}^{j-1}\mathbb{T}_j$ defined by equations [7.28], [7.29] and [7.30], in terms of the elastic and rigid velocities and positions as well as rigid accelerations;

ii) backward recurrence for $j = n,...1$ in order to calculate matrices and vectors, making it possible to express $\ddot{\mathbf{q}}_{ej}$ and ${}^{j}\mathbb{F}_j$ according to ${}^{j}\dot{\mathbb{V}}_j$;

iii) forward recurrence for $j = 1,...n$ in order to calculate $\ddot{\mathbf{q}}_{ej}$, ${}^{j}\mathbb{F}_j$ and τ_j.

The equations of the first and third recurrence are obtained directly from the relations developed in sections 7.2.6 and 7.2.7. Consequently, we will deal in detail with the first two iterations of the backward recurrence, and then we will give the complete algorithm.

First iteration: $j = n$

We will start from equations [7.28] and [7.30] when $j = n$:

$$\begin{pmatrix} \mathbb{J}_{rrn} & \mathbf{J}_{ren} \\ \mathbf{J}_{ern} & \mathbf{J}_{een} \end{pmatrix} \begin{Bmatrix} {}^{n}\dot{\mathbb{V}}_n \\ \ddot{\mathbf{q}}_{en} \end{Bmatrix} + \begin{Bmatrix} {}^{n}\mathbb{C}_n \\ \mathbf{c}_n \end{Bmatrix} = \begin{Bmatrix} {}^{n}\mathbb{F}_n \\ \mathbf{0} \end{Bmatrix} \qquad [7.34]$$

$${}^{n}\dot{\mathbb{V}}_n = {}^{n}\mathbb{T}_{n-1}\, {}^{n-1}\dot{\mathbb{V}}_{n-1} + {}^{n}\mathbb{R}_{n-1}\, {}^{n-1}\mathbf{\Phi}_{n-1}\ddot{\mathbf{q}}_{en-1} + {}^{n}\mathbb{H}_n \qquad [7.35]$$

We suppose that the end-effector does not have any contact with the environment, i.e.: ${}^{n+1}\mathbb{F}_{n+1} = \mathbb{O}$. Otherwise, we establish ${}^{n+1}\mathbb{F}_{n+1} = \mathbb{F}_{ext}$ and consequently we change ${}^{n}\mathbb{C}_n$ and $\mathbf{c}_n$. The elastic accelerations of the body n are obtained from equation [7.34]:

$$\ddot{\mathbf{q}}_{en} = \mathbf{J}_{een}^{-1}\left(-\mathbf{J}_{ern}\,{}^{n}\dot{\mathbb{V}}_n - \mathbf{c}_n\right) \tag{7.36}$$

If we report this vector in the higher part of equation [7.34], we obtain:

$${}^{n}\mathbb{F}_n = (\mathbb{J}_{rrn} - \mathbf{J}_{ren}\mathbf{J}_{een}^{-1}\mathbf{J}_{ern})\,{}^{n}\dot{\mathbb{V}}_n - \mathbf{J}_{ren}\mathbf{J}_{een}^{-1}\mathbf{c}_n + {}^{n}\mathbb{C}_n$$

which gives:

$${}^{n}\mathbb{F}_n = \mathbb{K}_n\,{}^{n}\dot{\mathbb{V}}_n + {}^{n}\mathbb{M}_n \tag{7.37}$$

with:

$$\mathbb{K}_n = \mathbb{J}_{rrn} - \mathbf{J}_{ren}\mathbf{J}_{een}^{-1}\mathbf{J}_{ern}, \quad {}^{n}\mathbb{M}_n = -\mathbf{J}_{ren}\mathbf{J}_{een}^{-1}\mathbf{c}_n + {}^{n}\mathbb{C}_n$$

By using equations [7.35] and [7.37] we obtain:

$${}^{n}\mathbb{F}_n = \mathbb{B}_n\,{}^{n-1}\dot{\mathbb{V}}_{n-1} + \mathbf{H}_n\ddot{\mathbf{q}}_{en-1} + {}^{n}\mathbb{P}_n \tag{7.38}$$

with:

$$\mathbb{B}_n = \mathbb{K}_n\,{}^{n}\mathbb{T}_{n-1},$$

$$\mathbf{H}_n = \mathbb{K}_n\,{}^{n}\mathbb{R}_{n-1}\,{}^{n-1}\mathbf{\Phi}_{n-1}, \quad {}^{n}\mathbb{P}_n = \mathbb{K}_n\,{}^{n}\mathbb{H}_n + {}^{n}\mathbb{M}_n$$

Second iteration: $j = n-1$

By using equation [7.38] in [7.28] as well as equation [7.29], we obtain for $j = n-1$:

$$\begin{pmatrix} \mathbb{J}^{*}_{rrn-1} & \mathbf{J}^{*}_{ren-1} \\ \mathbf{J}^{*}_{ern-1} & \mathbf{J}^{*}_{een-1} \end{pmatrix} \begin{Bmatrix} {}^{n-1}\dot{\mathbb{V}}_{n-1} \\ \ddot{\mathbf{q}}_{en-1} \end{Bmatrix} + \begin{Bmatrix} {}^{n-1}\mathbb{C}^{*}_{n-1} \\ \mathbf{c}^{*}_{n-1} \end{Bmatrix} = \begin{Bmatrix} {}^{n-1}\mathbb{F}_{n-1} \\ \mathbf{0} \end{Bmatrix} \tag{7.39}$$

$$ {}^{n-1}\dot{\mathbb{V}}_{n-1} = {}^{n-1}\mathbb{T}_{n-2}\,{}^{n-2}\dot{\mathbb{V}}_{n-2} + {}^{n-1}\mathbb{R}_{n-2}\,{}^{n-2}\mathbf{\Phi}_{n-2}\ddot{\mathbf{q}}_{en-2} + {}^{n-1}\mathbb{H}_{n-1} \quad [7.40] $$

with:

$$ \mathbb{J}^{*}_{rrn-1} = \mathbb{J}_{rrn-1} + {}^{n}\mathbb{T}^{T}_{n-1}\mathbb{B}_{n},\ \mathbf{J}^{*}_{ren-1} = \mathbf{J}_{ren-1} + {}^{n}\mathbb{T}^{T}_{n-1}\mathbf{H}_{n}, $$

$$ \mathbf{J}^{*}_{ern-1} = \mathbf{J}_{ern-1} + {}^{n-1}\mathbf{\Phi}^{T}_{n-1}\,{}^{n-1}\mathbb{R}_{n}\mathbb{B}_{n} $$

$$ \mathbf{J}^{*}_{een-1} = \mathbf{J}_{een-1} + {}^{n-1}\mathbf{\Phi}^{T}_{n-1}\,{}^{n-1}\mathbb{R}_{n}\mathbf{H}_{n},\ \ {}^{n-1}\mathbb{C}^{*}_{n-1} = {}^{n-1}\mathbb{C}_{n-1} + {}^{n}\mathbb{T}^{T}_{n-1}\,{}^{n}\mathbb{P}_{n} $$

$$ \mathbf{c}^{*}_{n-1} = \mathbf{c}_{n-1} + {}^{n-1}\mathbf{\Phi}^{T}_{n-1}\,{}^{n-1}\mathbb{R}_{n}\,{}^{n}\mathbb{P}_{n} \quad [7.41] $$

Equations [7.39] and [7.40] have the same form as [7.34] and [7.35]. We can thus calculate:

$$ \ddot{\mathbf{q}}_{en-1} = \mathbf{J}^{*-1}_{een-1}(-\mathbf{J}^{*}_{ern-1}\,{}^{n-1}\dot{\mathbb{V}}_{n-1} - \mathbf{c}^{*}_{n-1}) \quad [7.42] $$

$$ {}^{n-1}\mathbb{F}_{n-1} = \mathbb{B}_{n-1}\,{}^{n-2}\dot{\mathbb{V}}_{n-2} + \mathbf{H}_{n-1}\ddot{\mathbf{q}}_{en-2} + {}^{n-1}\mathbb{P}_{n-1} \quad [7.43] $$

with:

$$ \mathbb{K}_{n-1} = \mathbb{J}^{*}_{rrn-1} - \mathbf{J}^{*}_{ren-1}\mathbf{J}^{*-1}_{een-1}\mathbf{J}^{*}_{ern-1},\ \mathbb{B}_{n-1} = \mathbb{K}_{n-1}\,{}^{n-1}\mathbb{T}_{n-2} $$

$$ \mathbf{H}_{n-1} = \mathbb{K}_{n-1}\,{}^{n-1}\mathbb{R}_{n-2}\,{}^{n-2}\mathbf{\Phi}_{n-2} $$

$$ {}^{n-1}\mathbb{M}_{n-1} = -\mathbf{J}^{*}_{ren-1}\mathbf{J}^{*-1}_{een-1}\mathbf{c}^{*}_{n-1} + {}^{n-1}\mathbb{C}^{*}_{n-1} $$

$$ {}^{n-1}\mathbb{P}_{n-1} = \mathbb{K}_{n-1}\,{}^{n-1}\mathbb{H}_{n-1} + {}^{n-1}\mathbb{M}_{n-1} \quad [7.44] $$

We can repeat this procedure for $j = n-2, \ldots 1$. The last iteration of the recurrence gives ${}^{1}\mathbb{F}_{1}$ in terms of ${}^{0}\dot{\mathbb{V}}_{0}$ whose elements are known since ${}^{0}\dot{\mathbb{V}}_{0} = -\begin{Bmatrix} {}^{0}\mathbf{g} \\ \mathbf{0} \end{Bmatrix}$. This recurrence provides the elements $\mathbb{K}_{j}$, $\mathbb{B}_{j}$, $\mathbf{H}_{j}$, ${}^{j}\mathbb{P}_{j}$, ${}^{j}\mathbb{C}^{*}_{j}$, ${}^{j}\mathbb{M}_{j}$, $\mathbf{c}^{*}_{j}$, for $j = 1, \ldots, n$.

In short, the complete calculation algorithm of the inverse dynamic model is given by the three following recurrences:

i) for $j = 1$ to $j = n$, calculate:

$$^{j}\boldsymbol{\omega}_j = {}^{j}\mathbf{R}_{j-1}\,{}^{j-1}\boldsymbol{\omega}_{j-1} + {}^{j}\mathbf{R}_{j-1}\,{}^{j-1}\boldsymbol{\Phi}_{rj-1}\,\dot{\mathbf{q}}_{ej-1} + \dot{q}_{rj}\,{}^{j}\mathbf{a}_j$$

then calculate $^{j}\mathbb{H}_j$, $^{j}\mathbb{C}_j$, $\mathbf{c}_j$, $\mathbb{J}_{rrj}$, $\mathbf{J}_{rej}$, $\mathbf{J}_{erj}$, $^{j-1}\mathbb{R}_j$, $^{j-1}\mathbb{T}_j$.

If the base is fixed, we set $^{0}\boldsymbol{\omega}_0 = \mathbf{0}$;

ii) for $j = n$ to $j = 1$, calculate:

$$\mathbb{K}_j = \mathbb{J}^{*}_{rrj} - \mathbf{J}^{*}_{rej}\mathbf{J}^{*-1}_{eej}\mathbf{J}^{*}_{erj},$$

$$^{j}\mathbb{M}_j = -\mathbf{J}^{*}_{rej}\mathbf{J}^{*-1}_{eej}\mathbf{c}^{*}_j + {}^{j}\mathbb{C}^{*}_j,$$

if $j \neq n$ calculate:

$$\mathbb{J}^{*}_{rrj} = \mathbb{J}_{rrj} + {}^{j+1}\mathbb{T}^{T}_j\mathbb{B}_{j+1},$$

$$\mathbf{J}^{*}_{rej} = \mathbf{J}_{rej} + {}^{j+1}\mathbb{T}^{T}_j\mathbf{H}_{j+1},$$

$$\mathbf{J}^{*}_{erj} = \mathbf{J}^{*T}_{rej},$$

$$\mathbf{J}^{*}_{eej} = \mathbf{J}_{eej} + {}^{j}\boldsymbol{\Phi}^{T}_j\,{}^{j}\mathbb{R}_{j+1}\mathbf{H}_{j+1},$$

$$^{j}\mathbb{C}^{*}_j = {}^{j}\mathbb{C}_j + {}^{j+1}\mathbb{T}^{T}_j\,{}^{j+1}\mathbb{P}_{j+1},$$

$$\mathbf{c}^{*}_j = \mathbf{c}_j + {}^{j}\boldsymbol{\Phi}^{T}_j\,{}^{j}\mathbb{R}_{j+1}\,{}^{j+1}\mathbb{P}_{j+1}$$

if $j \neq 1$ calculate:

$$\mathbb{B}_j = \mathbb{K}_j\,{}^{j}\mathbb{T}_{j-1},$$

$$\mathbf{H}_j = \mathbb{K}_j \, {}^j\mathbb{R}_{j-1} \, {}^{j-1}\mathbf{\Phi}_{j-1},$$

$${}^j\mathbb{P}_j = \mathbb{K}_j \, {}^j\mathbb{H}_j + {}^j\mathbb{M}_j$$

This recurrence is initialized by: $\mathbb{J}^*_{rrn} = \mathbb{J}_{rrn}$, $\mathbf{J}^*_{ren} = \mathbf{J}_{ren}$, $\mathbf{J}^*_{een} = \mathbf{J}_{een}$, ${}^n\mathbb{C}^*_n = {}^n\mathbb{C}_n$, $\mathbf{c}^*_n = \mathbf{c}_n$;

iii) from equations [7.35] and [7.36], we obtain the equations of the third recurrence:

for $j = 1$ to $j = n$, calculate:

$${}^j\dot{\mathbb{V}}_j = {}^j\mathbb{T}_{j-1} \, {}^{j-1}\dot{\mathbb{V}}_{j-1} + {}^j\mathbb{R}_{j-1} \, {}^{j-1}\mathbf{\Phi}_{j-1}\ddot{\mathbf{q}}_{ej-1} + {}^j\mathbb{H}_j$$

$$\ddot{\mathbf{q}}_{ej} = \mathbf{J}^{*-1}_{eej}\left(-\mathbf{J}^*_{erj} \, {}^j\dot{\mathbb{V}}_j - \mathbf{c}^*_j\right)$$

The torque of joint j is obtained by projecting ${}^j\mathbb{F}_j$ on ${}^j\mathbb{A}_j$; using equation [7.37] we have:

$$\tau_j = {}^j\mathbb{A}_j^T(\mathbb{K}_j \, {}^j\dot{\mathbb{V}}_j + {}^j\mathbb{M}_j)$$

When the base is fixed, this calculation is initialized by ${}^0\dot{\mathbb{V}}_0 = -\begin{Bmatrix} {}^0\mathbf{g} \\ \mathbf{0} \end{Bmatrix}$, $\ddot{\mathbf{q}}_{e0} = \mathbf{0}$.

7.2.8.3. *Recursive algorithm of the direct dynamic model*

The calculation of the direct dynamic model through a recursive algorithm based on the Newton-Euler equations was first proposed by D'Eleuterio [DEL 92]. Its solution consists of gathering the rigid and elastic accelerations of each body in one vector and then applying the resolution procedure suggested for the rigid robots. We propose a different approach here, which initially calculates the elastic accelerations using equation [7.36].

The algorithm uses three recurrences:

i) forward recurrence for $j = 1,\ldots n$ in order to calculate the matrices ${}^j\mathbb{H}_j$, ${}^j\mathbb{C}_j$, $\mathbf{c}_j$, $\mathbb{J}_{rrj}$, $\mathbf{J}_{rej}$, $\mathbf{J}_{erj}$, ${}^{j-1}\mathbb{R}_j$, ${}^{j-1}\mathbb{T}_j$ which depend on the elastic and rigid positions and velocities;

ii) backward recurrence for $j = n,...1$ in order to calculate the matrices and vectors making it possible to calculate $\ddot{q}_{rj}$, $\ddot{\mathbf{q}}_{ej}$ and ${}^{j}\mathbb{F}_{j}$ according to ${}^{j}\dot{\mathbb{V}}_{j}$;

iii) forward recurrence for $j = 1,...n$ in order to calculate the $\ddot{q}_{rj}$ and $\ddot{\mathbf{q}}_{ej}$.

The equations of the first and third recurrences are obtained directly from the relations developed in sections 7.2.6 and 7.2.7. Further on, we will deal in detail with the first two iterations of the backward recurrence and then we will give the complete algorithm.

First iteration: $j = n$

Using [7.30] we replace ${}^{n}\mathbb{H}_{n}$ by ${}^{n}\mathbb{G}_{n} + \ddot{q}_{rn}\,{}^{n}\mathbb{A}_{n}$ in equation [7.35] in order to express the nth joint acceleration, which is unknown in the case of the direct dynamic model, we obtain:

$$ {}^{n}\dot{\mathbb{V}}_{n} = {}^{n}\mathbb{T}_{n-1}\,{}^{n-1}\dot{\mathbb{V}}_{n-1} + {}^{n}\mathbb{R}_{n-1}\,{}^{n-1}\mathbf{\Phi}_{n-1}\ddot{\mathbf{q}}_{en-1} + {}^{n}\mathbb{G}_{n} + \ddot{q}_{rn}\,{}^{n}\mathbb{A}_{n} \qquad [7.45] $$

As in the inverse model, based on equation [7.34] and [7.36], we have:

$$ {}^{n}\mathbb{F}_{n} = \mathbb{K}_{n}\,{}^{n}\dot{\mathbb{V}}_{n} + {}^{n}\mathbb{M}_{n} \qquad [7.46] $$

with:

$$ \mathbb{K}_{n} = \mathbb{J}_{rrn} - \mathbf{J}_{ren}\mathbf{J}_{een}^{-1}\mathbf{J}_{ern}, \quad {}^{n}\mathbb{M}_{n} = -\mathbf{J}_{ren}\mathbf{J}_{een}^{-1}\mathbf{c}_{n} + {}^{n}\mathbb{C}_{n} $$

To calculate $\ddot{q}_{rn}$, we pre-multiply [7.45] by ${}^{n}\mathbb{A}_{n}^{T}\mathbb{K}_{n}$, which gives:

$$ {}^{n}\mathbb{A}_{n}^{T}\mathbb{K}_{n}\,{}^{n}\dot{\mathbb{V}}_{n} = $$

$$ {}^{n}\mathbb{A}_{n}^{T}\mathbb{K}_{n}\,{}^{n}\mathbb{A}_{n}\ddot{q}_{rn} + {}^{n}\mathbb{A}_{n}^{T}\mathbb{K}_{n}[{}^{n}\mathbb{T}_{n-1}\,{}^{n-1}\dot{\mathbb{V}}_{n-1} + {}^{n}\mathbf{R}_{n-1}\,{}^{n-1}\mathbf{\Phi}_{n-1}\ddot{\mathbf{q}}_{en-1} + {}^{n}\mathbb{G}_{n}] \qquad [7.47] $$

Then by pre-multiplying [7.46] by ${}^{n}\mathbb{A}_{n}^{T}$, we obtain:

$$ \tau_{n} - {}^{n}\mathbb{A}_{n}^{T}\,{}^{n}\mathbb{M}_{n} = {}^{n}\mathbb{A}_{n}^{T}\mathbb{K}_{n}\,{}^{n}\dot{\mathbb{V}}_{n} $$

Finally, by using [7.47], we obtain:

$$\ddot{q}_{rn} = k_n^{-1}\{\tau_n - {}^n\mathbb{A}_n^T[{}^n\mathbb{M}_n + \mathbb{K}_n({}^n\mathbb{T}_{n-1}\,{}^{n-1}\dot{\mathbb{V}}_{n-1} + {}^n\mathbb{R}_{n-1}\,{}^{n-1}\mathbf{\Phi}_{n-1}\ddot{\mathbf{q}}_{en-1} + {}^n\mathbb{G}_n)]\} \quad [7.48]$$

where the scalar k_n is given by:

$$k_n = {}^n\mathbb{A}_n^T\mathbb{K}_n\,{}^n\mathbb{A}_n$$

We can now eliminate the acceleration $\ddot{q}_{rn}$ in equation [7.45]:

$${}^n\dot{\mathbb{V}}_n = \mathbb{E}_n\,{}^n\mathbb{T}_{n-1}\,{}^{n-1}\dot{\mathbb{V}}_{n-1} + \mathbb{E}_n\,{}^n\mathbb{R}_{n-1}\,{}^{n-1}\mathbf{\Phi}_{n-1}\ddot{\mathbf{q}}_{en-1} + \mathbb{E}_n\,{}^n\mathbb{G}_n + k_n^{-1}\,{}^n\mathbb{A}_n(\tau_n - {}^n\mathbb{A}_n^T\,{}^n\mathbb{M}_n)$$

with:

$$\mathbb{E}_n = \mathbb{I}_6 - k_n^{-1}\,{}^n\mathbb{A}_n\,{}^n\mathbb{A}_n^T\mathbb{K}_n$$

and $\mathbb{I}_6$ is the 6×6 identity matrix.

If we report this last expression of ${}^n\dot{\mathbb{V}}_n$ in [7.46], we obtain the interaction forces for the following iteration:

$${}^n\mathbb{F}_n = \mathbf{B}_n\,{}^{n-1}\dot{\mathbb{V}}_{n-1} + \mathbf{Y}_n\,\ddot{\mathbf{q}}_{en-1} + {}^n\mathbb{W}_n \quad [7.49]$$

with:

$$\mathbf{B}_n = \mathbb{K}_n\mathbb{E}_n\,{}^n\mathbb{T}_{n-1}$$

$$\mathbf{Y}_n = \mathbb{K}_n\mathbb{E}_n\,{}^n\mathbb{R}_{n-1}\,{}^{n-1}\mathbf{\Phi}_{n-1}$$

$${}^n\mathbb{W}_n = \mathbb{K}_n(\mathbb{E}_n\,{}^n\mathbb{G}_n + k_n^{-1}\,{}^n\mathbb{A}_n(\tau_n - {}^n\mathbb{A}_n^T\,{}^n\mathbb{M}_n)) + {}^n\mathbb{M}_n$$

Second iteration: $j = n-1$

The insertion of [7.49] in the balance equation of the body n–1 gives an equation similar to the one we started from, in order to "solve" the preceding iteration. And we have to consider the fact that the inertial matrices and forces are replaced by those of equation [7.41]. This enables us to obtain equations [7.50] and [7.51] in a way which is similar to the method used to calculate [7.48] and [7.49]:

$$\ddot{q}_{rn-1} = k_{n-1}^{-1}\{\tau_{n-1} - {}^{n-1}\mathbb{A}_{n-1}^{T}[{}^{n-1}\mathbb{M}_{n-1} + \mathbb{K}_{n-1}({}^{n-1}\mathbb{T}_{n-2}\,{}^{n-2}\dot{\mathbb{V}}_{n-2} +$$

$$+{}^{n-1}\mathbb{R}_{n-2}\,{}^{n-2}\mathbf{\Phi}_{n-2}\ddot{\mathbf{q}}_{en-2} + {}^{n-1}\mathbb{G}_{n-1})]\} \quad [7.50]$$

$${}^{n-1}\mathbb{F}_{n-1} = \mathbb{B}_{n-1}\,{}^{n-2}\dot{\mathbb{V}}_{n-2} + \mathbf{Y}_{n-1}\ddot{\mathbf{q}}_{en-2} + {}^{n-1}\mathbb{W}_{n-1} \quad [7.51]$$

with:

$$\mathbb{K}_{n-1} = \mathbb{J}_{rrn-1}^{*} - \mathbf{J}_{ren-1}^{*}\mathbf{J}_{een-1}^{*-1}\mathbf{J}_{ern-1}^{*},\quad {}^{n-1}\mathbb{M}_{n-1} = -\mathbf{J}_{ren-1}^{*}\mathbf{J}_{een-1}^{*-1}\mathbf{c}_{n-1}^{*} + {}^{n-1}\mathbb{C}_{n-1}^{*},$$

$$k_{n-1} = {}^{n-1}\mathbb{A}_{n-1}^{T}\mathbb{K}_{n-1}\,{}^{n-1}\mathbb{A}_{n-1}$$

$$\mathbb{E}_{n-1} = \mathbb{I}_{6} - k_{n-1}^{-1}\,{}^{n-1}\mathbb{A}_{n-1}\,{}^{n-1}\mathbb{A}_{n-1}^{T}\mathbb{K}_{n-1}$$

$$\mathbb{B}_{n-1} = \mathbb{K}_{n-1}\mathbb{E}_{n-1}\,{}^{n-1}\mathbb{T}_{n-2},$$

$$\mathbf{Y}_{n-1} = \mathbb{K}_{n-1}\mathbb{E}_{n-1}\,{}^{n-1}\mathbb{R}_{n-2}\,{}^{n-2}\mathbf{\Phi}_{n-2}$$

$${}^{n-1}\mathbb{W}_{n-1} = \mathbb{K}_{n-1}\left(\mathbb{E}_{n-1}\,{}^{n-1}\mathbb{G}_{n-1} + k_{n-1}^{-1}\,{}^{n-1}\mathbb{A}_{n-1}(\tau_{n-1} - {}^{n-1}\mathbb{A}_{n-1}^{T}\,{}^{n-1}\mathbb{M}_{n-1})\right) + {}^{n-1}\mathbb{M}_{n-1} \quad [7.52]$$

This procedure is repeated for $j = n - 2, \ldots 1$.

Finally, we obtain the rigid and elastic accelerations of the first body in terms of the known acceleration of the robot base, and the accelerations of the other bodies by a forward recurrence for $j = 1,\ldots n$.

The complete algorithm with its three recurrences is written:

i) This recurrence is similar to the first recurrence of the inverse dynamic model except for the calculation of ${}^{j}\mathbb{H}_{j}$ which is replaced by that of ${}^{j}\mathbb{G}_{j}$;

ii) the forward recurrence is given by:

For $j = n$ to $j = 1$, calculate:

$\mathbb{K}_{j}$ and ${}^{j}\mathbb{M}_{j}$: as for the inverse model,

$$k_{j} = {}^{j}\mathbb{A}_{j}^{T}\mathbb{K}_{j}\,{}^{j}\mathbb{A}_{j}$$

if $j \neq n$ calculate:

– calculation of: $\mathbb{J}^{*}_{rrj}, \mathbf{J}^{*}_{rej}, \mathbf{J}^{*}_{erj}, \mathbf{J}^{*}_{eej}, {}^{j}\mathbb{C}^{*}_{j}, \mathbf{c}^{*}_{j}$ (see the corresponding part in the inverse dynamic model),

if $j \neq 1$ calculate:

$$\mathbb{E}_j = \mathbb{I}_6 - k_j^{-1}\, {}^{j}\mathbb{A}_j\, {}^{j}\mathbb{A}_j^T \mathbb{K}_j$$

$$\mathbb{B}_j = \mathbb{K}_j \mathbb{E}_j\, {}^{j}\mathbb{T}_{j-1}$$

$$\mathrm{Y}_j = \mathbb{K}_j \mathbb{E}_j\, {}^{j}\mathbb{R}_{j-1}\, {}^{j-1}\mathbf{\Phi}_{j-1}$$

$${}^{j}\mathbb{P}_j = \mathbb{K}_j \left(\mathbb{E}_j\, {}^{j}\mathbb{G}_j + k_j^{-1}\, {}^{j}\mathbb{A}_j (\tau_j - {}^{j}\mathbb{A}_j^T\, {}^{j}\mathbb{M}_j) \right) + {}^{j}\mathbb{M}_j$$

This recurrence is initialized by: $\mathbb{J}^{*}_{rrn} = \mathbb{J}_{rrn}$, $\mathbf{J}^{*}_{ren} = \mathbf{J}_{ren}$, $\mathbf{J}^{*}_{een} = \mathbf{J}_{een}$, ${}^{n}\mathbb{C}^{*}_{n} = {}^{n}\mathbb{C}_{n}$, $\mathbf{c}^{*}_{n} = \mathbf{c}_{n}$;

iii) From equations [7.48], [7.45] and [7.36], we obtain:

For $j = 1$ to $j = n$, calculate:

$$\ddot{q}_{rj} = k_j^{-1} \left\{ \tau_j - {}^{j}\mathbb{A}_j^T \left[{}^{j}\mathbb{M}_j + \mathbb{K}_j \left({}^{j}\mathbb{T}_{j-1}\, {}^{j-1}\dot{\mathbb{V}}_{j-1} + {}^{j}\mathbb{R}_{j-1}\, {}^{j-1}\mathbf{\Phi}_{j-1} . \ddot{\mathbf{q}}_{ej-1} + {}^{j}\mathbb{G}_j \right) \right] \right\}$$

$${}^{j}\dot{\mathbb{V}}_j = {}^{j}\mathbb{T}_{j-1}\, {}^{j-1}\dot{\mathbb{V}}_{j-1} + {}^{j}\mathbb{R}_{j-1}\, {}^{j-1}\mathbf{\Phi}_{j-1} \ddot{\mathbf{q}}_{ej-1} + {}^{j}\mathbb{G}_j + \ddot{q}_{rj}\, {}^{j}\mathbb{A}_j$$

$$\ddot{\mathbf{q}}_{ej} = \mathbf{J}^{*-1}_{eej} \left(-\mathbf{J}^{*}_{erj}\, {}^{j}\dot{\mathbb{V}}_j - \mathbf{c}^{*}_j \right)$$

When the base is fixed, this algorithm is initialized by ${}^{0}\dot{\mathbb{V}}_0 = -\begin{Bmatrix} {}^{0}\mathbf{g} \\ \mathbf{0} \end{Bmatrix}$, $\ddot{\mathbf{q}}_{e0} = \mathbf{0}$.

We note that this algorithm requires only the inversion of scalars k_j and diagonal matrices $\mathbf{J}_{eej}$ of dimension $N_j \times N_j$, while the algorithm of [DEL 92] requires the inversion of the matrices of dimension $(6 + N_j) \times (6 + N_j)$.

7.2.8.4. *Iterative symbolic calculation*

The preceding algorithms can be calculated numerically. However, in order to reduce their number of operations, it is preferable to program them by using an iterative symbolic calculation which considers particular values of the geometric and dynamic parameters. The method is based on the development of the preceding equations without using loops and according to the following rules [KHA 87], [KHA 02]:

1) to create an intermediate variable each time that an element of a matrix or vector contains a mathematical operation;

2) not to create an intermediate variable to represent a result that does not contain an operation, but to use it as it is for the rest of the calculation;

3) to eliminate all the intermediate variables, which are not used in the calculation of the desired variables. Thus, once the model is generated, all the created variables are searched for in the equations and those which do not appear in the calculation of the desired variables are removed.

7.3. Control of flexible robot manipulators

7.3.1. *Introduction*

As regards the case of rigid robots, the control of flexible robot manipulators implies different problems because of the intrinsic difficulties due to the very nature of these systems. First of all, these systems are underactuated in the sense that the motorized joints are in a finite number, while the dynamics of the bodies are rigorously described by PDE (partial derivative equations) governing the evolution of an infinite number of degrees of freedom. This under-actuation is still present in the simplified models obtained by Rayleigh-Ritz discretization because then one rigid degree of freedom and at least one elastic mode are considered for each body. A second characteristic of flexible robots is that their dynamic in the operational space is of non-minimum phase and consequently the inversion techniques used in the rigid case are no longer adequate because of the instability of the zero dynamics, rendered unobservable by feedback [DEL 89].

In this context, we usually classify the control problems in order of increasing difficulty, based on the following objectives:

– regulation: placing the robot on a desired constant configuration;

– point-to-point in fixed time: in this case an additional constraint is added to the regulation problem, i.e. the time of the joint motion, which becomes a fixed parameter;

– trajectory tracking in the joint space: the robot is constrained to follow a joint time law, whilst canceling or minimizing the vibrations of the end-effector at the end of the desired motion;

– trajectory tracking in the operational (or Cartesian) space.

Later on we will present some control results, which made it possible to achieve the goals previously described. These methods are synthesized for the models exposed in section 7.2, whose notations we will discuss again below.

7.3.2. *Reminder of notations*

We recall below equation [7.31] of the dynamic model of a flexible robot:

$$\begin{pmatrix} \mathbf{M}_{rr} & \mathbf{M}_{re} \\ \mathbf{M}_{er} & \mathbf{M}_{ee} \end{pmatrix} \begin{pmatrix} \ddot{\mathbf{q}}_r \\ \ddot{\mathbf{q}}_e \end{pmatrix} = \begin{pmatrix} \mathbf{T}_r \\ \mathbf{T}_e \end{pmatrix} + \begin{pmatrix} \boldsymbol{\tau} \\ \mathbf{0} \end{pmatrix} \quad [7.53]$$

We can also use another form of this model, expressing the terms of Coriolis, centrifugal and gravitational. Hence, the equations of motion can be written as follows:

$$\mathbf{M}(\mathbf{q})\ddot{\mathbf{q}} + \mathbf{C}(\mathbf{q},\dot{\mathbf{q}}) + \mathbf{K}(\mathbf{q})\mathbf{q} + \mathbf{D}(\mathbf{q})\dot{\mathbf{q}} + \mathbf{G}(\mathbf{q}) = \boldsymbol{\tau} \quad [7.54]$$

Matrix **M** is the inertial matrix, matrix **C** represents the Coriolis and centrifugal effects, **G** represents the forces of gravity, **K** is the stiffness matrix and **D** is the damping matrix. These two last matrices will be considered constant and diagonals. Equation [7.54] can be rewritten by separating the rigid variables $\mathbf{q}_r$ from the elastic variables $\mathbf{q}_e$:

$$\begin{pmatrix} \mathbf{M}_{rr} & \mathbf{M}_{re} \\ \mathbf{M}_{er} & \mathbf{M}_{ee} \end{pmatrix} \begin{pmatrix} \ddot{\mathbf{q}}_r \\ \ddot{\mathbf{q}}_e \end{pmatrix} + \begin{pmatrix} \mathbf{C}_{rr} & \mathbf{C}_{re} \\ \mathbf{C}_{er} & \mathbf{C}_{ee} \end{pmatrix} \begin{pmatrix} \dot{\mathbf{q}}_r \\ \dot{\mathbf{q}}_e \end{pmatrix} + \begin{pmatrix} \mathbf{G}_r \\ \mathbf{G}_e \end{pmatrix} + \begin{pmatrix} \mathbf{0} \\ \mathbf{D}\dot{\mathbf{q}}_e + \mathbf{K}\mathbf{q}_e \end{pmatrix} = \begin{pmatrix} \boldsymbol{\tau} \\ \mathbf{0} \end{pmatrix} \quad [7.55]$$

where $\mathbf{q}^T = (\mathbf{q}_r^T \ \mathbf{q}_e^T)^T$. Equation [7.55] clearly shows that the control acts only on one part of dynamics and that the system is thus underactuated. However, in the case of planar robots, the indirectly actuated modes remain controllable via the actuated states [LOP 94].

7.3.3. *Control methods*

As in the case of rigid robots, in order to control flexible robots we firstly generate a time law translating the desired reference motion of the robot. Then we determine the control law which ensures its monitoring.

Further on, we will briefly recall some control results for each one of the objectives mentioned in section 7.3.1 by detailing certain recent methods.

7.3.3.1. *Regulation*

The goal here is to reach a point of desired equilibrium point which will be marked $\mathbf{q}_d = (\mathbf{q}_{rd}, \mathbf{q}_{ed})$. The linear case of a flexible axis was studied in the first works [SCH 85], where the author compared the performances of a PD regulator on the joint coordinates to that of LQ regulators. The passivity theory was also used in [LAN 92], [WAN 92], where the authors studied the case of a flexible axis, and demonstrated the passivity of the transfer between the joint velocity and the control torque, hence proving that it is possible to stabilize the joint position by a simple PD on these variables. Moreover, the authors underline that the observability of elastic variables, via the stabilized joint velocity output, implies the systematic regulation of elastic vibrations, and this without considering structural damping and without feedback on the elastic variables. The non-linear case of the flexible multi-axis robots was studied in [CAN 97], [DEL 93], where the authors show that a joint PD control, associated to a term of compensation for the gravity effects $\mathbf{G}_r$, makes it possible to fully and asymptotically stabilize the joint positions. This control is written as follows:

$$\boldsymbol{\tau} = \mathbf{K}_p(\mathbf{q}_{rd} - \mathbf{q}_r) - \mathbf{K}_d\, \dot{\mathbf{q}}_r + \mathbf{G}_r(\mathbf{q}_{rd}, \mathbf{q}_{ed}) \qquad [7.56]$$

where $\mathbf{K}_p$, $\mathbf{K}_d$ are positive definite diagonal matrices.

7.3.3.2. *Point-to-point movement in fixed time*

This problem was dealt with in [MEC 94] using a mass-spring model. The method consists of decomposing the control signal on the basis of sinusoidal and linear functions. The decomposition coefficients are calculated by minimizing the control energy during the movement and the elastic accelerations at the end of the desired movement. In [CHA 95], a method based on optimal rigid trajectory is suggested for a mass-spring model. The joint movements are planned in order to minimize the elastic velocities, accelerations and "jerk" (time derivatives of the accelerations). In [DEL 01a], for the linear model of a flexible axis, the authors build an output without zero dynamics (i.e. of degree relatively equal to the dimension of the state vector). This makes it possible to express all the states of the system according to this output. The joint movement planning, making it possible to reach the final joint position without oscillations, is then carried out easily. This idea is developed in [DEL 01b] for a non-linear manipulator with two axes, whose last axis is flexible and modeled by only one elastic mode. In [BEN 04a], the problem of point-to-point movement in fixed time, for the planar multi axis flexible arms, was solved by a calculated torque type control associated to a backward integration of elastic dynamics and a planning of optimal joint movements. To conclude, we note that the methods enabling a monitoring of the movements in the joint space or in the operational space can meet the objective of the point-to-point control in fixed time. Next, we present two methods applied to the one-link case.

7.3.3.2.1. Control of one-link flexible robot by operational movement planning [BEN 00a], [BEN 03]

This method is based on the stable inversion of the transfer between the joint torque and the operational location of the end-effector. This transfer function is at non-minimum phase [SCH 85] (i.e. a transfer function admitting zeros with positive real parts), so its direct inversion leads to divergent controls. To avoid this problem, an operational movement is planned so as to cancel the effect of the unstable zeros of the transfer. This approach is presented here for a linear model with two elastic modes. Equation [7.55] becomes in this case the following:

$$\begin{cases} M_{11}\ddot{q}_r + M_{12}\ddot{q}_1 + M_{13}\ddot{q}_2 = \tau \\ M_{21}\ddot{q}_r + M_{22}\ddot{q}_1 + K_1 q_1 = 0 \\ M_{31}\ddot{q}_r + M_{33}\ddot{q}_2 + K_2 q_2 = 0 \end{cases} \quad [7.57]$$

with the elastic vector $\mathbf{q}_e = (q_1, q_2)^T$.

Considering the operational output:

$$y_t = Lq_r(t) + \phi_1(L)q_1(t) + \phi_2(L)q_2(t) \quad [7.58]$$

where L indicates the length of the flexible arm and ϕ_i the ith shape function.

Equations [7.57] and [7.58] make it possible to write the differential equation:

$$\begin{aligned} a_5 y_t^{(6)}(t) + a_4\ y_t^{(5)}(t) + a_3\ y_t^{(4)}(t) + a_2\ y_t^{(3)}(t) + a_1\ y_t^{(2)}(t) + a_0 y_t(t) = \\ b_4\tau^{(4)}(t) + b_3\tau^{(3)}(t) + b_2\tau^{(2)}(t) + b_1\tau^{(1)}(t) + b_0\tau(t) \end{aligned} \quad [7.59]$$

where $y_t^{(i)}$ indicates the ith derivative of y_t.

Associated to the initial conditions:

$$\tau(0) = \tau^{(1)}(0) = \tau^{(2)}(0) = \tau^{(3)}(0) = 0$$

$$y_t(0) = y_t^{(1)}(0) = y_t^{(2)}(0) = y_t^{(3)}(0) = y_t^{(4)}(0) = y_t^{(5)}(0) = 0 \qquad [7.60]$$

The solution of the differential equation [7.59], where we substituted a desired trajectory y_{td} with y_t, writes:

$$\tau(t) = \tau_1(t) + \tau_2(t) \qquad [7.61]$$

where $\tau_1(t)$ is the general solution of the homogenous differential equation associated to [7.59] and $\tau_2(t)$ is the particular solution of the non-homogenous equation [7.59]. By setting for the output a polynomial trajectory as:

$$y_d(t) = \sum_{i=1}^{i=p} a_i t^i \qquad [7.62]$$

where the constant coefficients $a_i, i \in \{1...p\}$ will be determined so as to obtain a bounded solution and the degree p of the polynomial will depend on the initial and final conditions which we want to impose for the output and on the number of unstable zeros of the system (see below). The homogenious solution is then given by:

$$\tau_1(t) = \sum_{i=1}^{i=4} A_i(a_i, t_0, \tau(0), \tau^{(1)}(0), \tau^{(2)}(0), \tau^{(3)}(0))e^{r_i t} \qquad [7.63]$$

where the terms A_i are linear according to the coefficients a_i, this being due to the linearity of the differential equation. The non-homogenous solution is given by:

$$\tau_2(t) = \sum_{i=1}^{i=p} B_i(a_i)t^i \qquad [7.64]$$

where the terms B_i are linear in a_i and are obtained by the substitution of equation [7.64] in [7.59] and term by term identification. The coefficients a_i are adjusted so as to cancel the effect of zeros with positive real parts and zero real parts, by imposing the following equalities:

$$A_i(a_i, t_0, \tau(0), \tau^{(1)}(0), \tau^{(2)}(0), \tau^{(3)}(0)) = 0, \forall t \quad [7.65]$$

for the A_i associated in [7.63] to the terms $e^{z_i t}$, z_i being zeros with real positive or zero real parts.

The coefficients of the desired movement are then determined by solving the following linear system:

$$\begin{aligned}
&A_1(a_i, t_0, \tau(0), \tau^{(1)}(0), \tau^{(2)}(0), \tau^{(3)}(0)) = 0 \\
&A_2(a_i, t_0, \tau(0), \tau^{(1)}(0), \tau^{(2)}(0), \tau^{(3)}(0)) = 0 \\
&A_3(a_i, t_0, \tau(0), \tau^{(1)}(0), \tau^{(2)}(0), \tau^{(3)}(0)) = 0 \\
&y_{td}^{(i)}(0) = 0, \ i \in \{0,1,2,3,4,5\} \\
&y_{td}(t_f) = y_f \\
&y_{td}^{(i)}(t_f) = 0, \ i \in \{1,2,3,4,5\}
\end{aligned} \quad [7.66]$$

The limited nominal control is finally obtained by equation [7.61]. This closed-form feedforward control is then supplemented by a proportional-derivative regulator on the joint coordinates..

7.3.3.2.2. Control of a one-link flexible robot through model parameterization

The linear dynamic model of a one link arm, equation [7.57], can be written as a parameter in which all its variables are expressed as a function of one variable, marked λ, and of its derivatives $\lambda^{(i)}$. This parametric form can be obtained by various methods, the most direct being the calculation of the companion form for the control of the controllable linear system [7.53] [KAI 80]. Another approach based on the parameterization of linear differential operators was presented in [BEN 00b]. We also mention here [DEL 01a], where the parametric form is obtained by creating an auxiliary output without zeros. According to this idea, model [7.57] can be written in the following parametric form:

$$\begin{cases}
q_r(t) = A_1\lambda(t) + A_2\lambda^{(2)}(t) + A_6\lambda^{(4)}(t) \\
q_1(t) = A_3\lambda^{(2)}(t) + A_7\lambda^{(4)}(t) \\
q_2(t) = A_4\lambda^{(2)}(t) + A_8\lambda^{(4)}(t) \\
\tau(t) = A_5\lambda^{(2)}(t) + A_9\lambda^{(4)}(t) + A_{10}\lambda^{(6)}(t)
\end{cases} \quad [7.67]$$

where $A_i, i \in \{1 \dots 10\}$ are obtained in terms of the coefficients of the models [7.57]

Form [7.67] makes it possible to directly solve the problem of point-to-point movement in fixed time. Indeed, let us consider a desired movement, characterized by the time of movement t_f and the following initial and final constraints:

$$\begin{cases} q_r(t_0) = q_{r0} \\ \dot{q}_r(t_0) = \ddot{q}_r(t_0) = 0 \\ q_1(t_0) = q_2(t_0) = 0 \\ \dot{q}_1(t_0) = \dot{q}_2(t_0) = 0 \end{cases}, \quad \begin{cases} q_r(t_f) = q_{rf} \\ \dot{q}_r(t_f) = \ddot{q}_r(t_f) = 0 \\ q_1(t_f) = q_2(t_f) = 0 \\ \dot{q}_1(t_f) = \dot{q}_2(t_f) = 0 \end{cases} \qquad [7.68]$$

Based on equation [7.67], these conditions result in initial and final constraints on λ:

$$\begin{aligned} &\lambda(t_0) = \frac{q_{r0}}{A_1}, \lambda^{(i)}(t_0) = 0,\ i \in \{1,...,6\} \\ &\lambda(t_f) = \frac{q_{rf}}{A_1}, \lambda^{(i)}(t_f) = 0,\ i \in \{1,...,6\} \end{aligned} \qquad [7.69]$$

These constraints are then interpolated by the following polynomial function:

$$\lambda_d(t) = \sum_{i=0}^{i=13} a_i t^i \qquad [7.70]$$

with the coefficients:

$$\begin{aligned} &a_0 = q_{r0}\frac{M_{12}}{K_1}, a_i = 0,\ i \in \{1...6\},\ a_7 = 1{,}716\frac{q_{rf}}{A_1 t_f^{\,7}},\ a_8 = -9{,}009\frac{q_{rf}}{A_1 t_f^{\,8}} \\ &a_9 = 20{,}020\frac{q_{rf}}{A_1 t_f^{\,9}},\ a_{10} = -24{,}024\frac{q_{rf}}{A_1 t_f^{\,10}},\ a_{11} = 16{,}380\frac{q_{rf}}{A_1 t_f^{\,11}}, \\ &a_{12} = -6{,}006\frac{q_{rf}}{A_1 t_f^{\,12}},\ a_{13} = 924\frac{q_{rf}}{A_1 t_f^{\,13}} \end{aligned} \qquad [7.71]$$

By substituting [7.70] in [7.67], we obtain the desired movements and the feedforward control, which is associated to a joint regulator in order to realize the desired movement. The parametric form [7.67] can also be used to control the operational trajectories [BEN 02c].

7.3.3.3. *Trajectory tracking in the joint space*

One of the first studies dealing with this problem is the article [TRU 79], where an LQR control was used in order to stabilize the tangent linearized system along the desired trajectories. In [SIC 88] the method of singular disturbances was applied to the joint trajectory control for flexible robots, the idea being related to the concept of variables moving on various scales of time. The control is done in two steps: in the first instance, a slow control is calculated for the slow dynamic (i.e. rigid variables), as if for a rigid robot. Then in the second instance, we calculate a control which stabilizes the fast system (i.e. elastic variables). In [SIN 86], [CHE 89], [YAN 97], the authors proposed a non-linear control decoupling the joint coordinates from the elastic coordinates. This control, i.e. inversion or computed torque, requires the measurement or the estimation of elastic states. In [CHE 00] a control based on the passivity of the system was proposed. The results presented show a compromise between the precision of the joint trajectory tracking and the damping of residual oscillations. Below we will present in detail thc computed torque method, as well as the approach suggested in [BEN 02a], which makes it possible to carry out a precise joint trajectory tracking for flexible multi-axis planar robots, while damping the oscillations of the end-effector at the end of the rigid movement; and this without elastic state feedback.

7.3.3.3.1. The computed torque method

On the basis of equation [7.53], we can formally express the accelerations of elastic variables:

$$\ddot{\mathbf{q}}_e = \mathbf{M}_{ee}^{-1}\ (\mathbf{T}_e - \mathbf{M}_{er}\ddot{\mathbf{q}}_r) \qquad [7.72]$$

which makes it possible to rewrite the joint dynamics according to:

$$(\mathbf{M}_{rr} - \mathbf{M}_{re}\mathbf{M}_{ee}^{-1}\mathbf{M}_{er})\ddot{\mathbf{q}}_r = -\mathbf{M}_{re}\mathbf{M}_{ee}^{-1}\mathbf{T}_e + \mathbf{T}_r + \boldsymbol{\tau} \qquad [7.73]$$

Matrix $\mathbf{M}_{rr} - \mathbf{M}_{re}\mathbf{M}_{ee}^{-1}\mathbf{M}_{er}$ is called “decoupling matrix”. If we substitute the joint accelerations vector with the vector **W**, we can calculate the corresponding linearizing control $\boldsymbol{\tau}_d$ by:

$$\boldsymbol{\tau}_d = (\mathbf{M}_{rr} - \mathbf{M}_{re}\mathbf{M}_{ee}^{-1}\mathbf{M}_{er})\mathbf{W} + \mathbf{M}_{re}\mathbf{M}_{ee}^{-1}\mathbf{T}_e - \mathbf{T}_r \qquad [7.74]$$

and thus the following linear system is obtained:

$$\ddot{\mathbf{q}}_r = \mathbf{W} \qquad [7.75]$$

To give some robustness to the feedforward controller, we add a co-located joint PD, which gives finally:

$$\mathbf{W} = \ddot{\mathbf{q}}_{rd} + \mathbf{K}_p(\mathbf{q}_{rd} - \mathbf{q}_r) + \mathbf{K}_d(\dot{\mathbf{q}}_{rd} - \dot{\mathbf{q}}_r) \quad [7.76]$$

where the gain matrices $\mathbf{K}_p$, $\mathbf{K}_d$ are diagonal positive definite. The control [7.74] and [7.76] ensure the tracking of the desired joint movements. However, no action ensures the damping of the elastic variables controlled by the dynamic [7.72], even if this dynamic, which is identified with the zeros dynamics of the closed-loop system, is stable [CAN 97]. However, linear feedback terms on the elastic variables can be added to [7.76] in order to minimize the structural vibrations. Lastly, let us note that the control law [7.74] can be calculated online by the recursive algorithm for the calculation of the inverse dynamic model presented in section 7.2.8.2 [BOY 98].

7.3.3.3.2. Joint trajectory tracking by backward integration of elastic dynamics

This method of the type open loop "computed torque" is valid for a planar robot with multiple axes and with n flexible links [BEN 02a]. For a vector of a bounded desired joint movements:

$$\mathbf{q}_{rd}(t) = (q_{r1d}(t), \ldots, q_{rnd}(t))^T, \; t \in [t_0 \;\; t_f] \quad [7.77]$$

We try to find the control vector such that:

$$\begin{cases} \mathbf{q}_r(t) = \mathbf{q}_{rd}(t), \text{ for } t \in [t_0 \;\; t_f] \\ \mathbf{q}_r(t) = \mathbf{q}_{rd}(t_f), \text{ for } t \geq t_f \\ \mathbf{q}_e(t_0) = \mathbf{0}_{n_e \times 1}, \; \mathbf{q}_e(t) = \mathbf{0}_{n_e \times 1}, \text{ for } t \geq t_f \\ \dot{\mathbf{q}}_e(t_0) = \mathbf{0}_{n_e \times 1}, \; \dot{\mathbf{q}}_e(t) = \mathbf{0}_{n_e \times 1}, \text{ for } t \geq t_f \end{cases} \quad [7.78]$$

where $\mathbf{0}_{n \times m}$ indicates the zero matrix of dimension $(n \times m)$ and n_e is the dimension of the elastic vector $\mathbf{q}_e$. The control suggested, of the computed torque type, is calculated offline from equation [7.55]:

$$\boldsymbol{\tau}_{bo} = \mathbf{M}_{rr}^d \ddot{\mathbf{q}}_{rd} + \mathbf{M}_{re}^d \ddot{\mathbf{q}}_{ed} + \mathbf{C}_{rd}(\mathbf{q}_{rd}, \mathbf{q}_{ed}, \dot{\mathbf{q}}_{rd}, \dot{\mathbf{q}}_{ed}) \quad [7.79]$$

where

$$\begin{aligned} \mathbf{M}_{rr}^{d} &= \mathbf{M}_{rr}(\mathbf{q}_{rd}, \mathbf{q}_{ed}) \\ \mathbf{M}_{re}^{d} &= \mathbf{M}_{re}(\mathbf{q}_{rd}, \mathbf{q}_{ed}) \\ \mathbf{C}_{rd} &= \mathbf{C}_{rr}^{d}\dot{\mathbf{q}}_{rd} + \mathbf{C}_{re}^{d}\dot{\mathbf{q}}_{ed} \end{aligned} \qquad [7.80]$$

and the vector of the desired elastic coordinates is the solution of:

$$(\mathbf{M}_{er}^{d})^{T}\ddot{\mathbf{q}}_{rd} + \mathbf{M}_{ee}^{d}\ddot{\mathbf{q}}_{ed} + \mathbf{C}_{ed}(\mathbf{q}_{rd}, \mathbf{q}_{ed}, \dot{\mathbf{q}}_{rd}, \dot{\mathbf{q}}_{ed}) + \mathbf{K}\mathbf{q}_{ed} = \mathbf{0} \qquad [7.81]$$

with

$$\begin{aligned} \mathbf{M}_{ee}^{d} &= \mathbf{M}_{ee}(\mathbf{q}_{rd}, \mathbf{q}_{ed}) \\ \mathbf{C}_{ed} &= \mathbf{C}_{ee}(\mathbf{q}_{d}, \dot{\mathbf{q}}_{d})\dot{\mathbf{q}}_{rd} + \mathbf{C}_{er}(\mathbf{q}_{d}, \dot{\mathbf{q}}_{d})\dot{\mathbf{q}}_{ed} \end{aligned} \qquad [7.82]$$

The dynamic [7.81] is then integrated backward in time, based on the desired elastic final conditions:

$$\mathbf{q}_{ed}(t_f) = \mathbf{0}_{ne\times 1}, \quad \dot{\mathbf{q}}_{ed}(t_f) = \mathbf{0}_{ne\times 1} \qquad [7.83]$$

The initial elastic set points obtained, $\mathbf{q}_{ed}(t_o)$, $\dot{\mathbf{q}}_{ed}(t_o)$, can be non-zero. This can lead to errors on the initial elastic states. This difference is then corrected by a joint PD regulator of variable gain, which is added to the feedforward control [7.79], in order to ensure a local exponential convergence of the error dynamic toward zero [BEN 02a], [BEN 02b]. However, it should be stressed that because of the local nature of the obtained stability, in the case of the big initial errors on the elastic coordinates, the stability of the error dynamics equilibrium point can no longer be guaranteed. This situation arises in the case of the fast joint movements, for which the elastic displacements can be considerable. In this case, the causal nominal control [7.79] is associated to a non-causal feedforward control which is applied before the control of the desired movement in order to pre-set the elastic variables to pre-calculated desired values. The non-causal control is obtained as follows:

1) Planning of non-causal joint trajectories $\tilde{\mathbf{q}}_{rd}(t) = (\tilde{q}_{rd1}, \ldots, \tilde{q}_{rdn})^{T}$ like:

$$\tilde{q}_{rdk}(t) = \sum_{i=0}^{i=5} a_{ik}(t/t_0)^{i} + \sum_{j=1}^{j=m} b_{jk}(t/t_0)^{(j+5)}, \; k \in \{1 \ldots n\}, \; m \in N^{*} \qquad [7.84]$$

for $t \in [-\delta t, t_0]$. The coefficients a_{ik} are calculated as a function of b_{ik} by solving the constraints:

$$\begin{aligned} &\tilde{\mathbf{q}}_{rd}(-\delta t) = \mathbf{0}_{n\times 1},\ \dot{\tilde{\mathbf{q}}}_{rd}(-\delta t) = \mathbf{0}_{n\times 1},\ \ddot{\tilde{\mathbf{q}}}_{rd}(-\delta t) = \mathbf{0}_{n\times 1} \\ &\tilde{\mathbf{q}}_{rd}(t_0) = \mathbf{0}_{n\times 1},\ \dot{\tilde{\mathbf{q}}}_{rd}(t_0) = \mathbf{0}_{n\times 1},\ \ddot{\tilde{\mathbf{q}}}_{rd}(t_0) = \mathbf{0}_{n\times 1} \end{aligned} \qquad [7.85]$$

The coefficients b_{ik} are then obtained by minimizing the quadratic criterion (see [BEN 02b] for a proof of the solutions existence problem):

$$\begin{aligned} &\underset{\mathbf{b}}{Min}\left(\frac{1}{2}\boldsymbol{\varepsilon}_e(t_o)^T \mathbf{K}_1 \boldsymbol{\varepsilon}_e(t_o) + \frac{1}{2}\dot{\boldsymbol{\varepsilon}}_e(t_o)^T \mathbf{K}_2 \dot{\boldsymbol{\varepsilon}}_e(t_o)\right) \\ &\boldsymbol{\varepsilon}_e(t_o) = \tilde{\mathbf{q}}_{ed}(t_o) - \mathbf{q}_{ed}(t_o),\ \mathbf{K}_1, \mathbf{K}_2 > 0 \end{aligned} \qquad [7.86]$$

where $\tilde{\mathbf{q}}_{ed}$ verifies equation [7.81], in which $\mathbf{q}_{rd}$ is replaced by $\tilde{\mathbf{q}}_{rd}$ and $\mathbf{b} = (b_{11}, \ldots b_{mn})^T$.

2) Calculating the non-causal control $\boldsymbol{\tau}_{bo}^{nc}$ associated to $\tilde{\mathbf{q}}_{rd}$ and $\tilde{\mathbf{q}}_{ed}$, by using equation [7.79].

3) The final nominal control is then given by:

$$\begin{cases} \boldsymbol{\tau}_{bo}^{nc},\ t \in [-\delta t,\ t_0] \\ \boldsymbol{\tau}_{bo}^{c},\ t \in [t_0,\ t_f] \end{cases} \qquad [7.87]$$

This controller is added to a joint regulator based on the movement $\tilde{\mathbf{q}}_{rd}$ for the non-causal part and on $\mathbf{q}_{rd}$ for the causal part of the control.

7.3.3.4. *Trajectory tracking in the operational space*

This objective is the most difficult of all the control problems for the flexible robot manipulators, because of the non-minimum phase property which characterizes the models of flexible robots when the outputs are defined in the operational space. In this case, the computed torque control schemes, as applied to the rigid robots, leads to radially unbounded torques and states. Several control methods were suggested to solve this problem; we mention for example the approaches of so-called "output redefinition". The principle of these methods is the redefinition of the outputs to be controlled, so that the new dynamic is at minimum phase. In [WAN 91], it is shown that for a one-link flexible arm, the symmetrical position of the end-effector:

$$y_s = q_r L - \sum_{i=1}^{i=n_e} \phi_i q_i \qquad [7.88]$$

(see equation [7.58] for the notations) leads to a passive system (transfer function at a minimum phase), if the joint-motor hub-inertia is sufficiently important. The same idea was developed in [DAM 95], [CHR 00], where the authors show that the characteristic of passivity remains true for the multi-link case, if the end-effector load has a mass much larger than that of the manipulator. We will also mention the methods of the "stable inversion" type, where the control is obtained by model inversion, via non-causal operators, necessary to obtain a bounded control. Based on this approach, the case of a flexible axis robot is dealt with in [BAY 87]. A non-causal bounded control torque is obtained by applying the non-causal inverse Fourier transform to the inverse transfer function. This approach was extended by an iterative algorithm in the case of the multi-link planar robots [BAY 88]. Inversion methods were also proposed in the temporal field. In [BAY 89a], the stable inversion is obtained by using the non-causal impulse response. In [KWO 94], the inverse model is decomposed into two dynamics, one stable and the other one unstable, under the assumption that the inverse system is hyperbolic i.e. "does not admit imaginary poles". The stable sub-system is then directly integrated in time, whereas the unstable sub-system is integrated backward in time. The two integrations lead to bounded states, implying a bounded control. This approach was applied in [ZHA 98] to a robot with two planar flexible links with small elastic displacements. We present in detail below this method, which requires a rather difficult offline iterative calculation. The authors consider the model given by equation [7.55], where $\mathbf{q}_r \in \mathbb{R}^2$ and $\mathbf{q}_e \in \mathbb{R}^4$ (two elastic modes per segment). The operational output considered is given by:

$$\mathbf{Y} = (y_1, y_2)^T = (q_{r1}, q_{r2})^T + \left(arctg(\frac{v_1(L_1,t)}{L_1}), arctg(\frac{v_2(L_2,t)}{L_2}) \right)^T \qquad [7.89]$$

where y_1, y_2, represent the absolute angles of the tangents at the joint extremities of segments 1 and 2 respectively. As for L_1, L_2, and $v_1(L_1,t)$, $v_2(L_2,t)$, they indicate the lengths of the segments and the elastic displacements at the end of the first and second segments. We note that with this choice it is necessary to first establish a model that links these operational outputs to the "true" Cartesian coordinates of the tool in the operational plan. This is carried out via the inversion of the geometric model of a "virtual" rigid robot, whose segments would be supported at every moment by straight lines joining the extreme points of each real segment. In any event, such a construction is valid only for small elastic displacements, i.e. when **Y** can be written:

$$\mathbf{Y} = \mathbf{q}_r + \boldsymbol{\Gamma}\mathbf{q}_e \qquad [7.90]$$

with

$$\boldsymbol{\Gamma} = \begin{bmatrix} \frac{\phi_{11}}{L_1} & \frac{\phi_{12}}{L_1} & 0 & 0 \\ 0 & 0 & \frac{\phi_{21}}{L_2} & \frac{\phi_{22}}{L_2} \end{bmatrix}$$

By substituting $\mathbf{q}_r$ of equation [7.90] in equation [7.55], we obtain the elastic dynamics of reference:

$$\mathbf{M}_{er}\ddot{\mathbf{Y}}_d + (\mathbf{M}_{ee} - \mathbf{M}_{er}\boldsymbol{\Gamma})\ddot{\mathbf{q}}_e + \mathbf{C}_e + \mathbf{K}\mathbf{q}_e + \mathbf{D}\dot{\mathbf{q}}_e = \mathbf{0} \qquad [7.91]$$

where $\mathbf{C}_e$ is defined in equation [7.82]. This dynamic is linearized along the movements in order to obtain the linearized tangent model:

$$\mathbf{L}_1\ddot{\mathbf{q}}_e + \mathbf{L}_2\dot{\mathbf{q}}_e + \mathbf{L}_3\mathbf{q}_e = \mathbf{L}_4 \qquad [7.92]$$

Where:

$$\mathbf{L}_1 = \mathbf{M}^0_{ee} - \mathbf{M}^0_{er}\boldsymbol{\Gamma}, \quad \mathbf{L}_2 = \mathbf{D} + \partial^0_{\dot{\mathbf{q}}_e}\mathbf{C}_e,$$

$$\mathbf{L}_3 = \partial^0_{\mathbf{q}_e}(\mathbf{M}_{ee} - \mathbf{M}_{er}\boldsymbol{\Gamma})\ddot{\mathbf{q}}_{e0} + \mathbf{K} + \partial^0_{\mathbf{q}_e}\mathbf{C}_e + \partial^0_{\mathbf{q}_e}\mathbf{M}_{er}\ddot{\mathbf{Y}}_d$$

$$\mathbf{L}_4 = -\mathbf{M}^0_{re}\ddot{\mathbf{Y}}_d + \left[\partial^0_{\mathbf{q}_e}\mathbf{M}_{re}\ddot{\mathbf{Y}}_d\right]\mathbf{q}_{e0} + \left[\partial^0_{\mathbf{q}_e}(\mathbf{M}_{ee} - \mathbf{M}_{er}\boldsymbol{\Gamma})\ddot{\mathbf{q}}_{e0}\right]\mathbf{q}_{e0} +$$
$$\partial^0_{\mathbf{q}_e}\mathbf{C}_e\mathbf{q}_{e0} + \partial^0_{\dot{\mathbf{q}}_e}\mathbf{C}_e\dot{\mathbf{q}}_{e0} - \mathbf{C}^0_e$$

$\partial^0_{\mathbf{q}_e}\mathbf{C}_e$, $\partial^0_{\dot{\mathbf{q}}_e}\mathbf{C}_e$ indicate respectively the derivative of $\mathbf{C}_e$ compared to $\mathbf{q}_e$ and $\dot{\mathbf{q}}_e$ (the superscript 0 indicates that the derivatives are calculated along ($\mathbf{q}_{e0}, \dot{\mathbf{q}}_{e0}$), solutions of the preceding iteration). By defining the vector of state $\boldsymbol{\eta} = (\mathbf{q}_e^T, \dot{\mathbf{q}}_e^T)^T$, equation [7.92] will have the following form:

$$\dot{\boldsymbol{\eta}} = \mathbf{A}(t)\boldsymbol{\eta} + \mathbf{B}(t) \qquad [7.93]$$

with:

$$\dot{\boldsymbol{\eta}} = \mathbf{A}(t)\boldsymbol{\eta} + \mathbf{B}(t),$$

$$\mathbf{A}(t) = \begin{bmatrix} \mathbf{0} & \mathbf{I} \\ -\mathbf{L}_1^{-1}\mathbf{L}_3 & -\mathbf{L}_1^{-1}\mathbf{L}_2 \end{bmatrix}, \quad \mathbf{B}(t) = \begin{bmatrix} \mathbf{0} \\ -\mathbf{L}_1^{-1}\mathbf{L}_4 \end{bmatrix}$$

A solution of [7.93] is obtained by considering the initial/final conditions:

$$\boldsymbol{\eta}(t_0) \in E^u, \boldsymbol{\eta}(t_f) \in E^s \qquad [7.94]$$

where E^s, E^u are, respectively, the stable and unstable vector sub-spaces of the linear system [7.93] [WIG 90]. The condition $\boldsymbol{\eta}(t_0) \in E^u$ corresponds to equation:

$$\mathbf{C}_s \boldsymbol{\eta}(t_0) = \mathbf{0} \qquad [7.95]$$

and the final condition $\boldsymbol{\eta}(t_f) \in E^s$ corresponds to equation:

$$\mathbf{C}_u \boldsymbol{\eta}(t_f) = \mathbf{0} \qquad [7.96]$$

where $\mathbf{C}_s = \mathbf{Y}_s(t_0)\mathbf{A}(t_0)$, $\mathbf{C}_u = \mathbf{Y}_u(t_f)\mathbf{A}(t_f)$, $\mathbf{Y}_s$, $\mathbf{Y}_u$ being respectively the matrix of the eigenvectors associated to the eigenvalues of $\mathbf{A}$ with negative and positive real parts. To solve the linear problem at the two extremities [7.93], [7.95], [7.96], the authors propose the following change of variables:

$$\boldsymbol{\xi} = (\xi_1, \xi_2)^T = (\mathbf{C}_s\boldsymbol{\eta}, \mathbf{C}_u\boldsymbol{\eta})^T \qquad [7.97]$$

And thus:

$$\boldsymbol{\eta} = \begin{bmatrix} \mathbf{C}_s \\ \mathbf{C}_u \end{bmatrix}^{-1} \boldsymbol{\xi} \qquad [7.98]$$

Hence, equation [7.93] will be written:

$$\begin{cases} \dot{\xi}_1 = \mathbf{A}_{11}(t)\xi_1 + \mathbf{A}_{12}(t)\xi_2 + \mathbf{B}_1(t) \\ \dot{\xi}_2 = \mathbf{A}_{21}(t)\xi_1 + \mathbf{A}_{22}(t)\xi_2 + \mathbf{B}_2(t) \end{cases} \qquad [7.99]$$

where the first equation of the system [7.99] carries out a problem with the initial values: $\xi_1(t_0) = \mathbf{0}$,. while the second equation of the system [7.99] defines a problem with the final values: $\xi_2(t_f) = \mathbf{0}$. Moreover, these two problems being coupled linearly, their solutions are thus bound by a relation of type:

$$\xi_2(t) = \mathbf{s}(t)\xi_1(t) + \mathbf{v}(t) \quad [7.100]$$

with:

$$\mathbf{s}(t_f) = \mathbf{0},\ \mathbf{v}(t_f) = \mathbf{0}$$

Finally, considering [7.100] in the system [7.99] we obtain:

$$\begin{aligned} \dot{\mathbf{s}}(t) &= \mathbf{A}_{21}(t) + \mathbf{A}_{22}(t)\mathbf{s}(t) - \mathbf{s}(t)\mathbf{A}_{11}(t) - \mathbf{s}(t)\mathbf{A}_{12}(t)\mathbf{s}(t) \\ \dot{\mathbf{v}}(t) &= (\mathbf{A}_{22}(t) - \mathbf{s}(t)\mathbf{A}_{12}(t))\mathbf{v}(t) + (\mathbf{B}_2(t) - \mathbf{s}(t)\mathbf{B}_1(t)) \end{aligned} \quad [7.101]$$

Matrix $\mathbf{s}(t)$ is determined by backward integration in the time of the first equation [7.101]. This solution is then replaced in the second equation of [7.101] which makes it possible to obtain the vector $\mathbf{v}(t)$, there again, by backward integration. Having $\mathbf{s}(t)$, $\mathbf{v}(t)$ and substituting [7.100] in the equation of the first equation of [7.99], we obtain a differential equation in ξ_1 which is integrated in the direct sense (i.e. starting from $\xi_1(t_0) = \mathbf{0}$); ξ_2 is then calculated by [7.100].

Finally, the solution $\boldsymbol{\eta}$ of the current iteration is obtained by [7.98], then used to re-evaluate the linearized tangent [7.92] (the first iteration is calculated along the zero trajectory of the elastic dynamics) and the resolution is repeated until the difference between two successive solutions is below a given threshold. Once the limited elastic set point is obtained, the associated joint movement $\mathbf{q}_{rd}$ is obtained using [7.90] and the correspondent nominal control torque is calculated by equation [7.79]. The nominal control torque thus obtained is added to a joint regulator. It is necessary to stress that the presence of segment structural damping ensures a hyperbolic equilibrium point for the elastic dynamics. Moreover, in order to ensure the algorithm convergence based on the tangent linearized model, the position set points must be of low amplitude. The assumptions imposed by this method were relaxed in [BEN 01], [BEN 02b], where a stable inversion approach was introduced, which was based on a formulation of a boundary value problem. This method leads to causal controls without hyperbolicity constraints or restriction regarding the amplitude of the trajectories to follow. However, the controls are also calculated offline, and have discontinuities at the initial/final instants of the movement. The faster the desired movements are, the more significant these discontinuities.

7.4. Conclusion

In this chapter we presented a coherent approach to the geometric, kinematic and dynamic modeling of a flexible robot. The approach is based on the floating frame method. The deformation fields are reduced through "clamped-free" modes. The dynamic model suggested performs a generalization of the Newton-Euler model for the rigid robot manipulators. This generalization is conceptually guided by the concept of formalism of description of a motion applied as defined in the continuous medium mechanics.

This concept, contrary to the current practices in rigid robotics, is dissociated from the means of obtaining the dynamic equations (Newton-Euler theorems, Lagrange equations, Hamel equations, etc.). After considering this point, the principle of virtual powers proved to be best adapted to an Eulerian-Lagrangian mixed description, as imposed by the generalization of the Newton-Euler models within the framework of the floating frame. Based on this model, we developed two algorithms for the calculation of the direct and inverse dynamic models for a deformable robot manipulator. These two algorithms are in $o(n)$, and can be numerically or symbolically calculated.

The inverse dynamic algorithm is an essential element of the control laws such as they are studied thereinafter. On this subject, we adopted a classification of the control objectives for flexible robot manipulators by separating the point-to-point positioning tasks from the ones concerning the trajectory tracking. Then we detailed some control methods, which made it possible to achieve these goals. The general case of space movements with large elastic displacements remains, from our point of view, an open problem. Indeed, in the case of 3D space movements, the controllability problems may occur [LOP 94], making the objective of trajectory tracking more difficult.

Moreover, the models of the flexible multi-link manipulators with large elastic displacements are strongly non-linear [BOY 00]. Another problem that deserves attention is that of flexible robots with closed-loop structures [BAY 89b], [DAM 00]. Additional references on the control of flexible robots are indicated in the paper [BEN 04b], where a recent state of the art method is presented on the control of flexible arm manipulators. We will also mention the recent results related to the synthesis of robust control, namely the robust regulation of the end-effector based on the input-shaping method [PAO 00], [PAO 03], [PAO 04], [SIN 04], as well as the regulation controls and robust trajectory tracking, based on approaches through neuron networks [TAL 05], [JNI 05].

7.5. Bibliography

[ARM 79] ARMSTRONG W.W., "Recursive solution to the equations of an n-link manipulator", *Proc. 5th World Congress on Theory of Machines and Mechanisms*, Montreal, p. 1343-1346, 1979.

[BAY 87] BAYO E., "A finite element approach to control the end-point motion of a single-link flexible robot", *J. of Robotic Systems*, p. 63-75, 1987.

[BAY 88] BAYO E., "Computed torque for the position control of open-chain flexible robots", *Proc. IEEE Int. Conf. on Robotics and Automation*, Philadelphia, p. 316-321, 1988.

[BAY 89a] BAYO E., MOULIN H., "An efficient computation of the inverse dynamics of flexible manipulators in the time domain", *Proc. IEEE Int. Conf. on Robotics and Automation*, Scottsdale, p. 710-715, 1989.

[BAY 89b] BAYO E., PAPADOPOULOS P., STUBBE J., SERNA M.A, "Inverse dynamics and kinematics of multi-link elastic robots: an iterative frequency domain approach", *J. of Robotic Research*, Vol. 8, No. 6, p. 49-62, 1989.

[BEN 00a] BENOSMAN M., LE VEY G., "End effector motion planning for a one link flexible robot", *Proc. Symp. in Robotics Control*, SYROCO, Vienna, Austria, p. 561-566, 2000.

[BEN 00b] BENOSMAN M., LE VEY G., "Exact articular trajectory tracking for a one-link flexible manipulator: an approach through the parametrization of differential operators", *31st International Symposium on Robotics*, Montreal, Canada, p. 150-155, 2000.

[BEN 01] BENOSMAN M., LE VEY G., "Model inversion for a particular class of non-linear non-minimum phase systems: an application to the two-link flexible manipulator", *Proc. IEEE 40th Conference on Decision and Control*, Orlando, Florida, 2001.

[BEN 02a] BENOSMAN M., LE VEY G., "Joint trajectory tracking for a planar multi-link flexible manipulators: simulation and experiments for a two-link flexible manipulator", *Proc. IEEE Int. Conf. on Robotics and Automation*, p. 2461-2466, 2002.

[BEN 02b] BENOSMAN M., "Commande des bras manipulateurs souples et extensions aux systèmes à non minimum de phase", PhD Thesis, Ecole Centrale of Nantes/Nantes University, 2002.

[BEN 02c] BENOSMAN M., BOYER F., LE VEY G., PRIMAULT D., "Flexible link manipulators: from modelling to control", *J. of Intelligent and Robotic Systems*, Vol. 34, p. 381-414, 2002.

[BEN 03] BENOSMAN M., LE VEY G., "Stable inversion of SISO non-minimum phase linear dynamic through output planning: an experimental application to the one-link flexible manipulator", *IEEE Trans. on Control Systems Technology*, Vol. 11, No. 4, p. 588-597, 2003.

[BEN 04a] BENOSMAN M., LE VEY G., LANARI L., DE LUCA A., "Rest to rest motion for planar multi-link flexible manipulator", *ASME, J. of Dyn. Syst., Measurements and Control*, Vol. 126, p. 115-123, 2004.

[BEN 04b] BENOSMAN M., LE VEY G., "Control of flexible manipulators: a survey", *Robotica,* vol. 22, p. 533-545, 2004.

[BOO 84] BOOK J.W, "Recursive lagrangian dynamics of flexible manipulators arms", *Int. J. of Robotic Research*, Vol. 3, p. 87-101, 1984.

[BOY 94] BOYER F., "Contribution à la modélisation et à la commande des robots flexibles", PhD Thesis, Paris VI University, 1994.

[BOY 95] BOYER F., KHALIL W., "Recursive solution of inverse and forward dynamics of flexible manipulators", *Proc. 3rd European Control Conference*, Rome, Italy, p. 2696-2701, 1995.

[BOY 96a] BOYER F., COIFFET P., "Generalisation of Newton-Euler model for flexible manipulators", *J. of Robotic Systems*, Vol. 13, No. 1, p. 11-24, 1996.

[BOY 96b] BOYER F., COIFFET P., "Symbolic modelling of a flexible manipulator via assembling of its generalized Newton-Euler model", *Journal of Mechanism and Machine Theory*, Vol. 31, p. 45-56, 1996.

[BOY 98] BOYER F., KHALIL W., "An efficient calculation of flexible manipulator inverse dynamics", *J.of Robotic Research*, Vol. 17, No. 3, p. 282-293, 1998.

[BOY 99a] BOYER F., KHALIL W., "Kinematic model of a multi-beam structure undergoing large elastic displacement and rotations. Part one: model of an isolated beam", *Journ. of Mechanism and Machine Theory*, Vol. 34, p. 205-222, 1999.

[BOY 99b] BOYER F., KHALIL W., "Kinematic model of a multi-beam structure undergoing large elastic displacement and rotations. Part two: kinematic model of an open chain", *J. of Mechanism and Machine Theory*, Vol. 34, p. 223-242, 1999.

[BOY 00] BOYER F., GLANDAIS N., KHALIL W., "Flexible multibody dynamics based on non-linear Euler-Bernoulli kinematics", *Int. J. of Numerical Methods in Engineering*, 2001, Vol. 54, p. 27-59.

[BOY 04] BOYER F., PRIMAULT D., "Finite element of slender beams in finite transformations: a geometrically exact approach", *Int. J. of Numerical Methods in Engineering*, 2004, Vol. 59, p. 669-702.

[BOY 05] BOYER F., PRIMAULT D., "Finite element of non-linear cables: applications to Robotics", *Far East Journal of Applied Mathematics*, A special volume devoted to the applications of numerical methods in the partial differential equations, 2005, Vol. 19, p. 1-34.

[BRA 86] BRANDL H., JOHANNI R., OTTER M., "A very efficient algorithm for the simulation of robots and multibody systems without inversion of the mass matrix", *Proc. IFAC Symp. on Theory of Robots*, Vienna, Austria, p. 365-370, 1986.

[BRE 97] BREMER H., "Fast moving flexible robot dynamics. *Symposium on Robot Control", Proc. Symp. in Robotics Control, SYROCO,* Nantes, France, p. 45-52, 1997.

[CAN 77] CANAVIN J.R, LIKINS P.W, "Floating reference frames for flexible spacecraft", *AIAA 15th Aerospace Sciences Meeting*, Los Angeles, CA, p. 77-66, 1977.

[CAN 97] CANUDAS DE WIT C., SICILIANO B., BASTIN G., *Theory of Robot Control*, Springer-Verlag, 2nd edition, 1997.

[CAR 88] CARDONA A., GERADIN M., "A beam finite element non-linear theory with finite rotations", *Int. J. Numer. Methods. in Engin.*, Vol. 26, p. 2403-2438, 1988.

[CHA 95] CHAN T.M., STELSON K.A., "Point to point motion commands that eliminate residual vibration", *Proc. American Control Conference*, Washington, p. 909-913, 1995.

[CHE 89] CHEDMAIL P., KHALIL W., "Non-linear decoupling control of flexible robots", *Proc. ICAR 89*, Columbus, p.138-145, 1988.

[CHE 90] CHEDMAIL P., "Synthèse de robots et de sites robotisés, modélisation de robots souples", Thesis, ENSM, Nantes, 1990.

[CHE 00] CHEONG J., CHUNG W., YOUM Y., "Bandwidth modulation of rigid subsystem for the class of flexible robots", *Proc. IEEE Int. Conf. on Robotics and Automation*, San Francisco, p. 1478-1483, 2000.

[CHR 00] CHRISTOFOTOUA E.G., DAMAREN C.J., "The control of flexible-link robots manipulating large payloads: theory and experiments", *Int. J. of Robotic Systems*, Vol. 17, No. 5, p. 255-271, 2000.

[DAM 95] DAMAREN C.J., "Passivity analysis for flexible multilink space manipulators", *J. of Guidance, Control, and Dynamics*, Vol. 18, No. 2, p. 272-279, March-April 1995.

[DAM 00] DAMAREN C.J., "On the dynamics and control of flexible multibody systems with closed loops", *J. of Robotic Research*, Vol. 19, No. 3, p. 1-16, 2000.

[DEL 92] D'ELEUTERIO G.M.T, "Dynamics of an elastic multibody chain. Part C-recusive dynamics", *Dyn. Stab. of Syst.*, Vol. 7, No. 2, p. 61-89, 1992.

[DEL 89] DE LUCA A., SICILIANO B., "Trajectory control of a non-linear one link flexible arm", *Int. J. of Control*, Vol. 50(5), p. 1699-1715, 1989.

[DEL 93] DE LUCA A., SICILIANO B., "Regulation of flexible arms under gravity", *IEEE Trans. on Robotics and Automation*, vol. 50, No. 5, p. 463-467, 1993.

[DEL 01a] DE LUCA A., DI GIOVANNI G., "Rest to rest motion of a one-link flexible arm", *IEEE/ASME International Conference on Advanced Intelligent Mechatronics*, Como, Italy, p. 923-928, July 2001.

[DEL 01b] DE LUCA A., DI GIOVANNI G., "Rest to rest motion of a two-link flexible robot with a flexible forearm", *IEEE/ASME International Conference on Advanced Intelligent Mechatronics*, Como, Italy, p. 929-935, July 2001.

[FEA 83] FEATHERSTONE R., "The calculation of robot dynamics using articulated-body inertias", *Int. Journ. of Robotics Research*, Vol. 2, No. 1, p. 13-30, 1983.

[FIS 97] FISSETE P., JOHNSON D.A., SAMIN J.C., "A fully symbolic generation of the equations of motion of multibody systems containing flexible beams", *Comput. Methods Appl. Mech. Engrg.*, Vol. 142, p. 123-152, 1997.

[GER 86] GERMAIN P., *Cours de mécanique de l'école Polytechnique*, Vol. 1, Editions Ellipses, Paris, 1986.

[GLA 99] GLANDAIS N., "Modélisation et simulation des robots souples: extension de la méthode du repère flottant au domaine des grands déplacements élastiques", PhD Thesis, Ecole Centrale, Nantes, 1999.

[HUG 78] HUGHES T.J.R., PISTER K.S., "Consistent linearization in mechanics of solid and structures", *J. Computer & Structures*, Vol. 8, p. 391-397, 1978.

[HUG 89] HUGHES P.C., SINCARSIN G.B., "Dynamics of elastic multibody chains: part B-Global dynamics", *Dynamics and Stability of Systems*, Vol. 4, p. 227-243, 1989.

[IBR 98] IBRAHIMBEGOVIC A., AL MIKDAD M., "Finite rotations in dynamics of beams and implicit time-stepping schemes", *Int. J. for Numeric. Meth. in Engineer.*, Vol. 41, p. 781-814, 1998.

[JNI 05] JNIFENE A., ANDREWS W., "Experimental study on active control of a single-link manipulator fuzzy logic and neural networks", *IEEE Transactions on Instrumentation and Measurement,* Vol. 54, p. 1200-1208, 2005.

[KAI 80] KAILATH T., *Linear Systems*, Prentice-Hall, 1980.

[KHA 87] KHALIL W., KLEINFINGER J.-F, "Minimum operations and minimum parameters of the dynamic models of tree structure robots", *IEEE J. Robotics and Autom.* p. 517-526, 1987.

[KHA 02] KHALIL W., DOMBRE E., *Modelisation et commande des robots,* Hermès, Paris, 2002

[KIM 88] KIM S.S., HAUG D.S., "A recursive formulation for flexible multibody dynamics: part I. open loop systems", *Comp. Methods Appl. Mech. Eng.*, Vol. 71, p. 293-314, 1988.

[KWO 94] KWON D.S., BOOK W.J., "A time-domain inverse dynamic tracking control of a single-link flexible manipulator", *ASME, J. of Dyn. Syst., Measurements and control*, Vol. 116, p. 193-200, 1994.

[LAN 92] LANARI L., WEN J.T., " Asymtotically stable set point control laws for flexible robots ", *Systems and Control Letters*, Vol. 19, p. 119-129, 1992.

[LOP 94] LOPEZ-LINARES S.A., KONNO A., UCHIYAMA M., "Controllability of flexible manipulators", *Proc. Symp. in Robotics Control, SYROCO*, Capri, Italy, p. 509-516, 1994.

[LUH 80] LUH J.Y.S., WALKER M.W., PAUL R., "On-line computational scheme for mechanical manipulators", *ASME, J. of Dyn. Syst., Measurements and Control*, Vol. 102, No. 2, p. 69-76, 1980.

[MEC 94] MECKL P.H., "Robust motion control of flexible systems using feedforward forcing functions", *IEEE Trans. on Control Systems Technology*, Vol. 2, No. 3, p. 245-253, 1994.

[MEI 89] MEIROVITCH L., *Dynamics and Control of Structures,* John Wiley & Sons, New York, 1989.

[MEI 91] MEIROVITCH L. "Hybrid state equations of motion for flexible bodies in terms of quasi-coordinates", *J. Guidance*, Vol. 14, p. 1008-1013, 1991.

[PAO 00] PAO L.Y., LAU M.A., "Robust input shaper control for parameter in flexible structures", *ASME, J. of Dyn. Syst., Measurements and Control*, vol. 122, p. 63-70, 2000.

[PAO 03] PAO L.Y., CUTFORTH C.F., "On frequency-domain and time-domain input shaping for multi-mode flexible structures", *ASME, J. of Dyn. Syst., Measurements and Control*, Vol. 125, p. 494-497, 2003.

[PAO 04] PAO L.Y., CUTFORTH C.F., "Shaped time-optimal feedback controllers for flexible structures", *ASME, J. of Dyn. Syst., Measurements and Control*, vol. 126, p. 173-186, 2004.

[SCH 85] SCHMITZ E., "Experiments on the end-point position control of a very flexible one-link manipulator" PhD Thesis, Stanford University, June 1985.

[SCH 99] SCHWERTASSEK R., WALLRAPP O., SHABANA A.A., "Flexible multibody simulation and choice of shape functions", *Non-linear Dynamics*, Vol. 20, p. 361-380, 1999.

[SER 89] SERNA M.A., BAYO E., "A simple and efficient computational approach for the forward dynamics of elastic robots", *J. of Robotic Systems*, Vol. 6, No. 4, p. 363-382, 1989.

[SHA 90] SHABANA A.A., "Dynamics of flexible bodies using generalized Newton-Euler equations", *J. of Dyn. Syst. Meas. Cont.*, Vol. 112, p. 496-503, 1990.

[SHA 92] SHARF I., DAMAREN C., "Simulation of flexible-link manipulators: basis functions and non-linear terms in the motion equations", *Proc. IEEE Int. Conf. on Robotics and Automation*, Nice, France, p. 1956-1962, 1992.

[SHA 95] SHARF I., "A survey of geometric stiffening in multibody dynamics formulations", *Wave Motion, Intelligent Structures and Non-linear Mechanics*, Editions World Scientific, p. 239-279, 1995.

[SIC 88] SICILIANO B., BOOK W.J., "A singular perturbation approach to control of lightweight flexible manipulators", *Int. J. of Robotics Research*, Vol. 7, No. 4, p. 79-90, 1988.

[SIM 86] SIMO J.C., VU-QUOC L., "On the dynamics of flexible beams under large overall motions – the plane case: Part I", *J. Appl. Mech.*, Vol. 53, p. 849-854, 1986.

[SIM 88] SIMO J.C., VU-QUOC L., "On the dynamics in space of rods undergoing large motions – a geometrically exact approach", *Computer Methods in Applied Mech. and Engin.*, Vol. 66, p. 125-161, 1988.

[SIM 91] SIMO J.C., "Unconditionally stable algorithms for rigid body dynamics that exactly preserve energy and momentum", *International Journal for Numerical Methods in Engineering*, Vol. 31, p. 19-52, 1991.

[SIN 84] SINGH R.P., VANDER VOORT R.J., LIKINS R.J., "Dynamics of flexible bodies in tree topology – a computer oriented approach", *Proc. of the AIAA Dynamics Specialist Conference*, Palm Springs, California, p. 327-337, 1984.

[SIN 86] SINGH S.N., SCHY A.A., "Control of elastic robotic systems by non-linear inversion and modal damping", *ASME, J. of Dyn. Syst., Measurements and Control*, Vol. 108, No. 2, p. 180-189, September 1986.

[SIN 04] SINGHOSE W., BIEDIGER E.O., CHEN Y., "Reference command shaping using specified-negative-amplitude input shapers for vibrations reduction", *ASME, J. of Dyn. Syst., Measurements and Control*, Vol. 126, p. 210-214, 2004.

[TAL 05] TALBI H.A., PATEL R.V., KHORASANI K., "A neural network controller for a class of non-linear non-minimum phase systems with applications to the flexible-link manipulator", *ASME, J. of Dyn. Syst., Measurements and Control*, Vol. 127, p. 289-294, 2005.

[TRU 79] TRUCKENBRODT A., "Dynamics and control methods for moving flexible structures and their application to industrial robots", *Proc. of the 5th World Congress on Theory of Machines and Mechanisms*, p. 831-834, 1979.

[VER 94] VERLINDEN O., DEHOMBREUX P., CONTI C., BOUCHER S., "A New formulation for the direct dynamic simulation of flexible mechanisms based on the Newton-Euler inverse methods", *Int. Journ. for Numerical Methods in Engineering*, Vol. 37, p. 3363-3387, 1994.

[WAN 91] WANG D., VIDYASAGAR M., "Transfer function for a flexible link", *Int. J. of Robotic Research*, Vol. 10, No. 5, p. 540-549, October 1991.

[WAN 92] WANG D., VIDYASAGAR M., "Passive control of a stiff flexible link", *Int. J. of Robotic Research*, Vol. 11, No. 6, p. 572-578, December 1992.

[WIG 90] WIGGINS S., "Introduction to applied non-linear dynamical systems and chaos", *Texts in Applied Mathematics 2*, Springer-Verlag, 1990.

[YAN 97] YANG H., LIAN F.L., FU L.C., "Non-linear adaptive control for flexible-link manipulators", *IEEE J. Robotics and Autom.*, Vol. 13, No. 1, p. 140-148, 1997.

[ZHA 98] ZHAO H., CHEN D., "Tip trajectory tracking for multilink flexible manipulators using stable inversion", *J. of Guidance, Control, and Dynamics*, Vol. 21, No. 2, p. 314-320, March-April 1998.

List of Authors

Frédéric Boyer
IRCCyN
Ecole des Mines de Nantes
France

Mouhacine Benosman
IRCCyN
Ecole Centrale de Nantes
France

François Chaumette
IRISA
Rennes
France

Taha Chettibi
Laboratory of Structure Mechanics
Military Polytechnical School
Algiers
Algeria

Pierre Dauchez
LESIA
Toulouse
France

Etienne Dombre
LIRMM
University of Montpellier II-CNRS
France

Philippe Fraisse
LIRMM
University of Montpellier II-CNRS
France

Moussa Haddad
Laboratory of Structure Mechanics
Military Polytechnical School
Algiers
Algeria

Wisama Khalil
IRCCyN
Ecole Centrale de Nantes
France

Georges Le Vey
IRCCyN
Ecole des Mines de Nantes
France

Halim Lehtihet
Laboratory of Structure Mechanics
Military Polytechnical School
Algiers
Algeria

Jean-Pierre Merlet
INRIA
Sophia-Antipolis
France

François Pierrot
LIRMM
University of Montpellier II-CNRS
France

Philippe Wenger
IRCCyN
Ecole Centrale de Nantes
France

Index

A

accessibility 141, 143-146, 152, 163, 165, 183
angles
    Euler 7

C

calibration 26-39
    geometric 26, 30-35
cameras
    mounted 280, 282, 326
chain
    control 243, 245
configurations
    eye-in-hand 282-283, 285
    eye-to-hand 283-284, 285
    singular 190

D

dexterity 141, 174, 181, 183

H

hand-eye calibration 282

K

kinematic screw 281-285, 300, 303

M

matrix
    inertia 205, 243, 249, 253
    Jacobian 242, 249, 264, 270, 287
    transformation 267, 282, 284, 302, 319
model
    calibration 30-31, 32, 35-36, 37
    direct dynamic 40
    direct kinematic 14-20
    energetic 60, 69
    inverse dynamic 40, 50, 66, 69
    inverse kinematic 1, 21-25

N

Newton-Euler equations 50-53

P

parameters
    base 32, 53, 58, 62
    base inertial 50, 53-59

dynamic 59-70, 246, 248, 249, 256, 373
geometric 2, 3, 5, 26-39, 222, 224
identifiable 31, 32-33, 53, 62-63, 70

proximity sensor 279, 320, 325

R

robots
parallel 81-134
redundant 169, 170-172

S

space
configuration 146
operational 147, 162, 163, 171, 191, 206, 229, 373, 383-387
primary 148, 149-151, 182
secondary 148, 149-151, 181

structure
parallel 263-268

T

task function 301-325